T0143604

Computational Number Theory and Digital Signal Processing

Fast Algorithms and Error Control Techniques

Hari Krishna, Ph.D.
Syracuse University
Syracuse, New York

Bal Krishna, Ph.D.
University of Roorkee
Roorkee, India

Kuo-Yu Lin, Ph.D.
GVC Corporation
Taiwan

Jenn-Dong Sun, Ph.D.
Chung-Shan Institute of Science and Technology
Taiwan

CRC Press
Boca Raton   Ann Arbor   London   Tokyo

Catalog record is available from the Library of Congress
ISBN: 0-8493-7177-5

No claim to original U.S. Government works
International Standard Book Number 0-8493-7177-5
Printed in the United States of America   1   2   3   4   5   6   7   8   9   0
Printed on acid-free paper

INSPIRED BY AND DEDICATED TO
THE GENIUS IN THE ANCIENT CHINESE

HK
BK
K-YL
J-DS

Preface

There exist a number of books on the topic of fast algorithms for digital signal processing of data sequences. However, they all deal with data sequences defined over complete number systems termed as the fields. The most frequently encountered number systems in this regard are the rational, real and complex number systems, and Galois fields. Also one find a number of books on the topic of error control techniques for data sequences defined over finite fields. The polynomial algebra so widely used in all these instances is well known and well understood. For example, the number of complex roots of a polynomial with complex coefficients is equal to its degree. The unique polynomial factorization property holds and provides a fundamental foundation to many of the results.

The most distinctive feature of this book is its focus on techniques for processing data sequences defined over finite integer rings. Such sequences occur naturally in many systems. In other instances they can equivalently be converted to one. The main idea is to design techniques which retain the infinite precision present in the arithmetic associated with finite integer rings. The lack of a field presents a major mathematical challenge to the framework as it makes most of the polynomial algebra results invalid. For example, a degree 3 polynomial may have 6 roots over a finite integer ring. The unique polynomial factorization property does not hold either.

In this book we turn to number theory to find answers to many of the fundamental questions that surround the behavior of polynomial algebra over finite integer rings and their complex extensions. This is done with a view to (1) establish new computationally efficient algorithms for digital signal processing, (2) establish a mathematical framework for error control techniques, and (3) obtain new computationally efficient algorithms for error control in residue number systems.

A common thread present throughout the book is the Chinese Remainder Theorem, a technique that has its origins in the ancient Chinese writing. We have developed new versions of the theorem based on the classical factorization method known as the Hensel's theorem. It is to the Genius in the ancient Chinese that we dedicate this book.

We have explored the topic of design of error control techniques for residue number systems from a coding theory point of view. It has resulted in a complete mathematical framework and computationally efficient algorithms. But, perhaps most importantly, it has led to immense satisfaction on our part for we have always believed in taking a broader approach to problem solving.

An attempt has been made to make this book as self-contained as possible in terms of its coverage of elementary number theory and other related topics. The readers are assumed to be familiar with the basics of error control coding for latter chapters. The Chinese Remainder Theorem is proven for all pertinent cases as it plays a vital role in this book.

We hope that the readers will enjoy the material and find it useful in their work.

Hari Krishna
Syracuse, NY, USA

Acknowledgments

This book is an outgrowth of our research work of the past six years. The work has been exciting, intriguing and intellectually rewarding. Many individuals have contributed to this book in a direct and indirect manner.

HK has fond memories of the times that he spent at IIT Delhi as an undergraduate student and at Concordia University as a graduate student. Professors S.I. Ahson, D.N. Chaudhary, H.M. Gupta, S.C. Gupta, S. Prasad, S.C.D. Roy, P.S. Satsangi and S.N. Tandon of IIT Delhi, and A. Antoniou, J.F. Hayes, R.V. Patel, V. Ramachandran and M.N.S. Swamy of Concordia University have influenced HK by virtue of their excellence in teaching and research. Two professors, Vijay K. Bhargava and Salvatore D. Morgera have played crucial roles in the graduate education of HK. They provided the opportunity and guidance not only as advisors, but also as friends. Thanks VJ. Thanks SAL.

In the past decade, Professor Norman Balabanian has supported the academic activities of HK and given a focus to his research work. HK owes the biggest debt of gratitude to Norm. Professors K. Jabbour, B.J. Strait, P.K. Varshney and D.D. Weiner of Syracuse University have always encouraged HK to strive for excellence in all his endeavors. HK has enjoyed stimulating discussions with professors Varshney and Weiner on numerous occasions. HK also thanks Professor E.J. Coyle of Purdue University for his valuable assistance from time to time.

This book was written in major parts during the sabbatical leave that HK spent at IIT Delhi. HK expresses his sincere thanks to Drs. N.C. Nigam, Director, IIT Delhi and C.S. Indulkar, Head, Department of Electrical Engineering, for making it possible for him to visit IIT Delhi. Dr. Nigam is a great inspiration to anyone he comes in contact with. It is a pleasure to know him.

Section 9.5 is taken from the reviews of one of our papers. The reviewer was anonymous at the time. He agreed to come forward upon mutual agreement after the paper was accepted for publication. He is B. Musicus of Bolt, Beranek and Newman. Thanks Bruce. We were fortunate to have you as a reviewer.

Two persons who have put this book together are Ms. D. Tysco and Mrs. M. Marano. All of us are thankful to them. They are genuine experts in their work. Mr. K. York is to be thanked for his technical support. We express our sincere thanks to Ms. N'Quavah R. Velazquez and her competent staff at the CRC Press for making our dream a reality.

The love, affection and care that we always receive from our families is gratefully acknowledged. Thanks Sarojini Agarwal, Ritu Krishna, Anuradha Aggarwal, Ravi Krishna Garg, Arun Kumar Aggarwal, Nalini Garg, Raj Krishna, Rajat Krishna, Tzu-Hsun Lin, Chien-Chuan Sun, Chen-Lan Hsieh, Dong-Shian Sun, Hsiu-Shian Lu, Shiuh-Ying Hsiao and Yu-Kai Sun for being there for us.

List of Abbreviations and Important Symbols

NTT:	Number-Theoretic-Transform
CNTT:	Complex NTT
CRT:	Chinese Remainder Theorem
CRT-I:	The Chinese Remainder Theorem for Integers
RNS:	Residue Number Systems
MRC:	Mixed Radix Conversion
BEX:	Base Extension
CRT-P:	The Chinese Remainder Theorem for Polynomials
AICE-CRT:	American Indian Chinese Extension of the Chinese Remainder Theorem for Polynomials
WSLC:	Word Sequence Length Constraint Problem
RRNS:	Redundant Residue Number Systems
RNS-PC:	Residue Number System Product Codes
gcd:	greatest common divisor
iff:	if and only if
ℓi:	linearly independent
ℓd:	linearly dependent
ℓhs :	left hand side
rhs :	right hand side
MNT:	Mersenne Number Transform
FNT:	Fermat Number Transform
LI:	Legrange Interpolation
DFT:	Discrete Fourier Transform
MDS:	Maximum-Distance-Separable
MULT:	Multiplication
ADD:	Addition
MMULT:	Modular Multiplication
MADD:	Modular Addition
ROM:	Read-only Memory
COMP:	Comparator
MUX:	Multiplexers
VSLI:	Very Large Scale Integration

Contents

List of Figures

List of Tables

Chapter 1

Introduction

Design of fast computationally efficient algorithms has been a major focus of research activity in digital signal processing. Such algorithms are based on the mathematical structure inherent in the computational task. In many instances, these algorithms also lead to suitable architectures for hardware implementation. The emphasis in most cases is on designing the algorithms and their respective implementations in a way so as to perform the required computations in the least amount of time. In this regard, parallel processing has also received much attention in the research community.

It is clear that number theory plays a fundamental role in the design of fast algorithms for performing some of the most computationally intensive tasks arising in digital signal processing. Examples of such tasks include cyclic and acyclic convolution, correlation, discrete Fourier transformation, and solving a system of linear equations. A flurry of research activity has resulted in numerous algorithms which exploit the number-theoretic properties of the discrete-time data and the indices of such data.

Residue number systems are closely related to number theory via the much celebrated Chinese Remainder Theorem, a method used to partition a large size computational task into a number of smaller but independent computational subtasks which may be performed in parallel. Residue number systems have also found numerous applications in digital signal processing. Both number theory and residue number systems are based on the mathematical properties and algebraic structure of the integers.

Along other lines, coding theory has flourished as an art and science to protect numeric data from errors, thereby improving the reliability and overall performance of data processing systems. However, by and large, with certain exceptions, coding theory is developed and studied as it applies to digital communication systems. This is evident from a large number

1

of texts on coding theory as it relates to the disciplines of information theory and communication theory. Exceptions to this are the applications of coding theory concepts to fault tolerance in computing systems. Some research has also focused on techniques for error correction in residue number systems where extensive use is made of the Chinese Remainder Theorem.

1.1 Outline

In this book, we are primarily interested in the design of fast computationally efficient algorithms for processing integers and integer valued data sequences. Specifically, we describe number theory based algorithms for computing the cyclic and acyclic convolution of sequences defined over a ring of integers. The problem of error control in residue number systems is studied from a coding theory view point with the objective of deriving fast algorithms and the respective hardware realizations for error correction and detection. A complete mathematical framework for error control in data vectors defined over a ring of integers is described. Certain known but rather old results in the theory of numbers dealing with linear congruences are also presented using mathematical terminology that the modern day researchers are expected to be familiar with.

A common thread present throughout this book is the Chinese Remainder Theorem in integer and polynomial forms. We expect the relationship between the Chinese Remainder Theorem and the various algorithms for processing integers and integer valued sequences to become widely known and used as a result of this effort.

1.2 The Organization

Chapter 2 is on elementary number theory for integers as it relates to the computational problems that arise in digital signal processing. This chapter is a collection of results most of which are already available in the literature. The focus in this chapter is on mathematical properties of integers crucial to the understanding of computational problems such as solving a system of linear equations, number-theoretic-transforms (NTTs), and cyclic and acyclic convolution of integer sequences.

The definitions and other material related to groups, rings and fields are presented. The Chinese Remainder Theorem for integers (CRT-I) is established. The CRT-I constitutes the heart of residue number systems (RNS) and we describe its various applications for simplifying computations. Other residue operations such as the mixed radix conversion (MRC) and base extension (BEX) and algorithms for performing them are also described. The problem of error control in RNS is stated.

Chapter 3 is on the algebra of polynomials defined on a field or a ring of integers. The Chinese Remainder Theorem for polynomials (CRT-P) is established. The NTTs are presented as a computational scheme for computing the convolution of integer valued sequences. Two special cases of NTTs, namely, the Mersenne number transforms (MNTs) and Fermat number transforms (FNTs) are also described briefly.

In Chapter 4, the CRT-P is extended for the case of polynomials defined on a ring of integers. We term this extension as the American-Indian-Chinese extension of the CRT-P (AICE-CRT) in honor of the genius in the ancient Chinese. Once again, AIC indicates the origin of the authors and the place where this work was completed. The AICE-CRT leads to fast algorithms for computing the cyclic and acyclic convolution of integer sequences. The NTTs may be interpreted as a special case of the AICE-CRT. Also, AICE-CRT solves the word sequence length constraint problem (WSLC) associated with the NTTs.

In Chapter 5, the AICE-CRT is described for processing data sequences defined over complex integer rings. Once again, the complex number-theoretic-transforms (CNTTs) may be interpreted as a special case of this AICE-CRT. The complex version of the AICE-CRT also solves the WSLC associated with CNTTs.

In Chapter 6 we bring together the number theory and the coding theory with the objective of developing a mathematical framework for error control in vectors defined over a ring of integers. The impetus for this work is provided by the numerous applications of number theory in certain computational problems arising in digital signal processing. Lack of an underlying Galois field presents a unique challenge to this framework. We note here that the most of the coding theory is developed in context to digital communication systems where the data is defined on a Galois field. We develop the theory and algorithms for single as well as multiple error correction and detection. Interestingly, NTTs play a fundamental role in the design and implementation of these error control techniques.

Chapters 7-11 are based on results derived recently by the authors in their research on a coding theory based mathematical framework for error control in RNS. Both techniques for error control, namely, redundant residue number systems (RRNS) and residue number system product codes (RNS-PC), are studied in depth. These chapters are a departure from the conventional way in which error control techniques for RNS are described in the literature. They provide new insights into the algebraic structure of these techniques as well as lead to new computationally efficient algorithms for decoding. These algorithms reduce the computational complexity of the previously known algorithms by at least an order of magnitude. It is worthwhile to mention here that with some exceptions, the focus of the entire technical literature on error control techniques for RNS is on single error correction while we consider these techniques in the more general sense of multiple error/erasure correction and multiple error detection.

PART I

COMPUTATIONAL NUMBER THEORY

Scalar and Polynomial Algebra

Chapter 2

Computational Number Theory

There is a fundamental need to define a number system over which all the computations are to be performed. In the present context we are interested in the study of integer based number systems and the computational algorithms that employ integer arithmetic. In this chapter, we define such number systems and the affiliated algebra of interest.

2.1 Groups, Rings, and Fields

Consider a set $\mathcal{G}$ containing M elements and a single valued arithmetic operation o defined between any two elements of $\mathcal{G}$. The set $\mathcal{G}$ is called a *group* if it satisfies the following properties:

A Closure: For arbitrarily chosen $a, b \in \mathcal{G}$, if $c = aob$, then $c \in \mathcal{G}$.

B Associative law: $(aob)oc = ao(boc)$ for all $a, b, c \in \mathcal{G}$.

C Identity: There exists an element $e \in \mathcal{G}$ such that $aoe = a$, for all $a \in \mathcal{G}$.

D Inverse: For every $a \in \mathcal{G}$, there exists $b \in \mathcal{G}$, such that $aob = e$.

Thus a group is a number system with one arithmetic operation. A group $\mathcal{G}$ that satisfies the commutative law ($aob = boa$) is called a commutative (or an Abelian) group. The number of elements, M, in a group $\mathcal{G}$ is called the order of $\mathcal{G}$. The arithmetic operation can be either addition (+) or multiplication ($\cdot$).

A set $\mathcal{R}$ containing M elements is called a *ring* if it satisfies the following properties:

A $\mathcal{R}$ is a commutative group under $+$. The additive identity (called zero) and the additive inverse of a $\in \mathcal{R}$ are denoted by 0 and $-a$ respectively.

B Closure: For arbitrarily chosen $a, b \in \mathcal{R}$, if $c = a \cdot b$, then $c \in \mathcal{R}$.

C Associative law: $(a \cdot b) \cdot c = a \cdot (b \cdot c)$ for all $a, b, c \in \mathcal{R}$.

D Distributive law: $a \cdot (b + c) = a \cdot b + a \cdot c$ for all $a, b, c \in \mathcal{R}$.

A ring $\mathcal{R}$ that satisfies the commutative law under '$\cdot$' is called a commutative ring. The number of elements, M, in a ring $\mathcal{R}$ is called the order of $\mathcal{R}$. Clearly, a ring is a number system with two arithmetic operations namely '$+$' and '$\cdot$'.

A field $\mathcal{F}$ is a complete number system over which all the arithmetic operations are defined. A field $\mathcal{F}$ is defined as a commutative ring that contains a multiplicative identity and a multiplicative inverse for every non-zero element in $\mathcal{F}$. The multiplicative identity (called one or unity) and the multiplicative inverse of $a \in \mathcal{F}$ are denoted by 1 and a^{-1} respectively. The number of elements, M, in a field $\mathcal{F}$ is called the order of $\mathcal{F}$.

In this book, the focus is on algorithms for processing integers and integer valued sequences. Consequently, we will limit our study to rings and fields having integers or polynomials having integer coefficients as their elements. Based on the above definitions, it is straightforward to verify that the set of integers is a commutative ring (hereafter called simply a ring), and the sets of rational, real, and complex numbers are fields. The order in each case is infinite. This is summarized in Table 2.1.

2.2   Elements of Number Theory

Given two positive integers A and M, we may divide A by M and write

$$A = QM + R, \ 0 \leq R < M. \tag{2.2.1}$$

Here the integers Q and R are called the quotient and the remainder respectively. Without any loss in generality we assume throughout this chapter that all the integers are non-negative in order to simplify our description and analysis. If $R = 0$, then M divides A or M is a factor of A. This is denoted by $M|A$. We also write (2.2.1) as

$$R \equiv A \mod M, \tag{2.2.2}$$

Table 2.1: Properties of Some Sets of Numbers

Set of Numbers	Ring	Field	Order
Integers	$\checkmark$		∞
Rational		$\checkmark$	∞
Real		$\checkmark$	∞
Complex		$\checkmark$	∞
$Z(M), M$: composite	$\checkmark$		M
$GF(p), p$: prime		$\checkmark$	p

or

$$R = <A>_M, \qquad (2.2.3)$$

and read it as "A is equivalent or congruent to R mod (for modulo) M." The integers M and R are also called modulus and residue respectively. The mod operation satisfies certain basic properties listed in the following:

A Given three integers A_1, A_2, and M such that $<A_1>_M = <A_2>_M$, then $M \mid (A_1 - A_2)$,

B $<A_1 + A_2>_M = <<A_1>_M + <A_2>_M>_M$,

C $<A_1 \cdot A_2>_M = <<A_1>_M \cdot <A_2>_M>_M$.

Given two integers A and M, the largest integer D that divides both the integers A and M is called the *greatest common divisor* (*gcd*) and is denoted by

$$D = (A, M). \qquad (2.2.4)$$

It may also be denoted by $gcd(A, M)$. If $D = 1$, then the integers A and M have no factors in common and are said to be *relatively prime* or *coprime*. The algorithm for determining D is called the Euclid's algorithm. It consists in repeated application of (2.2.1) as follows. Since $D|A$ and $D|M$, we have $D|R$ from (2.2.1). In other words,

$$(A, M) = (M, R), \ 0 \le R < M. \qquad (2.2.5)$$

Applying (2.2.1) to M and R, we get

$$M = Q_1 R + R_1, \ 0 \le R_1 < R. \qquad (2.2.6)$$

Once again $(M, R) = (R, R_1)$. We repeat this process and obtain a series of decreasing positive integers $R_2, R_3, \ldots$, such that

$$
\begin{aligned}
R &= Q_2 R_1 + R_2 & 0 \le R_2 < R_1 \\
R_1 &= Q_3 R_2 + R_3 & 0 \le R_3 < R_2 \\
&\;\;\vdots \qquad \cdots \\
R_{\ell-2} &= Q_\ell R_{\ell-1} + R_\ell & 0 \le R_\ell < R_{\ell-1} \\
R_{\ell-1} &= Q_{\ell+1} R_\ell.
\end{aligned}
\tag{2.2.7}
$$

Since $R > R_1 > R_2 \ldots \ge 0$, this process must terminate, say at the ℓth step, $R_{\ell+1} = 0$. Equations (2.2.1), (2.2.5)-(2.2.7) imply that $D = (A, M) = (M, R_1) = (R_1, R_2) = \ldots = (R_{\ell-1}, R_\ell) = R_\ell$. Also, (2.2.1), (2.2.6), and (2.2.7) may be written as,

$$
\begin{aligned}
R &= A - QM \\
R_1 &= M - Q_1 R \\
R_2 &= R - Q_2 R_1 \\
&\;\;\vdots \\
R_\ell &= R_{\ell-2} - Q_\ell R_{\ell-1}.
\end{aligned}
$$

Using the first two expressions from above, we may express R as a linear combination of A and M. Similarly, using the second and the third expressions, we may express R_2 as a linear combination of A and M and so on. Finally, $R_\ell (= D)$ can be written as

$$
D = R_\ell = \alpha A + \beta M.
\tag{2.2.8}
$$

The fact that *gcd* of two integers can be expressed as their linear combination will be used very frequently in solving systems of linear equations. If A and M are coprime, then (2.2.8) becomes

$$
1 = \alpha A + \beta M.
\tag{2.2.9}
$$

The definition of *gcd* and the above analysis can be extended to more than two integers, say $A_1, A_2, \ldots, A_N$. Once again,

$$
D = (A_1, A_2, \ldots, A_N),
\tag{2.2.10}
$$

and we can express D as

$$
D = \sum_{i=1}^{n} \alpha_i A_i.
\tag{2.2.11}
$$

Consider the linear congruence in an unknown X as

$$
AX \equiv C \mod M.
\tag{2.2.12}
$$

Equivalently, we wish to find two integers X and Y such that

$$AX + MY = C. \qquad (2.2.13)$$

Since $D|A$ and $D|M$, it divides the sum $AX + MY$ for all X and Y. Therefore, for a solution to (2.2.12) to exist, a necessary condition is $D|C$. On the other hand, if $D|C$ ($C = C_1 D$) then using (2.2.8), we may write

$$C = (C_1 \alpha)A + (C_1 \beta)M,$$

and the solution to (2.2.12) is obtained as $X = C_1 \alpha$. It may be easily verified that if $C_1 \alpha$ is a solution to (2.2.12), then so is $C_1 \alpha + (M/D)\theta, \theta = 0, \ldots, D - 1$. The D integers $C_1 \alpha + (M/D)\theta, \theta = 0, 1, \ldots, D - 1$ are incongruent mod M. In summary, the congruence in (2.2.12) has a solution if and only if (*iff*) $D|C, D = (A, M)$. Also, if $D|C$, then there are D distinct or incongruent solutions. Thus if $D = 1$, that is, the integers A and M are coprime, then a solution to (2.2.12) always exists and more over it is unique. Also, if $D = 1$ and $C = 1$, then X is called the inverse of A and is denoted by A^{-1}. We emphasize that A^{-1} exists whenever $(A, M) = 1$ and is unique.

An integer is called *prime* if it has only 1 and itself as its factors. Prime integers are denoted by p in this work. An integer is called *composite* if it is not prime. The fundamental theorem of arithmetic states that a composite integer M can be uniquely factored as (standard factorization of M),

$$M = \prod_{i=1}^{t} p_i^{\alpha_i} = \prod_{i=1}^{t} q_i, \qquad (2.2.14)$$

where p_i is the ith prime factor, α_i is a positive integer exponent and $q_i = p_i^{\alpha_i}$. For a prime $p, (A, p) = 1$ for $0 < A < p$ and therefore, a solution to the congruence

$$AX \equiv C \mod p$$

will always exist and be unique, $0 < A, C < p$.

2.2.1 Integer Rings and Fields

Consider the set of integers, $\mathcal{G} = \{0, 1, 2, \ldots, M - 1\}$ and let the arithmetic operations of '+' and '·' between two elements of $\mathcal{G}$ be defined mod M, that is, if $C = A + B$, then

$$C = <A + B>_M,$$

and if $C = A \cdot B$, then

$$C = <A \cdot B>_M .$$

Using the properties of the modulo operation, it can be verified that $\mathcal{G}$ forms a ring. In order to compute the multiplicative inverse of $A \in \mathcal{G}$, $A \neq 0$, we have to compute $X \in \mathcal{G}$ that satisfies

$$AX \equiv 1 \quad \mathrm{mod} \ M.$$

Based on the analysis for the linear congruence in (2.2.12), we see that X exists and is unique *iff* $(A, M) = 1$. This condition is trivially satisfied for all $A \in \mathcal{G}$ if $M = p$, a prime integer. However, for composite M, only those elements of $\mathcal{G}$ will have a multiplicative inverse that are coprime to M.

We conclude this discussion by stating that the set $\mathcal{G} = \{0, 1, \ldots, M - 1\}$ forms an integer ring, $Z(M)$, for composite M and a field, $GF(p)$, for $M = p$, a prime. The order of $Z(M)$ and $GF(p)$ is M and p respectively which are finite. Therefore, $Z(M)$ and $GF(p)$ are termed a finite integer ring and a finite integer field respectively. This is summarized in Table 2.1. The notation GF stands for 'Galois field'. We reaffirm that the arithmetic operations of '+' and '·' are defined mod M for $Z(M)$ and mod p for $GF(p)$.

2.3   Linear Congruences Over $Z(M)$

In its most general form, a system of linear congruences deals with the computation of a solution vector $\underline{X} = [X_1 X_2 \ldots X_k]$ having k unknowns to n congruences specified by

$$\sum_{i=1}^{k} A_{ij} X_i \equiv C_j \quad \mathrm{mod} \ M, \quad j = 1, 2, \ldots, n, \qquad (2.3.1)$$

or

$$\underline{X}\mathbf{A} \equiv \underline{C} \quad \mathrm{mod} \ M. \qquad (2.3.2)$$

Algebraically, there are two important parts in the study of the above system. They are

1. Does a solution to the system exist?

2. If it does, then how many incongruent solutions are there?

In this section, we study the system of linear congruences in (2.3.2) and answer the following questions:

1. Does a solution exist?

2. If it does, then do there exist more than (or exactly equal to) one solution?

Two matrices $\mathbf{A}$ and $\mathbf{B}$ are said to be equivalent, $\mathbf{A} \sim \mathbf{B}$, if there exist two unimodular matrices $\mathbf{J}$ and $\mathbf{K}$ having dimensions $(k \times k)$ and $(n \times n)$ respectively such that

$$\mathbf{B} = \mathbf{JAK}. \tag{2.3.3}$$

Since unimodular matrices can be expressed as products of elementary matrices, the mathematical properties associated with $\mathbf{A}$ remain unaltered upon performing the equivalence transformation in (2.3.3). Given the system of linear congruences in (2.3.2), we can equivalently solve the system

$$\underline{Y}\mathbf{B} \equiv \underline{C}^* \bmod M, \tag{2.3.4}$$

where $\underline{X} = \underline{Y}\mathbf{J}$ and $\underline{C}^* = \underline{C}\mathbf{K}$. The systems in (2.3.2) and (2.3.4) are equivalent in all respects, that is, every noncongruent solution to (2.3.4) gives rise to a noncongruent solution to (2.3.2) and vice-versa.

It is well known in number theory that every $(k \times n)$ integer matrix $\mathbf{A}$ is equivalent to a matrix of the form

$$\begin{bmatrix} D_1 & 0 & 0 & 0 & & 0 & 0 \\ 0 & D_1 D_2 & & & & & \\ & 0 & D_1 D_2 D_3 & & & & \\ \vdots & \vdots & \cdots & & & \ddots & \vdots \\ 0 & 0 & 0 & D_1 D_2 \cdots D_k & & 0 & 0 \end{bmatrix} (k \leq n), \tag{2.3.5}$$

$$\begin{bmatrix} D_1 & 0 & & 0 \\ 0 & D_1 D_2 & & 0 \\ 0 & & D_1 D_2 \cdots D_n & \\ 0 & \vdots & \cdots & 0 \\ 0 & & & 0 \end{bmatrix} (n \leq k), \tag{2.3.6}$$

where $D_i \geq 0$. The integers $D_i, i = 1, 2, \ldots$, satisfy the important property that the integer H_i defined as

$$H_i = D_1 \cdot (D_1 D_2) \cdots (D_1 D_2 \cdots D_i), \tag{2.3.7}$$

is the *gcd* of the determinant of all the minors of order i which appertain to $\mathbf{A}$. The integer H_i is called the invariant factor of $\mathbf{A}$ of order $i, i = 1, 2, \ldots$ The matrices of the form (2.3.5) or (2.3.6) are called the *normal forms of Smith*. The normal form of Smith for a matrix is unique. It is assumed that the matrix $\mathbf{A}$ is asyzygetic, that is, the determinants of the minor of the highest order are not all equal to zero.

Let us consider the case $n = k$. As stated above, there exist two unimodular matrices $\mathbf{J}$ and $\mathbf{K}$ of order n such that

$$\mathbf{JAK} = \begin{bmatrix} D_1 & & 0 \\ & D_1 D_2 & \\ 0 & & D_1 D_2 \cdots D_n \end{bmatrix} = \mathbf{D}. \qquad (2.3.8)$$

We now let $\underline{X} \equiv \underline{Y}\mathbf{J}$ and $\underline{C}^* = \underline{C}\mathbf{K}$. This leads to an equivalent but decoupled system of congruences

$$\underline{Y}\mathbf{D} = \underline{C}^* \bmod M \qquad (2.3.9)$$

or

$$D_1 D_2 \cdots D_i Y_i \equiv C_i^* \bmod M, \ i = 1, 2, \ldots, n. \qquad (2.3.10)$$

A solution to (2.3.9) or equivalently a solution to (2.3.2) exists *iff*

$$(D_1 D_2 \cdots D_i, M) \mid C_i^*, \ i = 1, 2, \ldots, n. \qquad (2.3.11)$$

The number of incongruent solutions is given by

$$D = \prod_{i=1}^{n} (D_1 D_2 \cdots D_i, M). \qquad (2.3.12)$$

Since $\mathbf{A}$ is a square matrix, the *gcd* of the determinant of all minors of order n (there is only one minor of order n, the matrix $\mathbf{A}$ itself) is the determinant of $\mathbf{A}$. Therefore, in this case,

$$det(\mathbf{A}) = \Delta = H_n = \prod_{i=1}^{n} D_1 D_2 \ldots D_i. \qquad (2.3.13)$$

If $(\Delta, M) = 1$, that is, Δ and M have no factors in common, then from (2.3.13), we have

$$(\prod_{i=1}^{n} D_1 D_2 \cdots D_i, M) = 1,$$

or

$$(D_1 D_2 \cdots D_i, M) = 1, i = 1, 2, \ldots, n. \qquad (2.3.14)$$

In such a case, (2.3.11) is satisfied trivially and the number of incongruent solutions from (2.3.12) is given by $D = 1$. Therefore, we can state that, if $(\Delta, M) = 1$, then a *solution* to the system of linear congruences in (2.3.1) *always exists and is unique.*

Also, for a unique solution to exist, that is, for $D = 1$, we have

$$(\prod_{i=1}^{n} D_1 D_2 \cdots D_i, M) = 1$$

or

$$(D_1 D_2 \cdots D_i, M) = 1, \quad i = 1, 2, \ldots, n. \tag{2.3.15}$$

Once again (2.3.11) is satisfied trivially. Combining (2.3.15) with (2.3.13), we get

$$(\Delta, M) = 1. \tag{2.3.16}$$

We can alternatively argue that if $(\Delta, M) > 1$, then there is at least one integer say D_j such that $(D_j, M) > 1$. Clearly, in this case either no solution will exist if (2.3.11) is not satisfied or the number of incongruent solutions $D > 1$ (refer to (2.3.12)) if (2.3.11) is satisfied. This discussion is summarized in the form of the following theorem.

THEOREM 2.1 *A unique solution to an $(n \times n)$ system of linear congruences $\underline{X} \mathbf{A} \equiv \underline{C}$ modulo M exists iff the determinant of the coefficient matrix $\mathbf{A}, \Delta$, satisfies $(\Delta, M) = 1$.*

We further observe that $\mathbf{A}^{-1}$ exists and is unique *iff* $(\Delta, M) = 1$.

2.3.1 Linear Independence and Vector Spaces Over $Z(M)$

Consider a collection of k vectors $\underline{V}_1, \underline{V}_2, \ldots \underline{V}_k$ each having n components in $Z(M)$. Vectors $\underline{V}_1, \underline{V}_2, \ldots, \underline{V}_k$ are said to be linearly independent (ℓi) over $Z(M)$, if there do not exist scalars $A_1, A_2, \ldots, A_k$ in $Z(M)$, not all zeros, such that

$$\sum_{i=1}^{k} A_i \underline{V}_i \equiv \underline{0} \mod M. \tag{2.3.17}$$

Otherwise, they are said to be linearly dependent (ℓd) over $Z(M)$. Here, addition of vectors and multiplication by a scalar are defined in the usual sense. Clearly $k \leq n$.

Given k ℓi vectors, a vector space over $Z(M)$ is defined as the set of vectors

$$V = \left\{ \underline{V} : \underline{V} = \sum_{i=1}^{k} A_i \underline{V}_i, \forall A_i \in Z(M), i = 1, 2 \ldots, k \right\}. \tag{2.3.18}$$

The vectors $\underline{V}_1, \underline{V}_2, \ldots, \underline{V}_k$ are said to span $\mathcal{V}$. The integer k is called the dimension of $\mathcal{V}$. This definition of a vector space over $Z(M)$ satisfies all properties of a vector space. A method to test ℓi of vectors $\underline{V}_1, \underline{V}_2, \ldots, \underline{V}_k$ can be derived by creating a $(k \times n)$ matrix $\mathbf{A}$ as

$$\mathbf{A} = [\underline{V}_1^t \ \underline{V}_2^t \cdots \underline{V}_k^t]^t, \qquad (2.3.19)$$

consisting of vectors $\underline{V}_1, \underline{V}_2, \ldots, \underline{V}_k$ as its rows. The superscript t denotes the operation of transposition. The vectors are ℓd *iff* there exists a non-trivial solution to the linear system

$$\underline{X}\mathbf{A} \equiv \underline{0} \bmod M. \qquad (2.3.20)$$

Otherwise, they are ℓi. Clearly $\underline{X} = \underline{0}$ is a trivial solution to (2.3.20). In this case, the equivalent system of congruences becomes

$$\underline{Y}\mathbf{D} \equiv \underline{0} \bmod M. \qquad (2.3.21)$$

Clearly, (2.3.21) is satisfied trivially as $\underline{X} = \underline{0}$ is a solution to these congruences. Therefore, for the vectors to be ℓi, we must have $D = 1$ and for the vectors to be ℓd, $D > 1$. The normal form of Smith, $\mathbf{D}$, is the one shown in (2.3.5) and the decoupled system is

$$D_1 D_2 \cdots D_i Y_i \equiv 0 \bmod M, i = 1, 2, \ldots, k. \qquad (2.3.22)$$

In this case, $D > 1$ *iff* $(D_1 D_2 \cdots D_i, M) > 1$ for at least one value of i. Let us say that for $i \leq j, (D_1 D_2 \cdots D_i, M) = 1$ while for $i > j, (D_1 D_2 \cdots D_i, M) > 1$. Then the invariant factors H_i are such that $(H_i, M) = 1, i \leq j$, and $(H_i, M) > 1, i > j$, thereby implying that all minors of $\mathbf{A}$ of order $j + 1$ or higher have *gcd* which are not coprime to M. The integer j will be termed as the *rank of the matrix* $\mathbf{A}$. For $D = 1$, we must have $(H_k, M) = 1$, that is, all minors of $\mathbf{A}$ of order k have *gcd* that are coprime to M. In other words, rank$(\mathbf{A}) = k$. If there is one square submatrix in $\mathbf{A}$ of order i having determinant Δ_i such that $(\Delta_i, M) = 1$, then $(H_i, M) = 1$. This discussion leads to the following lemma.

LEMMA 2.1 *For a given $(k \times n)$ matrix $\mathbf{A}, k \leq n$, if there is one square submatrix in $\mathbf{A}$ of order k having determinant Δ_k such that $(\Delta_k, M) = 1$, then rank$(\mathbf{A}) = k$ and the row vectors that constitute $\mathbf{A}$ are ℓi.*

We note that $(H_k, M) = 1$ is a necessary and sufficient condition for rank$(\mathbf{A}) = k$, while existence of a square submatrix in $\mathbf{A}$ of order k having determinant Δ_k such that $(\Delta_k, M) = 1$ is only a sufficient condition for rank$(\mathbf{A}) = k$.

2.3.2 Orthogonality and Null Spaces Over $Z(M)$

Two vectors $\underline{U}$ and $\underline{V}$ are said to be *orthogonal* over $Z(M)$ *iff* their inner product is 0,

$$\underline{V}\underline{U}^t \equiv 0 \mod M. \tag{2.3.23}$$

Given a k-dimensional vector space $\mathcal{V}$ over $Z(M)$ spanned by the vectors $\underline{V}_1, \underline{V}_2, \ldots, \underline{V}_k$ of length n, $k \leq n$, consider all those length n vectors $\underline{X}$ that are orthogonal to any arbitrarily chosen vector $\underline{V} \in \mathcal{V}$. Such a vector can be computed as a solution to the system of linear congruences,

$$\underline{X}\mathbf{A} \equiv \underline{0} \mod M,$$

where $\mathbf{A}$ is a $(n \times k)$ matrix obtained by arranging the given k ℓi vectors as its columns. Once again the above equation is satisfied trivially as $\underline{X} = \underline{0}$ is a solution to these congruences. The normal form of Smith, $\mathbf{D}$, is the one shown in (2.3.6) and the decoupled system is given by

$$D_1 D_2 \cdots D_i Y_i \equiv 0 \mod M, \; i = 1, 2, \ldots, k. \tag{2.3.24}$$

Since the row vectors that constitute $\mathbf{A}$ are ℓi, rank$(\mathbf{A}) = k$ or the invariant factor H_k satisfies $(H_k, M) = 1$. Since $(H_k, M) = 1$, we have $(D_1 D_2 \cdots D_i, M) = 1, i = 1, 2, \ldots, k$. Thus, the system of linear congruences in (2.3.24) has the solution

$$Y_i = 0, i = 1, 2, \ldots, k. \tag{2.3.25}$$

Clearly, the unknowns $Y_i, i = k+1, \ldots, n$ can take any one of M values, thereby, leading to the total number of incongruent solutions to the system of linear congruences in (2.3.24) to be $D = M^{n-k}$. It is easy to establish that these M^{n-k} vectors that are orthogonal to the k ℓi vectors that span $\mathcal{V}$ in turn form a vector space $\mathcal{U}$ of dimension $n - k$ over $Z(M)$. The vector space $\mathcal{U}$ is called the null space of $\mathcal{V}$. We note that given $\mathcal{V}$, its null space $\mathcal{U}$ is unique and vice versa.

2.3.3 Rank of Matrices Over $Z(M)$

This section extends the various results in sections 2.3.1 and 2.3.2 as they pertain to the present work.

Consider a $(k \times n)$ matrix $\mathbf{A}$, having integer elements defined over the ring $Z(M), M = q_1 q_2 \cdots q_t$. If A_{ij} is the (i,j)th element of $\mathbf{A}$, then $0 \leq A_{ij} < M$. In order to determine rank$(\mathbf{A})$ over $Z(M)$, we seek a non-zero solution to the linear system

$$\underline{X}\mathbf{A} \equiv \underline{0} \mod M. \tag{2.3.26}$$

If such a solution does not exist then $\text{rank}(\mathbf{A}) = k$ over $Z(M)$ otherwise $\text{rank}(\mathbf{A}) < k$. The linear system is equivalent to the system

$$\underline{Y}\mathbf{D} \equiv \underline{0} \bmod M, \tag{2.3.27}$$

where $\mathbf{D}$ is as stated in (2.3.5). For $\text{rank}(\mathbf{A}) = k$ over $Z(M)$, we have the necessary and sufficient condition that $(H_k, M) = 1$, H_k being the kth invariant factor of $\mathbf{A}$. If $(H_k, M) \neq 1$, then $\text{rank}(\mathbf{A}) < k$.

Given a matrix $\mathbf{A}$ over $Z(M)$, consider the matrix $\mathbf{A}^{(e)}$ defined as

$$\mathbf{A}^{(e)} \equiv \mathbf{A} \bmod q_e. \tag{2.3.28}$$

Clearly, the (i,j)th element of $\mathbf{A}^{(e)}$, A_{ij}^e, can be expressed as

$$A_{ij}^e = A_{ij} + \alpha_{ij}q_e,$$

and the determinants of all the $(k \times k)$ submatrices of $\mathbf{A}^{(e)}$ can be expressed in terms of the determinants of the corresponding $(k \times k)$ submatrices of $\mathbf{A}$ as

$$\Delta_i^e = \Delta_i + L_i q_e, i = 1, 2, \ldots \tag{2.3.29}$$

The kth invariant factor of $\mathbf{A}^e$ is given by

$$H_k^e = (\Delta_1^e, \Delta_2^e, \ldots). \tag{2.3.30}$$

If $(H_k, M) = 1$, then $(H_k, q_j) = 1, j = 1, 2, \ldots, t$. If $(H_k, M) > 1$, then two cases follow, namely $(H_k, q_e) = 1$ or $(H_k, q_e) > 1$. In the following, we first pursue the case $(H_k, q_e) > 1$ to show that in this case $(H_k^e, q_e) > 1$ as well.

Case 1 $(H_k, q_e) > 1$.
In this case, H_k and q_e have a common factor, say F. Since $H_k = (\Delta_1, \Delta_2, \ldots)$, F is also a factor of $\Delta_1, \Delta_2, \ldots$. Since F is a factor of $\Delta_1, \Delta_2, \ldots$, and q_e, we see that F is also a factor of $\Delta_1^e, \Delta_2^e, \ldots$ from (2.3.29). Clearly, F is also a factor of H_k^e. Consequently, $(H_k^e, q_e) > 1$, or $\text{rank}(\mathbf{A}^{(e)}) < k$ over $Z(q_e)$.

Case 2 $(H_k, q_e) = 1$.
In this case, we can write

$$AH_k + Bq_e = 1. \tag{2.3.31}$$

Also, since $H_k = (\Delta_1, \Delta_2, \ldots)$, there exist integers $C_1, C_2, \ldots$, such that

$$C_1\Delta_1 + C_2\Delta_2 + \ldots = H_k. \tag{2.3.32}$$

Substituting for Δ_i from (2.3.29) into (2.3.32), we get

$$\sum_i C_i(\Delta_i^e - L_i q_e) = H_k. \tag{2.3.33}$$

Substituting (2.3.33) into (2.3.31) we get

$$A \sum_i C_i(\Delta_i^e - L_i q_e) + B q_e = 1.$$

Rearranging various terms, we get

$$\sum_i A C_i \Delta_i^e + B' q_e = 1$$

$$B' = -\sum_i A C_i L_i + B. \tag{2.3.34}$$

It is clear from (2.3.34) that

$$(\Delta_1^e, \Delta_2^e, \ldots, q_e) = 1,$$

or

$$((\Delta_1^e, \ldots), q_e) = 1,$$

or

$$(H_k^e, q_e) = 1. \tag{2.3.35}$$

The last expression implies that $\operatorname{rank}(\mathbf{A}^{(e)}) = k$ over $Z(q_e)$. This analysis can be summarized in the form of a lemma as follows

LEMMA 2.2 *If the rank of a $(k \times n)$ matrix $\mathbf{A}$ over $Z(M)$ is k, then the rank of the matrix $\mathbf{A}^{(j)} \equiv \mathbf{A}$ modulo q_j, $M = q_1 q_2 \cdots q_t$, $j = 1, 2, \ldots, t$, is k. Also, if $\operatorname{rank}(\mathbf{A}) < k$, then there is at least one value of j, say $j = e$, such that $\operatorname{rank}(\mathbf{A}^{(e)}) < k$.*

A special case of this lemma corresponds to $t = 1$, that is $M = q = p^\alpha$. For this special case, we restate the lemma as

LEMMA 2.3 *If the rank of a $(k \times n)$ matrix $\mathbf{A}$ over $Z(q), q = p^\alpha$ is k, then the rank of the matrix $\mathbf{A}^{(j)} \equiv \mathbf{A}$ mod p^j, $j \leq \alpha$ is k. Also, if $\operatorname{rank}(\mathbf{A}) < k$, then the $\operatorname{rank}(\mathbf{A}^{(j)}) < k$.*

Now, consider a $(k \times n)$ matrix $\mathbf{A}(k \leq n)$ having integer elements defined over the field $GF(p)$, that is, if A_{ij} is the (i, j)th element of $\mathbf{A}$, then $0 \leq$

$A_{ij} < p$. We establish the relationship between rank($\mathbf{A}$) over $GF(p)$ and rank($\mathbf{A}$) over $Z(q), q = p^\alpha$.

If rank($\mathbf{A}$) $= k$ over $GF(p)$, then the kth invariant factor of $\mathbf{A}$, H_k, is such that $(H_k, p) = 1$, or H_k has no factors in common with p. Clearly, in such a case, H_k has no factor in common with p^α either, thereby leading to $(H_k, p^\alpha) = 1$. As a result, rank($\mathbf{A}$) $= k$ over $Z(q)$.

If rank($\mathbf{A}$) $< k$ over $GF(p)$, then $(H_k, p) > 1$. Since p is prime $(H_k, p) > 1$ implies that $(H_k, p) = p$. Once again, $(H_k, p^\alpha) > 1$, or rank($\mathbf{A}$) $< k$ over $Z(q)$. In summary, we can state the following.

LEMMA 2.4 *If the rank of a $(k \times n)$ matrix over $\mathbf{A}$ over $GF(p)$ is k, then rank($\mathbf{A}$) $= k$ over $Z(q), q = p^\alpha$. Also, if rank($\mathbf{A}$) $< k$ over $GF(p)$, then rank($\mathbf{A}$) $< k$ over $Z(q)$ as well.*

The Lemma 2.4 may also be treated as a corollary to the Lemma 2.3.

2.4   The Chinese Remainder Theorem for Integers

The Chinese Remainder Theorem in two versions (one for integers and the second for polynomials) constitutes the heart and soul of the majority of fast algorithms in digital signal processing. In this section, the Chinese Remainder Theorem for integers (CRT-I) is stated and proved. The second version will be presented in Chapter 3.

Consider the following extension of the linear congruence in (2.2.12) in an unknown X,

$$X \equiv x_i \bmod m_i, \quad i = 1, 2, \ldots, k. \tag{2.4.1}$$

The integers $x_1, x_2, \ldots, x_k$ are known and the moduli $m_1, m_2, \ldots, m_k$ are pairwise coprime integers, that is $(m_i, m_j) = 1, i \neq j; i,j = 1,2,\ldots,k$.

THEOREM 2.2 *There exists a unique solution $X \in Z(M)$, $M = \prod_{i=1}^{k} m_i$ to the k congruences in (2.4.1). It can be computed as*

$$X \equiv \sum_{j=1}^{k} x_j T_j \left(\frac{M}{m_j} \right) \bmod M, \tag{2.4.2}$$

where the integers T_j are precomputed using the congruences

$$T_j \left(\frac{M}{m_j} \right) \equiv 1 \bmod m_j. \tag{2.4.3}$$

PROOF. Since the moduli $m_1, m_2, \ldots, m_k$ are pairwise relatively coprime and $(M/m_j) = \prod_{\substack{i=1 \\ i \neq j}}^{k} m_i$, we have $(M/m_j, m_j) = 1$. Therefore, the congruence in (2.4.3) can always be solved for T_j. The integers T_j are unique and have the property $(T_j, m_j) = 1, j = 1, 2, \ldots, k$. It is clear from the form of X in (2.4.2) that $X \in Z(M)$. Taking mod m_i on both sides of (2.4.2), we get

$$
\begin{aligned}
X \bmod m_i &\equiv \left(\sum_{j=1}^{k} x_j T_j \left(\frac{M}{m_j} \right) \right) \bmod m_i \\
&\equiv \left(\sum_{j=1}^{k} (x_j T_j \left(\frac{M}{m_j} \right)) \bmod m_i \right) \bmod m_i \\
&\equiv x_i T_i \left(\frac{M}{m_i} \right) \bmod m_i \\
&\equiv x_i \bmod m_i.
\end{aligned}
$$

The last step is based on (2.4.3). Thus there is one solution to the k congruences in (2.4.1) as expressed in (2.4.2). Now we prove the uniqueness of the solution. Suppose there are two solutions to (2.4.1) in $Z(M)$, say X and X'. Without any loss in generality, we assume that $X > X'$. Also, since X and X' belong to $Z(M)$, they satisfy $0 \leq X, X' < M$. Substituting X and X' in (2.4.1), we get

$$X \equiv x_i \bmod m_i \tag{2.4.4}$$

and

$$X' \equiv x_i \bmod m_i, \ i = 1, 2, \ldots, k \tag{2.4.5}$$

Subtracting (2.4.5) from (2.4.4) leads to

$$X - X' \equiv 0 \bmod m_i, \ i = 1, 2, \ldots, k.$$

This implies that $m_i | (X - X'), i = 1, 2, \ldots, k$. Since the moduli are pairwise relatively coprime, we have $(m_1 m_2 \cdots m_k) | (X - X')$ or $M | (X - X')$. This is not possible as $0 \leq X - X' < M$. Therefore, $X - X' = 0$ or $X = X'$, thereby implying that the solution is unique.

The CRT-I forms the basis for representing an integer $X \in Z(M)$ as a residue vector $\underline{X} = [x_1 x_2 \ldots x_k], x_i \in Z(m_i)$, where $x_i \equiv X \bmod m_i$. We express this relationship as

$$X \leftrightarrow \underline{X} = [x_1 x_2 \ldots x_k]. \tag{2.4.6}$$

Also, given two integers $X, Y \in Z(M)$ and their corresponding residue vectors $\underline{X} = [x_1 x_2 \ldots x_k]$, $\underline{Y} = [y_1 y_2 \ldots y_k]$, then

$$X + Y \leftrightarrow \underline{X} + \underline{Y} = [x_1 + y_1 \quad x_2 + y_2 \ldots x_k + y_k] \qquad (2.4.7)$$

$$X \cdot Y \leftrightarrow \underline{X} \cdot \underline{Y} = [x_1 \cdot y_1 \quad x_2 \cdot y_2 \ldots x_k \cdot y_k]. \qquad (2.4.8)$$

The above two equations must be interpreted carefully. The sum and the product $(X + Y)$ and $(X \cdot Y)$ on the left-hand-side (*lhs*) are defined over $Z(M)$, that is, they are performed mod M, while the sum and the product in the ith component on the right-hand-side (*rhs*) are defined over $Z(m_i)$. This follows directly from the properties of the mod operation and the definitions of $\underline{X}$ and $\underline{Y}$.

The CRT-I is a fundamental method of partitioning a large ring $Z(M)$ into a number of smaller but independent rings $Z(m_i)$ over which the various computations may be performed in parallel. This is the *direct sum property* of the CRT-I and is expressed as

$$Z(M) \approx Z(m_1) \oplus Z(m_2) \oplus \cdots \oplus Z(m_k). \qquad (2.4.9)$$

The CRT-I is used extensively to convert one-dimensional data sequences into two or more dimensional data sequences by applying its inherent mapping properties to the index of the sequence. The data sequence itself may or may not be integer valued. Consider the following example.

EXAMPLE 2.1 For a data sequence of length $n = 15$, the index i goes from 0 to 14, that is, $i \in Z(n)$. Since $15 = 3 \cdot 5$ and $(3,5) = 1$, let $n_1 = 3$ and $n_2 = 5$. We apply the CRT-I to convert the one-dimensional data sequence to a two-dimensional data sequence having indices (i_1, i_2) such that $i_1 \equiv i \bmod n_1$ and $i_2 \equiv i \bmod n_2$. The complete mapping is given by

$$
\begin{array}{lllll}
0 \to (0\ 0) & 1 \to (1\ 1) & 2 \to (2\ 2) & 3 \to (0\ 3) & 4 \to (1\ 4) \\
5 \to (2\ 0) & 6 \to (0\ 1) & 7 \to (1\ 2) & 8 \to (2\ 3) & 9 \to (0\ 4) \\
10 \to (1\ 0) & 11 \to (2\ 1) & 12 \to (0\ 2) & 13 \to (1\ 4) & 14 \to (2\ 4)
\end{array}
$$

In this book, we are primarily interested in the direct sum property of the CRT-I as stated in (2.4.9) and its applications to processing integer valued scalars and data sequences defined over $Z(M)$.

2.5   Residue Number Systems

The direct sum property of CRT-I in (2.4.9) and the properties of the residue representation in (2.4.7) and (2.4.8) provide means for realizing carry-free arithmetic over $Z(M)$. This is called a residue number system (RNS).

An RNS is defined by a set of k pairwise relatively prime integers $m_1, m_2, \ldots, m_k$. Let $M = \prod_{i=1}^{k} m_i$. The interval from 0 to $M - 1$, denoted by $[0, M)$ represents the useful range of the RNS. Any positive integer $X \in Z(M)$ (obviously, $0 \le X < M$) is uniquely represented by the k-tuple, $X \leftrightarrow \underline{X} = [x_1 x_2 \ldots x_k]$ of its residue with respect to the moduli $m_1, m_2, \ldots, m_k$, where $x_i \equiv X \bmod m_i$.

2.5.1 Other Residue Operations

Given a residue vector $\underline{X} = [x_1 x_2 \ldots x_k]$, a frequently used operation is to compute the residue $x_j \equiv X(\bmod\ m_j)$. Here, $(m_i, m_j) = 1$, $i = 1,2,\ldots,k$. This operation is called the base extension (BEX) operation. It consists in finding the residue digit for a new modulus, given the residue digits relative to a set of moduli. The CRT-I provides a method for BEX. However, the CRT-I based method for BEX operation may not be practical as it requires arithmetic operations modulo M on large valued integers. In order to avoid processing large valued integers, the mixed radix conversion (MRC) is an useful method for BEX. The MRC is an operation to represent the integer X in the form

$$X = \sum_{\ell=1}^{k} a_\ell \prod_{i=1}^{\ell-1} m_i, \qquad (2.5.1)$$

where $0 \le a_\ell < m_\ell, \ell = 1, 2, \ldots, k$, and $\prod_{i=1}^{0} m_i = 1$. The digits $a_1, a_2, \ldots, a_k$ are called the mixed radix digits and are computed using the congruences

$$
\begin{aligned}
a_1 &\equiv x_1(\bmod\ m_1), \\
a_2 &\equiv m_1^{-1}(x_2 - x_1)(\bmod\ m_2), \\
a_3 &\equiv m_2^{-1}(m_1^{-1}(x_3 - a_1) - a_2)(\bmod\ m_3) \\
&\vdots
\end{aligned}
$$

Given a residue vector $\underline{X}$ corresponding to an integer $X, X \in Z(M)$, the residue $x_j \equiv X(\bmod\ m_j)$ can be computed using

$$x_j \equiv \sum_{i=1}^{k} a_\ell \prod_{i=1}^{\ell-1} m_i\ (\bmod\ m_j). \qquad (2.5.2)$$

The advantage of using the MRC based method for BEX is that the computation of x_j can be overlapped with the computation of the mixed radix digits by making use of the mixed radix digits as soon as they become available.

2.6   The Problem of Error Control

Error detection and/or correction is required for the protection of the arithmetic and data transmission operations in digital processors as well as in the general purpose computers. An RNS is made fault-tolerant by introducing redundancy in its form. An RNS has two important properties especially useful for digital processing which may be exploited to provide fault tolerant capability. These properties are the carry-free arithmetic and the lack of ordered significance among residue digits which result from the nonweighted nature of the RNS. These two properties play a crucial role in the design of a fault tolerant RNS. The carry-free arithmetic implies that error generated in a certain residue digit during arithmetic operations or due to transmission noise, remains confined to its original digit position. The second property implies that any erroneous digit can be discarded without affecting the result, provided that enough redundancy is added to the original RNS. Therefore, the residue-number-arithmetic based digital filters can be suitably implemented using multiprocessor systems with fault tolerant capability. In the following, we describe two methods for adding redundancy to an RNS.

Redundant Residue Number Systems
A redundant residue number system (RRNS) is obtained by adjoining a set of $n - k$ additional moduli $m_{k+1}, \ldots, m_n$, called the redundant moduli to form n pairwise relatively coprime moduli. An integer $X, X \in Z(M)$, is represented as the residue vector $\underline{X}, X \leftrightarrow \underline{X} = [x_1 x_2 \cdots x_k x_{k+1} \cdots x_n]$. The range [0,M) is called the *legitimate range* and the range $[M, M \cdot M_r)$ is called the *illegitimate range*, where

$$M_r = \prod_{i=k+1}^{n} m_i. \tag{2.6.1}$$

$[0, M \cdot M_r)$ is the total range of the RRNS. Let $M_T = M \cdot M_r$. Given an arbitrary residue vector $\underline{X} = [x_1 x_2 \ldots x_n]$, the corresponding integer $X \in Z(M_T)$ can be computed using the CRT-I as

$$\underline{X} \leftrightarrow X \equiv \sum_{i=1}^{n} x_i T_i'(M M_r / m_i) \bmod (M_T). \tag{2.6.2}$$

The integer T'_i are precomputed by solving the congruences

$$T'_i(M_T/m_i) \equiv 1 \bmod m_i, \ i = 1, 2, \ldots, n. \tag{2.6.3}$$

We define the RRNS as an (n, k) RRNS code. This code is a systematic code in which every code-integer $\underline{X}$ is divided into two parts, the information part and the parity part. The information part consists of the first k residue digits and the parity part consists of the remaining $n - k$ redundant parity check digits. The parity check digits are residue digits derived from the information residue digits.

Residue Number System Product Code

The residue number system product code (RNS-PC) is a nonsystematic code and is defined in the RNS as follows. All the code-integers X in RNS-PC are integers of the form $AX', 0 \leq AX' < M_T$, where A is a given positive integer termed the code generator. In other words, an integer $X \in Z(M_T)$ is a code-integer *iff* $X \equiv 0 \bmod A$. Associated with each code-integer X is the residue vector $\underline{X} = [x_1 x_2 \cdots x_n]$ corresponding to the moduli $m_1, m_2, \ldots, m_n$. It is clear that the integer X' is in the range $[0, R)$, $R = [M_T/A] + 1$, where $[\alpha]$ denotes the integral part of α. $[0, R)$ is the useful range of the RNS-PC.

In a multiprocessor environment, where each component of the code-integer is processed on one processor, if the ith processor is faulty, it will introduce an error in the ith component, x_i, of the code-integer $\underline{X}$. The output of the ith processor, therefore, will be y_i, where,

$$y_i \equiv x_i + e_i \ (\bmod \ m_i), 0 \leq e_i < m_i, \tag{2.6.4}$$

e_i being the error value. If $e_i = 0$, then the ith processor is fault free.

Given the mathematical structure of the technique used for error control, we will focus on fast algorithms for error detection and correction. Error correction consists of computing the location of the error (ith processor) and the error value e_i while error detection consists of detecting the presence of errors ($e_i = 0$ or not). This will be studied extensively in chapters 7-11.

NOTES

We have used several excellent texts in number theory to put together this chapter. The books by McClellan and Radar [1979] and Nussbaumer [1981] present number theory as it relates to the field of digital signal processing. We have used Peterson and Weldon [1972], Lin and Costello [1983] and MacWilliams and Sloane [1977] extensively to learn the algebra of finite fields and its applications in error control coding.

Among other books on number theory, the books by Dickson [1920], MacDuffee [1940], Hua [1988] and Shapiro [1983] come to our minds as very valuable sources of information. Section 2.3 on linear congruences over a finite integer ring is based on sections 14-5 and 14-6 of Hua [1982] and earlier works of Smith [1861,1873].

A word of caution is in order here for the material presented in sections 2.3.1, 2.3.2 and 2.3.3. Speaking from a strict mathematical point of view, the notions of linear independence, vector spaces, orthogonality and rank of a matrix exist only for vectors defined over a field. Analogous to vector spaces over a field, the term 'module' is used for vectors defined over a ring. However, in all of our analysis, a module may be treated as a natural extension of a vector space. Consequently, we have intentionally chosen to retain the nomenclature of the matrix algebra over a field for the matrix algebra over a ring. We hope that it does not confuse the readers. In fact, we hope that this choice will assist the readers in their understanding of this material.

The material on the residue number systems is based on Szabo and Tanaka [1967] and Soderstrand et al. [1986]. The Chinese Remainder Theorem for integers can be found in any text on number theory including the ones mentioned here. Our presentation was particularly influenced by Blahut [1984], McClellan and Radar [1979], Nussbaumer [1981] and Soderstrand et al. [1986].

2.7   Bibliography

[2.1] J.H. McClellan and C.M. Radar, *Number Theory in Digital Signal Processing*, Prentice Hall, Inc., 1979.

[2.2] H.J. Nussbaumer, *Fast Fourier Transform and Convolution Algorithms*, Springer Verlag, 1981.

[2.3] W.W. Peterson and E.J. Weldon, Jr., *Error Correcting Codes*, IInd Edition, MIT Press, 1972.

[2.4] S. Lin and D.J. Costello, Jr., *Error Control Coding: Fundamentals and Applications*, Prentice Hall, 1983.

[2.5] F.J. MacWilliams and N.J.A. Sloane, *The Theory of Error Correcting Codes*, North Holland Publishing Co., 1977.

[2.6] L.E. Dickson, *History of the Theory of Numbers*, vol. I-V, Carnegie Institution of Washington, 1920.

[2.7] C.C. MacDuffee, *An Introduction to Abstract Algebra,* John Wiley and Sons, Inc., 1940.

[2.8] L.-K. Hua, *Introduction to Number Theory,* Springer Verlag, 1982.

[2.9] H.N. Shapiro, *Introduction to the Theory of Numbers,* John Wiley and Sons, Inc., 1983.

[2.10] H.J.S. Smith, "On Systems of Linear Indeterminate Equations and Congruences," *Philosophy Transactions Royal Society of London,* A 151, pp. 293-326, 1861.

[2.11] H.J.S. Smith, "On the Arithmetical Invariant of a Rectangular Matrix of Which the Constituents are Integral Numbers," *Proceedings London Mathematical Society,* 4, pp. 236-249, 1873.

[2.12] N.S. Szabo and R.I. Tanaka, *Residue Arithmetic and its Applications to Computer Technology,* McGraw Hill Co., 1967.

[2.13] M.S. Soderstrand, W.K. Jenkins, G.A. Jullien, and F.J. Taylor, *Residue Number System Arithmetic: Modern Applications in Digital Signal Processing,* IEEE Press, 1986.

[2.14] R. E. Blahut, *Fast Algorithms for Digital Signal Processing,* Addison-Wesley Publishing Co., 1984.

Chapter 3

Polynomial Algebra

In many digital signal processing applications, one is required to process data sequences defined over a given number system. This number system may either be a field or a ring of finite or infinite order. In this chapter, we present some fundamental results in the theory of polynomial algebra. The emphasis, of course, is on data sequences and the corresponding polynomials defined over finite fields and finite rings. We observe here that all the results valid for polynomials defined over finite fields are also valid for polynomials defined over infinite fields. The order of the field does not play a very critical role in the polynomial algebra. The important distinction to be drawn is between the polynomials defined over fields and rings. Many basic results in the polynomial algebra over a field are no longer valid for the polynomial algebra over a ring.

The exposition in this chapter is similar to the exposition in the Chapter 2.

DEFINITION 3.1 Given a field $\mathcal{F}$ or a ring $\mathcal{R}$, a polynomial $A(u)$ is defined as

$$A(u) = A_0 + A_1 u + \ldots + A_n u^n, \tag{3.0.1}$$

where u denotes an indeterminate quantity over $\mathcal{F}$ or $\mathcal{R}$, and the coefficients $A_0, A_1, \ldots, A_n$ belong to either $\mathcal{F}$ (for a polynomial defined over $\mathcal{F}$) or $\mathcal{R}$ (for a polynomial defined over $\mathcal{R}$).

The degree of $A(u)$, $deg(A(u))$, is the largest integer i for which $A_i \neq 0$. All the elements of $\mathcal{F}$ or $\mathcal{R}$ can be expressed as polynomials of degree 0 and are termed as scalars. The arithmetic operations of addition (+) and

29

multiplication ($\cdot$) between two polynomials are defined as

$$A(u) + B(u) = \sum_i (A_i + B_i)u^i, \qquad (3.0.2)$$

and

$$A(u) \cdot B(u) = \sum_\ell (\sum_j A_j B_{\ell-j})u^\ell, \qquad (3.0.3)$$

respectively. Here, '+' and '$\cdot$' in the *rhs* of (3.0.2) and (3.0.3) are defined over $\mathcal{F}$ or $\mathcal{R}$.

Given a data sequence of length $n + 1$ expressed as a vector of length $n + 1$, that is, $\underline{A} = [A_0 \ A_1 \ \dots \ A_n]$, it can equivalently be represented as a polynomial $A(u)$ as in (3.0.1). This relationship is expressed as $\underline{A} \leftrightarrow A(u)$. The polynomial $A(u)$ is also known as the generating function of the sequence A_i, $i = 0, 1, \dots, n$.

One of the most widely occurring operations in digital signal processing is the computation of acyclic convolution of two data sequences. Given two data sequences $A_i, i = 0, 1, \dots, k - 1$ of length k and $B_j, j = 0, 1, \dots, d - 1$ of length d, the acyclic convolution consists in the computation of the sequence $C_\ell, \ell = 0, 1, \dots, n - 1$, $n = k + d - 1$, defined as

$$C_\ell = \sum_{\substack{\ell=i+j \\ 0 \le i \le k-1 \\ 0 \le j \le d-1}} A_i B_j, \quad \ell = 0, 1, \dots, n - 1. \qquad (3.0.4)$$

If $C(u)$ is the generating function of the sequence $C_\ell, \ell = 0, 1, \dots, n - 1$, then comparing (3.0.4) with (3.0.3), we may write

$$C(u) = A(u)B(u), \qquad (3.0.5)$$

where $A(u)$ and $B(u)$ are the generating functions of the sequences A_i, $i = 0, 1, \dots, k - 1$ and $B_j, j = 0, 1, \dots, d - 1$. The degree of the polynomials $A(u)$, $B(u)$, and $C(u)$ can be at most $k - 1$, $d - 1$, and $n - 1$ respectively; $n = k + d - 1$.

The fact that acyclic convolution of two sequences can equivalently be expressed as a polynomial product has led to a flurry of research activity to derive computationally efficient algorithms for computing the product of two polynomials. The main idea is to exploit the mathematical properties of the number system over which the indices of the sequences (integers) and the sequences themselves are defined. In this regard, the focus of this book is on sequences defined over $Z(M)$.

Before proceeding further with design of algorithms for computing the polynomial product, one has to understand the essential elements of the algebra of polynomials. These are established in the following.

Consider the set of all polynomials defined over a field $\mathcal{F}$ or a ring $\mathcal{R}$. It can be easily shown that under the arithmetic operations of '+' and '·' between two polynomials as defined in (3.0.2) and (3.0.3) respectively, this set forms a ring. In the next section, we describe the algebra of ring of polynomials defined over a field $\mathcal{F}$. The algebra of ring of polynomials defined over a ring will be described in Section 3.5.

3.1 Algebra of Polynomials Over a Field

All the polynomials in this section are defined over a field $\mathcal{F}$. Given two polynomials $A(u)$ and $B(u)$ of degree k and d respectively ($A_k \neq 0, B_d \neq 0$), if $C(u) = A(u) \cdot B(u)$, then $deg(C(u)) = k + d$ as $C_{k+d} = A_k \cdot B_d \neq 0$. This leads to the following theorem

THEOREM 3.1 *If $C(u) = A(u) \cdot B(u)$, then $deg(C(u)) = deg(A(u)) + deg(B(u))$.*

Given two polynomials $A(u)$ and $M(u)$, we may divide $A(u)$ by $M(u)$ and write

$$A(u) = Q(u)M(u) + R(u), \ 0 \leq deg(R(u)) < deg(M(u)). \qquad (3.1.1)$$

The polynomials $Q(u)$ and $R(u)$ are called the quotient and the remainder respectively. They are unique as if they are non-unique then there exist $Q_1(u), Q_2(u)$ such that $Q_1(u) \neq Q_2(u)$, and $R_1(u), R_2(u)$ such that $R_1(u) \neq R_2(u), 0 \leq deg(R_1(u)), deg(R_2(u)) < deg(M(u))$ that satisfy

$$A(u) = Q_1(u)M(u) + R_1(u) = Q_2(u)M(u) + R_2(u)$$

or

$$[Q_1(u) - Q_2(u)]M(u) = R_2(u) - R_1(u).$$

This is not possible as the degree of the non-zero polynomial in the *rhs* is less than $deg(M(u))$ while the degree of the non-zero polynomial in the *lhs* is greater than $deg(M(u))$.

If $R(u) = 0$ in (3.1.1), then $M(u)$ divides $A(u)$ or $M(u)$ is a factor of $A(u)$. This is denoted by $M(u)|A(u)$. We also write (3.1.1) as

$$R(u) \equiv A(u) \bmod M(u) \qquad (3.1.2)$$

or

$$R(u) = < A(u) >_{M(u)} \qquad (3.1.3)$$

and read it as "$A(u)$ is congruent to $R(u)$ mod $M(u)$." The polynomials $M(u)$ and $R(u)$ are also called the modulus and the residue polynomials respectively. The polynomial modulo operation satisfies certain basic properties listed in the following:

A Given three polynomials $A_1(u)$, $A_2(u)$, and $M(u)$ such that $< A_1(u) >_{M(u)} = < A_2(u) >_{M(u)}$, then $M(u) \mid (A_1(u)-A_2(u))$,

B $< A_1(u)+A_2(u) >_{M(u)} = << A_1(u) >_{M(u)} + < A_2(u) >_{M(u)} >_{M(u)}$,

C $< A_1(u) \cdot A_2(u) >_{M(u)} = << A_1(u)>_{M(u)} \cdot < A_2(u) >_{M(u)} >_{M(u)}$.

Given two polynomials $A(u)$ and $M(u)$, the largest degree polynomial $D(u)$ that divides both $A(u)$ and $M(u)$ is called the *gcd* and is denoted by

$$D(u) = (A(u), M(u)). \qquad (3.1.4)$$

If $D(u)|A(u)$, then $fD(u)|A(u)$, for all $f \in \mathcal{F}$, $f \neq 0$, and vice versa. Therefore, all the factor polynomials can be specified only within a scalar multiple of the field. Consequently, we will assume that the *gcd* polynomial is such that the coefficient of its highest degree is equal to 1. Polynomials that satisfy this property are called *monic*.

DEFINITION 3.2 A polynomial $A(u)$ is called *monic* if the coefficient of its highest degree is equal to 1.

If $D(u) = 1$, then $A(u)$ and $M(u)$ have no factors in common and are said to be *relatively prime* or *coprime*. The algorithm for determining $D(u)$ is called the Euclid's algorithm. It consists in repeated application of (3.1.1) as follows. Since $D(u)|A(u)$ and $D(u)|M(u)$, we have $D(u)|R(u)$ from (3.1.1). In other words,

$$(A(u), M(u)) = (M(u), R(u)), \; 0 \le deg(R(u)) < deg(M(u)). \qquad (3.1.5)$$

Applying (3.1.1) to $M(u)$ and $R(u)$, we get

$$M(u) = Q_1(u)R(u) + R_1(u), 0 \le deg(R_1(u)) < deg(R(u)). \qquad (3.1.6)$$

Once again, $(M(u), R(u)) = (R(u), R_1(u))$. We repeat this process and obtain a series of polynomials $R_2(u), R_3(u)$ with decreasing degrees such that

$$
\begin{aligned}
R(u) &= Q_2(u)R_1(u) + R_2(u), \; 0 \le deg(R_2(u)) < deg(R_1(u)), \\
R_1(u) &= Q_3(u)R_2(u) + R_3(u), \; 0 \le deg(R_3(u)) < deg(R_2(u)),
\end{aligned}
$$

$$\vdots$$

$$
\begin{aligned}
R_{\ell-2}(u) &= Q_\ell(u)R_{\ell-1}(u) + R_\ell(u),\ 0 \le deg(R_\ell(u)) < deg(R_{\ell-1}(u)), \\
R_{\ell-1}(u) &= Q_{\ell+1}(u)R_\ell(u). \qquad\qquad\qquad\qquad\qquad\qquad (3.1.7)
\end{aligned}
$$

Since $deg(R(u)) > deg(R(u)) > \ldots \ge 0$, this process must terminate, say at the ℓth step, $R_{\ell+1}(u) = 0$. Equations (3.1.1), (3.1.5)-(3.1.7) imply that $D(u) = (A(u), M(u)) = (M(u), R_1(u)) = (R_1(u), R_2(u)) = \ldots = (R_{\ell-1}(u), R_\ell(u)) = R_\ell(u)$. Also, (3.1.1), (3.1.6) and (3.1.7) may be written as,

$$
\begin{aligned}
R(u) &= A(u) - Q(u)M(u) \\
R_1(u) &= M(u) - Q_1(u)R(u) \\
R_2(u) &= R(u) - Q_2(u)R_1(u) \\
&\vdots \\
R_\ell(u) &= R_{\ell-2}(u) - Q_\ell(u)R_{\ell-1}(u).
\end{aligned}
$$

Using the first two expressions from above, we may express $R_1(u)$ as a linear combination of $A(u)$ and $M(u)$. Similarly, using the second and the third expression, we may express $R_2(u)$ as a linear combination of $A(u)$ and $M(u)$ and so on. Finally, $R_\ell(u)\ (= D(u))$ can be written as

$$D(u) = R_\ell(u) = \alpha(u)A(u) + \beta(u)M(u). \qquad (3.1.8)$$

The fact that *gcd* of two polynomials can be expressed as their linear combination will be used frequently in solving polynomial congruences. If $A(u)$ and $M(u)$ are coprime then (3.1.8) becomes

$$1 = \alpha(u)A(u) + \beta(u)M(u). \qquad (3.1.9)$$

Consider the polynomial congruence in an unknown polynomial $X(u)$ as,

$$A(u)X(u) \equiv C(u) \bmod M(u). \qquad (3.1.10)$$

Equivalently, we wish to find two polynomials $X(u)$ and $Y(u)$ such that

$$A(u)X(u) + M(u)Y(u) = C(u). \qquad (3.1.11)$$

Since $D(u)|A(u)$ and $D(u)|M(u)$, it divides the sum $A(u)X(u) + M(u)Y(u)$ for all $X(u)$ and $Y(u)$. Therefore, for a solution to (3.1.10) to exist, a necessary condition is $D(u)|C(u)$. On the other hand, if $D(u)|\ C(u)\ (C(u) = C_1(u)D(u))$, then using (3.1.8), we may write

$$C(u) = [C_1(u)\alpha(u)]A(u) + [C_1(u)\beta(u)]M(u),$$

and the solution to (3.1.10) is obtained as $X(u) = C_1(u)\alpha(u)$. It may be verified that if $C_1(u)\alpha(u)$ is a solution to (3.1.10), then so is $C_1(u)\alpha(u) + (M(u)/D(u))\cdot\theta(u)$, $0 \leq deg(\theta(u)) < deg(D(u))$. The polynomials $C_1(u)$ $\alpha(u) + (\frac{M(u)}{D(u)})\cdot\theta(u)$, $0 \leq deg(\theta(u)) < deg(D(u))$ are incongruent mod $M(u)$.

In summary, the congruence in (3.1.10) has a solution *iff* $D(u) \mid C(u)$, $D(u) = (A(u),M(u))$. Thus, if $D(u) = 1$, that is the polynomials $A(u)$ and $M(u)$ are coprime, then a solution to (3.1.10) always exists and more over it is unique. Also, if $D(u) = 1$ and $C(u) = 1$, then $X(u)$ is called the inverse of $A(u)$ and is denoted by $A^{-1}(u)$. We emphasize that $A^{-1}(u)$ exists whenever $(A(u),M(u)) = 1$ and is unique.

DEFINITION 3.3 A polynomial $M(u)$ is called *irreducible* if it has only a scalar and itself as its factors. Once again, *all irreducible polynomials will be assumed to be monic.* A polynomial that is not irreducible is termed as reducible.

Given two polynomials $M_1(u)$ and $M_2(u)$ such that $M_1(u) \mid A(u)$, $M_2(u) \mid A(u)$, and $(M_1(u),M_2(u)) = 1$, then $[M_1(u) \cdot M_2(u)] \mid A(u)$. This can be proved using (3.1.1) and the discussion on the congruence in (3.1.10). Finally, every polynomial $A(u)$ of degree n over a field can be uniquely factored as

$$A(u) = \sum_{i=0}^{n} A_i u^i = A_n \prod_{i=1}^{\ell} p_i^{\alpha_i}(u) = A_n \prod_{i=1}^{\ell} q_i(u), \qquad (3.1.12)$$

where $p_i(u)$ is the ith irreducible polynomial and α_i is a positive exponent. This is known as the *unique factorization property of polynomials.* Two properties immediately follow from the factorization in (3.1.12). They are

$$deg(A(u)) = n = \sum_{i=1}^{\ell} deg(q_i(u)) = \sum_{i=1}^{\ell} \alpha_i deg(p_i(u))$$

and

$$(q_i(u), q_j(u)) = 1, \; i, j = 1, 2, \ldots, \ell; \; i \neq j.$$

For an irreducible polynomial $P(u)$, $(A(u),P(u)) = 1$ for $0 \leq deg(A(u)) < deg(P(u))$ and therefore a solution to the congruence

$$A(u)X(u) \equiv C(u) \bmod P(u)$$

will always exist and be unique, $0 \leq deg(A(u)), deg(C(u)) < deg(P(u))$.

3.2 Roots of a Polynomial

Given $A(u)$ over a field $\mathcal{F}$ and a scalar $\alpha \in \mathcal{F}$, consider dividing $A(u)$ by $(u - \alpha)$. Using (3.1.1), we get

$$A(u) = Q(u)(u - \alpha) + R(u), \tag{3.2.1}$$

where $0 = deg(R(u)) < deg(u - \alpha) = 1$. Therefore, $R(u) = R$, $R \in \mathcal{F}$. Setting $u = \alpha$ in both sides of (3.2.1) we get

$$A(\alpha) = R. \tag{3.2.2}$$

The scalar α is said to be a root of the polynomial $A(u)$ if $A(\alpha) = 0$ or equivalently $R = 0$. From (3.2.1), we see that if α is a root of a polynomial $A(u)$, then $(u - \alpha)|A(u)$. If α_1 is a root of $A(u)$, the (3.2.1) becomes

$$A(u) = Q(u)(u - \alpha_1). \tag{3.2.3}$$

Let α_2 be a second root of $A(u)$, $\alpha_2 \neq \alpha_1$. Dividing $Q(u)$ by $(u - \alpha_2)$ we may write

$$Q(u) = Q_1(u)(u - \alpha_2) + R_2. \tag{3.2.4}$$

Substituting (3.2.4) in (3.2.3), $A(u)$ can be expressed as,

$$A(u) = Q_1(u)(u - \alpha_2)(u - \alpha_1) + R_2(u - \alpha_1). \tag{3.2.5}$$

Setting $u = \alpha_2$ gives

$$R_2(\alpha_2 - \alpha_1) = 0, \tag{3.2.6}$$

where $\alpha_2 - \alpha_1 \neq 0$ as $\alpha_2 \neq \alpha_1$. Over a field, the multiplicative inverse of $(\alpha_2 - \alpha_1)$ exists and (3.2.6) implies that $R_2 = 0$. From (3.2.5), we have $[(u - \alpha_1)(u - \alpha_2)]|A(u)$. This leads to the following theorem.

THEOREM 3.2 *A polynomial $A(u)$ of degree n can have at most n roots over a field.*

3.3 Polynomial Fields and Rings

Consider the set $\mathcal{G}$ of all polynomials of degrees up to $m - 1$, that is, $\mathcal{G} = \{A(u) = \sum_{i=0}^{m-1} A_i u^i, A_i \in \mathcal{F}\}$ and let the arithmetic operations of '+' and '·' between two elements of $\mathcal{G}$ be defined mod $M(u)$, $deg\,(M(u)) = m$, that is, if $C(u) = A(u) + B(u)$, then

$$C(u) = < A(u) + B(u) >_{M(u)}$$

and if $C(u) = A(u) \cdot B(u)$, then

$$C(u) = < A(u) \cdot B(u) >_{M(u)}, \quad A(u), B(u) \in \mathcal{G}.$$

Using the properties of the modulo operation, it can be verified that $\mathcal{G}$ forms a ring. In order to compute the multiplicative inverse of $A(u) \in \mathcal{G}$, $A(u) \neq 0$, we have to compute a polynomial $X(u) \in \mathcal{G}$ that satisfies

$$A(u)X(u) \equiv 1 \bmod M(u).$$

Based on the analysis of the congruence in (3.1.10), we see that $X(u)$ exists and is unique *iff* $(A(u), M(u)) = 1$. This condition is trivially satisfied for all $A(u) \in \mathcal{G}$ if $M(u) = P(u)$, an irreducible polynomial. However, for a reducible polynomial $M(u)$, only those elements of $\mathcal{G}$ will have a multiplicative inverse that are coprime to $M(u)$.

Let the field $\mathcal{F}$ be $GF(p)$. Then the set of all polynomials of degrees up to $m - 1$ over $GF(p)$ form a field if the arithmetic operations of '+' and '·' are defined mod $P(v)$, $P(v)$ being an irreducible polynomial over $GF(p)$, $deg(P(v)) = m$. The order of this field is p^m and it is denoted by $GF(p^m)$. $GF(p^m)$ is a polynomial extension of $GF(p)$. Note that we have changed the indeterminate from u to v in order to distinguish the polynomial representation of the elements of $GF(p^m)$ from the remainder of our discussion on polynomial algebra. Having made this distinction, we may now define and analyze a polynomial $A(u)$ with coefficients over $GF(p^m)$. In general, we use $GF(q)$ to denote a Galois field of order q.

Similarly, the set of all polynomials of degrees up to $m - 1$ over $GF(q)$ forms a ring if the arithmetic operations of '+' and '·' are defined mod $M(u)$, $M(u)$ being a reducible polynomial over $GF(q)$, $deg(M(u)) = m$. The order of this ring is q^m.

There are two ways to create a ring of order p^m. The first way was studied in Section 2.2.1. Such a ring consists of integers $0, 1, \ldots, p^m - 1$ and the arithmetic operations of '+' and '·' are defined mod p^m. The second way is based on the polynomials defined over $GF(p)$ and its extensions.

3.4   The CRT for Polynomials

As will be demonstrated in the next section, the Chinese Remainder Theorem for polynomials (CRT-P) is extensively used in digital signal processing for performing computations involving polynomial products. All the polynomials in this section are defined over a field $\mathcal{F}$.

Consider the following extension of the linear congruence in (3.1.10) in an unknown polynomial $X(u)$,

$$X(u) \equiv x_i(u) \bmod m_i(u), \quad i = 1, 2, \ldots, k. \tag{3.4.1}$$

The polynomials $x_i(u), i = 1, 2, \ldots, k$ are known and the moduli $m_i(u)$, $i = 1, 2, \ldots, k$ are pairwise coprime polynomials, that is, $(m_i(u), m_j(u)) = 1, i \neq j; i, j = 1, 2, \ldots, k$.

THEOREM 3.3 *There exists a unique solution $X(u)$, $deg(X(u)) < deg$ $(M(u))$, $M(u) = \prod_{i=1}^{k} m_i(u)$, to the k congruences in (3.4.1) and can be computed as,*

$$X(u) \equiv \sum_{j=1}^{k} x_j(u)T_j(u)(\frac{M(u)}{m_j(u)}) \bmod M(u), \qquad (3.4.2)$$

where the polynomials $T_j(u)$ are precomputed using the congruences

$$T_j(u)(\frac{M(u)}{m_j(u)}) \equiv 1 \bmod m_j(u). \qquad (3.4.3)$$

PROOF. Since the moduli $m_i(u)$, $i = 1, 2, \ldots, k$ are pairwise coprime and $\frac{M(u)}{m_j(u)} = \prod_{\substack{i=1 \\ i \neq j}}^{k} m_i(u)$, we have $(\frac{M(u)}{m_j(u)}, m_j(u)) = 1$. Therefore, the congruence in (3.4.3) can always be solved for $T_j(u)$. The polynomials $T_j(u)$, $deg(T_j(u)) < deg(m_j(u))$, are unique and have the property that $(T_j(u), m_j(u)) = 1$, $j = 1, 2, \ldots, k$. It is clear from the form of $X(u)$ in (3.4.2) that $deg(X(u)) < deg(M(u))$. Taking $\bmod m_i(u)$ on both sides of (3.4.2), we get

$$
\begin{aligned}
X(u) \bmod m_i(u) &\equiv \left(\sum_{j=1}^{k} x_j(u)T_j(u) \left(\frac{M(u)}{m_j(u)} \right) \right) \bmod m_i(u) \\
&\equiv x_i(u)T_i(u) \left(\frac{M(u)}{m_i(u)} \right) \bmod m_i(u) \\
&\equiv x_i(u) \bmod m_i(u).
\end{aligned}
$$

The last step is based on (3.4.3). Thus there is one solution to the k congruences in (3.4.1) as expressed in (3.4.2). Now, we prove the uniqueness of the solution. Suppose there are two solutions to (3.4.1), say $X(u)$ and $X'(u), 0 \leq deg(X(u)), deg(X'(u)) < deg(M(u))$. Substituting $X(u)$ and $X'(u)$ in (3.4.1), we get

$$X(u) \equiv x_i(u) \bmod m_i(u), \qquad (3.4.4)$$

$$X'(u) \equiv x_i(u) \bmod m_i(u), \ i = 1, 2, \ldots, k. \qquad (3.4.5)$$

Subtracting (3.4.5) from (3.4.4) leads to

$$X(u) - X'(u) \equiv 0 \bmod m_i(u), \ i = 1, 2, \ldots, k.$$

This implies that $m_i(u)|(X(u) - X'(u))$, $i = 1, 2, \ldots, k$. Since the moduli are pairwise coprime, we have $\prod_{i=1}^{k} m_i(u) \mid (X(u) - X'(u))$ or $M(u) \mid (X(u)-X'(u))$. This is not possible as $0 \leq deg(X(u)-X'(u)) < deg(M(u))$. Therefore, $X(u) - X'(u) = 0$ or $X(u) = X'(u)$. This implies that the solution is unique.

The CRT-P forms the basis of representing a polynomial $X(u)$, deg $(X(u)) < deg(M(u))$ as a vector $\underline{X}(u) = [x_1(u) \ldots x_k(u)]$, $deg(x_i(u)) < deg$ $(m_i(u))$, where $x_i(u) \equiv X(u)$ mod $m_i(u)$. We express this relationship as

$$X(u) \leftrightarrow \underline{X}(u) = [x_1(u)x_2(u)\ldots x_k(u)]. \qquad (3.4.6)$$

Also, given two polynomials $X(u)$, $Y(u), 0 \leq deg(X(u))$, $deg(Y(u)) < deg(M(u))$, and their corresponding residue vectors as $\underline{X}(u) = [x_1(u) \ldots x_k(u)]$ and $\underline{Y}(u) = [y_1(u) \ldots y_k(u)]$, then

$$\begin{aligned} X(u) + Y(u) \quad &\leftrightarrow \quad \underline{X}(u) + \underline{Y}(u) = [x_1(u) + y_1(u) \\ &\qquad x_2(u) + y_2(u) \ldots x_k(u) + y_k(u)], \qquad (3.4.7) \\ X(u) \cdot Y(u) \quad &\leftrightarrow \quad \underline{X}(u) \cdot \underline{Y}(u) = [x_1(u) \cdot y_1(u) \\ &\qquad x_2(u) \cdot y_2(u) \ldots x_k(u) \cdot y_k(u)]. \qquad (3.4.8) \end{aligned}$$

The sum $X(u) + Y(u)$ and the product $X(u) \cdot Y(u)$ in the *lhs* are defined mod $M(u)$ while the sum and the product in the ith component in the *rhs* are defined mod $m_i(u)$. This follows directly from the properties of the mod operation and the definition of $\underline{X}(u)$ and $\underline{Y}(u)$.

The CRT-P is a fundamental method of partitioning a large degree polynomial $X(u)$ into a number of smaller degree but independent polynomials $x_i(u)$ which may be processed in parallel. This is the direct sum property of the CRT-P and is expressed as

$$\begin{aligned} < X(u) >_{M(u)} \quad &\approx \quad < X(u) >_{m_1(u)} \oplus < X(u) >_{m_2(u)} \\ &\oplus \ldots \oplus < X(u) >_{m_k(u)}. \qquad (3.4.9) \end{aligned}$$

The CRT-P as described here will be a starting point of our mathematical analysis for deriving fast algorithms for computing the cyclic and acyclic convolution of sequences defined over $Z(M)$.

3.5 Polynomial Algebra Over a Ring

The set of all polynomials with coefficients defined over the ring $Z(M)$ constitutes a polynomial ring. Given a degree m polynomial $M(u)$ over $Z(M)$, the set of polynomials of degree up to $m - 1$ over $Z(M)$ also constitutes a polynomial ring. Once again, the operations of '+' and '·' between two polynomials in such a polynomial ring are defined modulo $M(u)$. A close examination reveals that in order to perform the modulo $M(u)$ operation, the multiplicative inverse of the leading coefficient of $M(u)$ must exist. In other words, if $M(u) = \sum_{i=0}^{m} M_i u^i$, then M_m^{-1} must exist and be unique in $Z(M)$. This requires that $(M_m, M) = 1$. Without any loss of generality, we may assume that $M_m = 1$. Thus, $M(u)$ is a *monic* polynomial. *Therefore, all the modulo polynomials $M(u)$ that characterize the polynomial rings under study will be assumed to be monic throughout this work.*

Consider again the ring of all polynomials of degrees up to $m - 1$ over $Z(M)$ with operations defined modulo $M(u)$. This ring can also be expressed as a direct sum of t polynomial subrings each defined over $Z(p_i^{\alpha_i})$, $i = 1, 2, \ldots, t$. Each of the t polynomial subrings is a polynomial ring of all polynomials of degrees up to $m - 1$ over $Z(p_i^{\alpha_i})$ with operations defined modulo $M^{(i)}(u)$, where $M^{(i)}(u) \equiv M(u) \pmod{p_i^{\alpha_i}}$. In other words, given a polynomial $A(u)$ with coefficients over $Z(M)$, $M = p_1^{\alpha_1} p_2^{\alpha_2} \cdots p_t^{\alpha_t}$, it can be represented as a direct sum of polynomials $A^{(i)}(u)$, $i = 1, 2, \ldots, t$ where

$$A^{(i)}(u) \equiv A(u) \pmod{p_i^{\alpha_i}}. \tag{3.5.1}$$

We can also write

$$A(u) \approx A^{(1)}(u) \oplus A^{(2)}(u) \oplus \ldots \oplus A^{(t)}(u). \tag{3.5.2}$$

The above statements follow directly from the CRT-I and the definition of a polynomial. Therefore, analysis for polynomials over the ring $Z(M)$ can be performed once the mathematical properties of polynomials over integer rings of the type $Z(p^\alpha)$ are known. In summary, we will focus our attention on the algebraic structure of the rings of polynomials of degrees up to $m-1$ over $Z(p^\alpha)$ where the operations are performed modulo a degree m polynomial $M(u)$ over $Z(p^\alpha)$.

Next, we present the p-adic form of a polynomial defined over $Z(p^\alpha)$.

DEFINITION 3.4 Let $A(u) = \sum_{i=0}^{n-1} A_i u^i, A_i \in Z(p^\alpha)$, a polynomial of degree $n - 1$ over $Z(p^\alpha)$. Then, the p-adic representation of $A(u)$ is

$$A(u) = \sum_{j=0}^{\alpha-1} A_j(u) p^j, \tag{3.5.3}$$

where

$$A_j(u) = \sum_{i=0}^{n-1} A_{i,j} u^i, A_{i,j} \in GF(p). \tag{3.5.4}$$

The above p-adic representation of $A(u)$ is obtained by noting that the coefficient $A_i \in Z(p^\alpha)$ and can be represented in its p-adic form as

$$A_i = \sum_{j=0}^{\alpha-1} A_{i,j} p^j, \; i = 0, 1, \ldots, n-1, A_{i,j} \in GF(p).$$

Substituting in $A(u)$ and interchanging the summations, we get

$$A(u) = \sum_{i=0}^{n-1} (\sum_{j=0}^{\alpha-1} A_{i,j} p^j) u^i$$

$$= \sum_{j=0}^{\alpha-1} (\sum_{i=0}^{n-1} A_{i,j} u^i) p^j.$$

In this work, we wish to express a ring of polynomials of degrees up to $n-1$ over $Z(p^\alpha)$ as a direct sum in a manner analogous to (3.4.9). The various issues to be studied are (i) factorization of the modulo polynomial $M(u)$ over $Z(p^\alpha)$, (ii) direct sum representation of the polynomial ring based on the factorization of $M(u)$, and (iii) computational methods to convert from the direct sum representation to the original polynomial representation and vice-versa. This leads to an extension of the CRT-P to be termed as the AICE-CRT. Chapter 4 of this work will focus on the above mentioned issues, the AICE-CRT and the application of AICE-CRT to deriving fast computationally efficient algorithms for cyclic and acyclic convolution of data sequences defined over $Z(M)$.

DEFINITION 3.5   Let $M(u)$ be a polynomial over $Z(M)$. A polynomial $A(u)$ over $Z(M)$ has $M(u)$ as its factor whenever $A(u)$ and $M(u)$ can be written as $A(u) = M(u)Q(u) \bmod M$, where $Q(u)$ is a polynomial over $Z(M)$. We may also write $M(u) \mid A(u)$.

One has to be careful in interpreting the above definition. Many basic results in polynomial algebra over a field are no longer valid over a ring. For example, $(2u^2 + 3u + 1)(2u^2 + 3u + 3) = u^2 + 3$ over $Z(4)$. This simple example demonstrates that the sum of the degrees of the factor polynomials may not be equal to the degree of the product. Even more interestingly, we cannot even test if $2u^2 + 3u + 1$ divides $u^2 + 3$ over $Z(4)$ or not.

3.5.1 Roots of a Polynomial Over $Z(M)$

Given $A(u)$ over the ring $Z(M)$ and a scalar $\alpha \in Z(M)$, consider dividing $A(u)$ by $(u - \alpha)$ to get

$$A(u) = Q(u)(u - \alpha) + R(u),$$

where $0 \leq deg(R(u)) < deg(u - \alpha) = 1$. Therefore, $R(u) = R, R \in Z(M)$. Setting $u = \alpha$ in both sides of the above equation, we get

$$A(\alpha) = R.$$

The scalar α is said to be a root of a polynomial $A(u)$ if $A(\alpha) = 0$ or equivalently $R = 0$. We see that if α is a root of $A(u)$, then $(u - \alpha) \mid A(u)$. Therefore, if α_1 and α_2 are two roots of a polynomial $A(u)$, $\alpha_1 \neq \alpha_2$, then $(u - \alpha_1) \mid A(u)$, and $(u - \alpha_2) \mid A(u)$ over $Z(M)$. However, unlike a field, it *does not* imply that $[(u - \alpha_1)(u - \alpha_2)] \mid A(u)$ over $Z(M)$. Consequently, Theorem 3.2 does not hold for polynomials defined over $Z(M)$. The number of roots of a polynomial over $Z(M)$ may exceed its degree. For example, the polynomial $u^2 + u$ has $0, 2, 3$ and 5 as roots over $Z(6)$.

Let us examine the behavior of these roots further. The expressions (3.2.3)-(3.2.6) remain valid except that they are defined mod M in this case. The congruence $R_2(\alpha_2 - \alpha_1) = 0 \mod M$ does not imply that $R_2 = 0$. However, if $(\alpha_2 - \alpha_1, M) = 1$, then $R_2 = 0$ and $[(u - \alpha_1)(u - \alpha_2)] \mid A(u)$ from (3.2.5). This analysis leads to the following theorem.

THEOREM 3.4 *If* $\alpha_1, \alpha_2, \ldots, \alpha_\theta$ *are roots of a polynomial* $A(u)$ *over* $Z(M)$ *and*

$$(\alpha_i - \alpha_j, M) = 1, 1 \leq i, j \leq \theta, i \neq j, \qquad (3.5.5)$$

then we have

$$[\prod_{i=1}^{\theta}(u - \alpha_i)] \mid A(u).$$

Note that $(\alpha_2 - \alpha_1, M) = 1$ is only a sufficient condition for $R_2(\alpha_2 - \alpha_1) = 0$ to imply that $R_2 = 0$. It is *not* a necessary condition. For example, the polynomial $u^3 + u^2 + 2u + 2$ has $\alpha_1 = 2$, and $\alpha_2 = 4$ as roots and $(u + 2)(u + 4)$ as a factor over $Z(6)$ even though $(\alpha_2 - \alpha_1, M) = (2, 6) = 2 \neq 1$.

Integers $\alpha_1, \alpha_2, \ldots, \alpha_\theta$ that satisfy (3.5.5) are said to be *pairwise differentially coprime* to M. We observe that the condition in (3.5.5) does not imply that $(\alpha_i, M) = 1$, $i = 1, 2, \ldots, \theta$, in any way. For example, if $M = 5 \cdot 7 = 35$, one can have $\alpha_1 = 3$, $\alpha_2 = 4$, $\alpha_3 = 5$, $\alpha_4 = 6$, and $\alpha_5 = 7$. Finally, given the standard factorization of M in (2.2.14), it is easy to establish that the largest number of integers that are pairwise differentially coprime to M is p, $p = min\{p_1, p_2, \ldots, p_t\}$.

3.6   Computing Convolution Over a Field

It was stated in the beginning of this chapter that the computation of acyclic convolution of two data sequences or equivalently the corresponding polynomial product is one of the most widely occurring operations in digital signal processing. An equally important computation is the cyclic convolution of two data sequences which is the focus of our attention in this section.

Given two data sequences A_i and B_i, $i = 0, 1, \ldots, n - 1$, each of length n, the cyclic convolution consists in the computation of the sequence C_ℓ, $\ell = 0, 1, \ldots, n - 1$, defined as,

$$C_\ell = \sum_{i=0}^{n-1} A_i B_{<\ell-i>_n}, \; \ell = 0, 1, \ldots, n - 1. \qquad (3.6.1)$$

Note that the index on the second term in (3.6.1) is defined mod n. If $A(u)$, $B(u)$ and $C(u)$ are the generating functions of the sequences A_i, B_i and C_i, $i = 0, 1, \ldots, n - 1$, then we may express (3.6.1) as,

$$C(u) \equiv A(u)B(u) \bmod (u^n - 1). \qquad (3.6.2)$$

We now describe CRT-P based algorithms for computing the cyclic and acyclic convolutions expressed as polynomial products in (3.6.2) and (3.0.5) respectively. Since the polynomial $C(u)$ in (3.0.5) has degree $n - 1$, it remains unchanged when it is defined mod $M(u)$, $deg(M(u)) \geq n$. In other words, the acyclic convolution can be expressed as,

$$C(u) \equiv A(u)B(u) \bmod M(u), \; deg(M(u)) \geq n. \qquad (3.6.3)$$

The choice of the modulo polynomial $M(u)$ in (3.6.3) is arbitrary as long as $deg(M(u)) \geq n$. This is perhaps *the only fundamental difference between cyclic and acyclic convolution*. Among all the choices possible for $M(u)$, $M(u)$ will be selected as the polynomial for which the computation of the acyclic convolution has the least cost. In most instances, the cost is measured in terms of the total number of arithmetic operations (MULT and ADD) or just the number of MULT. Arithmetic operations in general and MULT in particular tend to dominate the time required for the computational algorithms. The focus of this book is also on the computational efficiency of the various algorithms as measured by the total number of MULT and ADD.

3.6.1 Algorithms for Computing Convolutions Over a Field

The CRT-P and the direct sum property in (3.4.9) provide a very convenient way to express an acyclic convolution of length n into a number of acyclic convolutions of smaller lengths.

Given a polynomial $M(u)$ and its unique factorization ($M(u)$ is assumed monic without any loss in generality),

$$M(u) = \prod_{i=1}^{\ell} p_i^{\alpha_i}(u) = \prod_{i=1}^{\ell} q_i(u), \qquad (3.6.4)$$

the polynomial product $C(u) = A(u)B(u) \bmod M(u)$ can be computed as a three step process outlined below

1. Reduce $A(u)$ and $B(u)$ mod $q_i(u)$ to get
   $A_i(u) \equiv A(u) \bmod q_i(u)$
   $B_i(u) \equiv D(u) \bmod q_i(u)$, $i = 1, 2, \ldots, \ell$.

2. Compute $C_i(u) \equiv C(u) \bmod q_i(u)$ as
   $C_i(u) \equiv A_i(u)B_i(u) \bmod q_i(u)$, $i = 1, 2, \ldots, \ell$.

3. Reconstruct $C(u) \bmod M(u)$ from the congruences
   $C(u) \equiv C_i(u) \bmod q_i(u)$, $i = 1, 2, \ldots, \ell$,
   using the CRT-P.

Since $deg(q_i(u)) < deg(M(u))$, it is expected that the polynomial product in step (2) can be computed very efficiently. The CRT-P provides a vehicle for breaking up a large degree polynomial product into a number of independent smaller degree polynomial products which may be computed in parallel.

Since the choice of the modulo polynomial $M(u)$ for computing the acyclic convolution is non-unique, it is generally selected in a way that it is a product of as large a number of relatively coprime polynomials as possible. This also ensures that the degree of each of the factors is as small as possible. Such a choice is expected to result in an algorithm having the least computational complexity.

The algorithm for computing the cyclic convolution in (3.6.2) has the same structure except that the modulo polynomial is restricted to be

$$M(u) = u^n - 1. \qquad (3.6.5)$$

Given the factorization of $u^n - 1$ in terms of its irreducible polynomials over the field of interest, the above three step algorithm can also be used to compute the cyclic convolution. Of course, the factorization of $u^n - 1$ depends on the field. In the following, we focus on the factorization of $u^n - 1$ over the finite fields $GF(p)$ and $GF(p^m)$. This factorization will also be of value in Chapter 4.

3.7   On Factorization of $(u^n - 1)$ Over Finite Fields

In this section, we study factorization of $u^n - 1$ over the finite fields $GF(p)$ and $GF(p^m)$. It will require developing further properties of the polynomial algebra over a field. In that respect, this section may be treated as an extension of the Section 3.1.

For the case that n is a multiple of p, $n = cp$, we have

$$u^n - 1 = (u^c - 1)^p. \qquad (3.7.1)$$

If c is still multiple of p, then (3.7.1) can be repeated until they are relatively prime. Therefore, we need to focus only on the case $(n, p) = 1$, that is, n and p are coprime.

THEOREM 3.5 *If $(\alpha, \beta) = 1$, then there always exists an integer ϕ such that*

$$\beta | (\alpha^\phi - 1).$$

PROOF. Consider the ring $Z(\beta)$ and the set A of all those integers in $Z(\beta)$ that are coprime to β. Clearly, this set is non-empty. Let the elements in A be labeled as $A_1, A_2, \ldots, A_\phi$, $\phi > 1$. The integer ϕ is known as the *Euler's totient function* of β. Consider the set B of integers B_i, $i = 1, 2, \ldots, \phi$ defined as

$$B_i \equiv \alpha A_i \bmod \beta. \qquad (3.7.2)$$

As $(\alpha, \beta) = 1$ and $(A_i, \beta) = 1$, we have $(B_i, \beta) = 1$, $B_i \in Z(\beta)$. Also $B_i \neq B_j$. Thus, the sets A and B contain the same integers. Therefore,

$$\prod_{i=1}^{\phi} A_i \equiv \prod_{i=1}^{\phi} (\alpha A_i) \bmod \beta$$

$$\equiv \alpha^\phi \prod_{i=1}^{\phi} A_i \bmod \beta. \qquad (3.7.3)$$

Since $(A_i, \beta) = 1$, $i = 1, 2, \ldots, \phi$, we have $(\prod_{i=1}^{\phi} A_i, \beta) = 1$. Therefore, the inverse of $\prod_{i=1}^{\phi} A_i \bmod \beta$ exists and is unique. Multiplying both sides of (3.7.3) by $(\prod_{i=1}^{\phi} A_i)^{-1} \bmod \beta$, we get $\alpha^{\phi} \equiv 1 \bmod \beta$ or $(\alpha^{\phi} - 1) \equiv 0 \bmod \beta$ or $\beta | (\alpha^{\phi} - 1)$.

Since $(n, p) = 1$, Theorem 3.5 can be used to find an integer such that $n | (p^m - 1)$. We shall use the smallest value of m for which $n | (p^m - 1)$ in order to factorize $u^n - 1$ over $GF(p)$ and $GF(p^m)$. It is clear from Theorem 3.5 that $m | \phi$. Another important consequence of Theorem 3.5 is as follows.

COROLLARY 3.1 *Let* $(A, p) = 1$; *then*

$$A^{(p-1)p^{\alpha-1}} \equiv 1 \ \bmod \ p^{\alpha}.$$

PROOF. $(A, p) = 1$ also implies that $(A, p^{\alpha}) = 1$. In the ring $Z(p^{\alpha})$, only those integers that contain p as a factor are *not* coprime to p^{α}. There are $p^{\alpha-1}$ such integers in $Z(p^{\alpha})$. Therefore, there are $p^{\alpha} - p^{\alpha-1} = (p-1)p^{\alpha-1}$ integers in $Z(p^{\alpha})$ that are coprime to p^{α}. Applying Theorem 3.5 gives the desired result.

THEOREM 3.6 *Let* A *be any non-zero element in* $GF(q)$, $q = p$ *for* $GF(p)$ *and* $q = p^m$ *for* $GF(p^m)$. *Then*

$$A^{q-1} = 1. \tag{3.7.4}$$

PROOF. Once again, let $A_1, A_2, \ldots, A_{q-1}$ be the $(q-1)$ non-zero elements of $GF(q)$. Consider the elements in $GF(q)$ defined as

$$B_i = A \cdot A_i, \ A \neq 0, \ i = 1, 2, \ldots, q-1.$$

Clearly, $B_i \in GF(q), B_i \neq 0$ and $B_i \neq B_j, i \neq j$. Therefore, $B_1, B_2, \ldots, B_{q-1}$ are also the $(q-1)$ non-zero elements of $GF(q)$. Thus

$$\prod_{i=1}^{q-1} B_i = \prod_{i=1}^{q-1} (A \cdot A_i) = \prod_{i=1}^{q-1} A_i. \tag{3.7.5}$$

Multiplying both sides by $(\prod_{i=1}^{q-1} A_i)^{-1}$, we get the result.

THEOREM 3.7 *A unique factorization of* $u^{q-1} - 1$ *over* $GF(q)$ *is given by*

$$u^{q-1} - 1 = \prod_{i=1}^{q-1} (u - A_i), \ A_i \in GF(q), \ A_i \neq 0. \tag{3.7.6}$$

PROOF.   From Theorem 3.6, every non-zero element $A_i \in GF(q)$, is a root of $u^{q-1} - 1$. Therefore, $(u - A_i)$, $i = 1, 2, \ldots, q - 1$, is a factor of $u^{q-1} - 1$. Since there are $q - 1$ non-zero elements in $GF(q)$, we have $\prod_{i=1}^{q-1}(u - A_i) | (u^{q-1} - 1)$. Both the polynomials $\prod_{i=1}^{q-1}(u - A_i)$ and $u^{q-1} - 1$ are monic and of degree $q - 1$. The theorem follows.

Theorem 3.7 gives a complete factorization of the polynomial $u^{q-1} - 1$ over $GF(q)$, $q = p^m$. Given the above factorization over $GF(p^m)$, we will explore the following two issues associated with the factorization:

1. Factorization of $u^n - 1$ over $GF(p)$, $n|(p - 1)$, and

2. Factorization of $u^n - 1$ over $GF(p)$ using $GF(p^m)$, $n|(p^m - 1)$.

DEFINITION 3.6 Given a non-zero element A in $GF(q)$, the smallest non-zero value of n such that $A^n = 1$ is called the order of A. It is clear from Theorem 3.6 that $n \leq q - 1$.

THEOREM 3.8 *The order of every non-zero element in $GF(q)$ divides $q - 1$.*

PROOF. If n does not divide $q - 1$, then we write

$$q - 1 = \alpha n + r, \ 0 \leq r < n.$$

Using Theorem 3.6, we get

$$A^{q-1} = 1 = A^{\alpha n + r} = (A^n)^\alpha . A^r = A^r, \ 0 \leq r < n.$$

However, this contradicts the definition of the order of A. Therefore, the statement of the theorem is true.

THEOREM 3.9
$$(u^\alpha - 1, u^\beta - 1) = u^{(\alpha, \beta)} - 1. \tag{3.7.7}$$

PROOF.   Let $\alpha > \beta$. We may divide $u^\alpha - 1$ by $u^\beta - 1$ to get

$$u^\alpha - 1 = (u^{\alpha - \beta} + u^{\alpha - 2\beta} + \ldots + u^{\alpha - c\beta})(u^\beta - 1) + u^r - 1 \tag{3.7.8}$$

corresponding to the expression

$$\alpha = c\beta + r, \ 0 \leq r < \beta. \tag{3.7.9}$$

It is clear from (3.7.8) and (3.7.9) that

$$(u^\alpha - 1, u^\beta - 1) = (u^\beta - 1, u^r - 1), \tag{3.7.10}$$

where

$$(\alpha, \beta) = (\beta, r) \tag{3.7.11}$$

If $r = 0$, that is, $\beta | \alpha$, then $(u^\beta - 1)|(u^\alpha - 1)$. This statement along with a recursive application of (3.7.10) and (3.7.11) establishes the theorem.

A consequence of the above theorem is that for every factor d of $q - 1$, that is, $d|(q-1)$, $(u^d - 1)|(u^{q-1} - 1)$. Combining this with the factorization in Theorem 3.7, we get

$$u^{q-1} - 1 = (u^d - 1)C(u) = \prod_{i=1}^{q-1}(u - A_i),\ deg(C(u)) = q - 1 - d.$$

Since $deg(C(u)) = q - 1 - d$, it can have only $q - 1 - d$ roots in $GF(q)$. Therefore, there must be d elements among the $q-1$ non-zero elements A_1, A_2, ..., A_{q-1} in $GF(q)$ that are roots of $u^d - 1$. Let these d elements be B_1, B_2, ..., B_d. This gives the factorization of $u^d - 1$ for every $d|(q - 1)$ over $GF(q)$ as

$$u^d - 1 = \prod_{i=1}^{d}(u - B_i),\ B_i \in GF(q), B_i \neq 0. \tag{3.7.12}$$

The above factorization also implies that the element B_i has order equal to d or a factor of d. This is consistent with Theorem 3.8. There are exactly d elements that have d or a factor of d as their order. Now, consider the factorization of $q - 1$ as

$$q - 1 = p_1^{\alpha_1}p_2^{\alpha_2} \cdots p_\ell^{\alpha_\ell}$$

and let $d = 1$. There is one element in $GF(q)$ of order equal to d. The element is 1. Let $d = p_1$. From (3.7.12), we see that there are p_1 distinct elements in $GF(q)$ having order p_1 or a factor of p_1. Since p_1 is prime, there must be exactly $p_1 - 1$ elements in $GF(q)$ of order p_1. Similarly, there are $p_1^i - p_1^{i-1}$ elements in $GF(q)$ of order p_1^i, $i = 1,2,..., \alpha_1$. Let A_j denote an element in $GF(q)$ of order $q_j = p_j^{\alpha_j}$, $j = 1, 2, ...,\ell$. A fundamental property of these elements is stated in the following.

THEOREM 3.10 *In $GF(q)$, there is an element of order $q - 1$.*

PROOF. Consider the element $A = A_1A_2$. Since $A_1^{q_1} = 1$, $A_2^{q_2} = 1$, we have $A_1^{q_1q_2} = 1$, and $A_2^{q_1q_2} = 1$. Therefore $(A_1A_2)^{q_1q_2} = 1$ or $A^{q_1q_2} = 1$. Let n be the order of the element A, that is, $A^n = 1$. Since $A^{q_1q_2} = 1$, we have $n|(q_1q_2)$. $A^n = 1$ also leads to $A^{nq_1} = 1$ and $A^{nq_2} = 1$. Replacing A

by $A_1 \cdot A_2$ in these equations and realizing that $A_1^{q_1} = 1$ and $A_2^{q_2} = 1$, we get

$$A_2^n = 1,$$

and

$$A_1^n = 1.$$

The above equations imply that $q_1|n$ and $q_2|n$. Since $(q_1, q_2) = 1$, we have $(q_1 \cdot q_2)|n$. The two relations $n|(q_1 q_2)$ and $(q_1 q_2)|n$ lead to $n = q_1 q_2$. Since the element $A_1 A_2$ has order $q_1 q_2$, by induction, the element $A = A_1 A_2 \cdots A_\ell$ has order $\prod_{i=1}^{\ell} p_i^{\alpha_i} = q - 1$.

The above theorem will play a critical role in the process of factorizing $u^n - 1$. An element A in $GF(q)$ of order $q - 1$ is called a *primitive element* of the field. Theorem 3.10 guarantees that every $GF(q)$ has at least one such element. Since A has order $q - 1$, every non-zero element A_i in $GF(q)$ can be expressed as $A_i = A^i$, $i = 1, 2, \ldots, q-1$. Based on (3.7.6), this leads to the factorization of $u^{q-1} - 1$ over $GF(q)$ as

$$u^{q-1} - 1 = \prod_{i=1}^{q-1}(u - A^i). \tag{3.7.13}$$

In terms of the element A, the factorization of $u^d - 1$, $d|(q-1)$ over $GF(q)$ can be expressed as

$$u^d - 1 = \prod_{i=1}^{d}(u - B^i), \tag{3.7.14}$$

where

$$B = A^{(q-1)/d}. \tag{3.7.15}$$

In digital signal processing, the emphasis is on processing polynomials defined over either $GF(p)$ or $Z(M)$. Therefore, we need to further analyze the factorization in (3.7.14) and (3.7.15). This is done in the following.

3.7.1   Factorization of $(u^n - 1)$ Over $GF(p)$

A direct consequence of (3.7.13)-(3.7.15) is that the factorization of $u^{p-1} - 1$ over $GF(p)$ or $GF(p^m)$ is given by

$$u^{p-1} - 1 = \prod_{i=1}^{p-1}(u - A^i), \tag{3.7.16}$$

where A is a primitive element in $GF(p)$ or equivalently an element of order $(q-1)/(p-1)$ in $GF(p^m)$. Also, for $d|(p-1)$, the factorization of $u^d - 1$

over $GF(p)$ or $GF(p^m)$ is given by

$$u^d - 1 = \prod_{i=1}^{d}(u - B^i), \qquad (3.7.17)$$

$$B = A^{(p-1)/d}. \qquad (3.7.18)$$

Given the factorization of $u^{q-1} - 1$ over $GF(q)$, $q = p^m$ in (3.7.13), we partition the set $\mho = \{1, 2, \ldots, q-1\}$ containing the powers of the non-zero elements in $GF(q)$, into subsets $\mho_{j_1}, \mho_{j_2}, \ldots$. A *cyclotomic set* $\mho_j$ begins with j, where j is the smallest power of A not included in the preceding subsets. Other elements in the subset $\mho_j$ are obtained as

$$\mho_j = \{j, jp, jp^2, jp^3, \ldots\}. \qquad (3.7.19)$$

Since $A^{p^m-1} = 1$, the powers of A are defined mod $(p^m - 1)$. Also, $A^{p^m-1} = 1$ implies that $A^{jp^m} = A^j$. Therefore, there are at most m elements in each $\mho_j$. It is easy to verify that no elements in two different cyclotomic sets are equal. Let $\mathcal{I}$ be the set of indices $j_1, j_2, \ldots$. Based on this partitioning and (3.7.13), we write the factorization of $u^{q-1} - 1$ as

$$
\begin{aligned}
u^{q-1} - 1 &= \prod_{j \in \mathcal{I}} \left\{ \prod_{\theta \in \mho_j} (u - A^\theta) \right\} \\
&= \prod_{j \in \mathcal{I}} Q_j(u) \qquad (3.7.20)
\end{aligned}
$$

The polynomial $Q_j(u)$ is defined as

$$Q_j(u) = (u - A^j)(u - A^{jp})(u - A^{jp^2}) \cdots (u - A^{jp^{\ell-1}}) \qquad (3.7.21)$$

such that

$$jp^\ell \equiv j \bmod (p^m - 1). \qquad (3.7.22)$$

The last step in this factorization process is to show that $Q_j(u)$ has coefficients over $GF(p)$ and that it is irreducible. This is established in the following.

THEOREM 3.11 *The polynomial $Q_j(u)$ as defined in (3.7.21) has coefficients over $GF(p)$.*

PROOF. Consider the polynomial

$$Q_j^p(u) = \left\{ \prod_{i=0}^{\ell-1} (u - A^{jp^i}) \right\}^p$$

$$= \prod_{i=0}^{\ell-1} (u - A^{jp^i})^p$$

$$= \prod_{i=0}^{\ell-1} (u^p - A^{jp^{i+1}})$$

$$= \left\{ \prod_{i=0}^{\ell-2} (u^p - A^{jp^{i+1}}) \right\} (u^p - A^{jp^\ell}).$$

Recalling (3.7.22), the above equation simplifies to

$$\left\{ \prod_{i=1}^{\ell-1} (u^p - A^{jp^i}) \right\} (u^p - A^j) = \prod_{i=0}^{\ell-1} (u^p - A^{jp^i}).$$

Thus, $Q_j^p(u) = Q_j(u^p)$. If $Q_j(u) = F_0 + F_1 u + \cdots + F_\ell u^\ell$, then

$$Q_j^p(u) = \left\{ \sum_{i=0}^{\ell} F_i u^i \right\}^p$$

$$= \sum_{i=0}^{\ell} F_i^p u^{ip},$$

$$Q_j(u^p) = \sum_{i=0}^{\ell} F_i u^{ip}.$$

The condition $Q_j^p(u) = Q_j(u^p)$ and the above expressions lead to the important property that if F_i is the ith coefficient of the polynomial $Q_j(u)$ defined over $GF(p^m)$, then $F_i^p = F_i$. In other words, $F_i = 0$ or F_i is a root of the polynomial $u^{p-1} - 1$. The polynomial $u^{p-1} - 1$ has $p - 1$ non-zero elements of $GF(p)$ as its roots. Consequently $F_i \in GF(p)$.

A result useful in establishing the irreducibility of the $Q_j(u)$ polynomials is as follows.

LEMMA 3.1 *If a polynomial $Q(u)$ defined over $GF(p)$ has $\beta \in GF(p^m)$ as a root, then β^p is also a root of $Q(u)$.*

PROOF. If β is a root of the polynomial $Q(u) = \sum_{i=0}^{\ell} F_i u^i$, $F_i \in GF(p)$, then $Q(\beta) = 0$ or

$$\sum_{i=0}^{\ell} F_i \beta^i = 0.$$

Raising both sides to the power p and recalling that since $F_i \in GF(p)$, $F_i^p = F_i$, we get

$$
\begin{aligned}
0 &= \left\{ \sum_{i=0}^{\ell} F_i \beta^i \right\}^p \\
&= \sum_{i=0}^{\ell} F_i^p \beta^{pi} = \sum_{i=0}^{\ell} F_i (\beta^p)^i,
\end{aligned}
$$

or β^p is a root of $Q(u)$.

Applying this lemma recursively we see that if a polynomial defined over $GF(p)$ has $\beta \in GF(p^m)$ as a root, then it also has β^p, β^{p^2}, ... as roots. Since the polynomial $Q_j(u)$ as defined in (3.7.21) has only A^j, A^{jp}, A^{jp^2},... as its roots and has coefficients in $GF(p)$ (Theorem 3.11), it is the polynomial of the lowest degree in $GF(p)$ having A^j as its root. Therefore, it must be irreducible over $GF(p)$.

This completes factorization of $u^n - 1$ over $GF(p)$, $n = p^m - 1$. Finally, if $d|(p^m - 1)$, then a factorization of $u^d - 1$ over $GF(p)$ begins with the factorization in (3.7.14) and (3.7.15) reproduced below:

$$
\begin{aligned}
u^d - 1 &= \prod_{i=1}^{d} (u - B^i) \\
B &= A^{(q-1)/d}.
\end{aligned}
\tag{3.7.23}
$$

Clearly, $B^d = 1$. The process of partitioning the set $\mho = \{1, 2,\ldots, d\}$ into cyclotomic sets remains the same as before. The cyclotomic subset $\mho_j$ is obtained as

$$\mho_j = \{j, jp, jp^2 \ldots\}. \tag{3.7.24}$$

In this case, $B^d = 1$ and therefore the powers of B are defined mod d. Once again, there are at most m elements in each $\mho_j$. The rest of the procedure for computing the irreducible factors of $u^d - 1$ over $GF(p)$ remains the same as before.

We summarize the procedure for factorizing $u^n - 1$ over $GF(p)$ in the Tables 3.1 and 3.2 along with the salient features of the factor polynomials.

Table 3.1: Factorization of $u^n - 1$ over $GF(p)$

	Let A be a primitive element in $GF(p)$.
$n = p - 1$	$u^n - 1 = \prod_{i=1}^{n}(u - A^i)$
$n\|(p-1)$	$u^n - 1 = \prod_{i=1}^{n}(u - B^i)$ $B = A^{\frac{p-1}{n}}$
	Note: A primitive integer in $GF(p)$ always exists.

3.7.2 Primitive Polynomial Over $GF(p)$

The structure of the field $GF(p^m)$ was described in Section 3.7. It consists
of all the polynomials of degrees up to $m - 1$. The arithmetic operations of
'+' and '·' in $GF(p^m)$ are defined mod $P(v)$, $P(v)$ being a monic degree m
irreducible polynomial over $GF(p)$. All the computation over $GF(p^m)$ can
be performed once $P(v)$ is specified. However, we encounter the primitive
elements of $GF(p^m)$ repeatedly in our factorization procedures. In this
section, we further study the algebraic properties of the primitive elements
of $GF(p^m)$.

A primitive element A of $GF(p^m)$ has order $q-1$, $q = p^m$. Thus, $A^i \neq 1$,
$i < q - 1$ and $A^{q-1} = 1$. Also, from (3.7.20), we see that A is a root of
$Q_1(v)$, $deg(Q_1(v)) = m$. Note that $Q_1(v)$ is an irreducible polynomial over
$GF(p)$ and is given by

$$Q_1(v) = \prod_{\theta=0}^{m-1}(v - A^{p^\theta}).$$

Definition 3.7 A degree m irreducible polynomial over $GF(p)$ having a
primitive element A of $GF(p^m)$ as its root is termed as a primitive polynomial.

By definition, every primitive polynomial is irreducible while the con-
verse is not true. Since A is a root of $v^{q-1} - 1$ and $Q_1(v)$, $Q_1(v)$ being the
smallest degree polynomial having A as a root, $Q_1(v)|(v^{q-1} - 1)$. In fact,
the smallest value of i for which $Q_1(v)|(v^i - 1)$ is given by $i = q - 1$ or else
A is not a primitive element.

Table 3.2: Factorization of $u^n - 1$ Over $GF(p)$ Using $GF(p^m)$

	Let A be a primitive element in $GF(p^m)$.
$n = (q - 1)$ $q = p^m$	$u^n - 1 = \prod_{j \in \mathcal{I}} Q_j(u)$ $Q_j(u) = \prod_{i=0}^{l-1}(u - A^{jp^i}), l \leq m$ $\mathcal{I}$ is the set of indices of the cyclotomic sets $\mho_j$, $\mho_j = \{j, jp, jp^2, \cdots\}$. Here, j is the smallest integer not appearing in $\mho_1, \cdots$, $\mho_{j-1}$. The elements of $\mho_j$ are defined modulo $(p^m - 1)$.
$n \mid (q - 1)$	Let $B = A^{(q-1)/n}$. $u^n - 1 = \prod_{j \in \mathcal{I}} Q_j(u)$ $Q_j(u) = \prod_{i=0}^{l-1}(u - B^{jp^i}), l \leq m$ $\mathcal{I}$ is the set of the indices of the cyclotomic set $\mho_j$, $\mho_j = \{j, jp, jp^2, \cdots\}$. Here, j is the smallest integer not appearing in $\mho_1, \cdots$, $\mho_{j-1}$. The elements of $\mho_j$ are defined modulo n.
$(n, p) = 1$	Find the smallest integer m such that $n \mid (p^m - 1)$. Apply the previous case to factorize $u^n - 1$.
$n = cp$ $(c, p) = 1$	$u^n - 1 = (u^c - 1)^p$ Factorize $u^c - 1$ using the techniques described above.
	Note: A primitive element in $GF(p^m)$ always exists. The polynomials $Q_j(u)$ are irreducible.

All the non-zero elements of $GF(p^m)$ can be represented as A^i, $i = 1, 2, \ldots, q-1$. The polynomial representation of A^i is obtained as A^i mod $Q_1(A)$. Thus

$$A^i \longleftrightarrow A^i \text{ mod } Q_1(A), \ i = 1, 2, \ldots, q-1. \qquad (3.7.25)$$

In *rhs* of (3.7.25), the elements of $GF(p^m)$ are expressed as polynomials in A degrees up to $m-1$ with the arithmetic operations defined mod $Q_1(A)$. The *lhs* is the corresponding representation as a power of A. In general, it is easier to perform '+' when two elements in $GF(p^m)$ are expressed in polynomial form. Similarly, it is easier to perform '·' when two elements in $GF(p^m)$ are expressed as a power of A.

We observe that primitive polynomials exist for all values of p and m. Furthermore, their choice is non-unique in the sense that there exists more than one primitive polynomial for all values of p and m. In the following, we present primitive polynomials over $GF(p)$ for $p = 2, 3, 5$ and 7 and $m = 2, 3, 4$ and 5,

$p = 2$

$\qquad m = 2, \quad Q_1(A) = 1 + A + A^2$

$\qquad m = 3, \quad Q_1(A) = 1 + A + A^3$

$\qquad m = 4, \quad Q_1(A) = 1 + A^3 + A^4$

$\qquad m = 5, \quad Q_1(A) = 1 + A^3 + A^5,$

$p = 3$

$\qquad m = 2, \quad Q_1(A) = 2 + A + A^2$

$\qquad m = 3, \quad Q_1(A) = 1 + 2A^2 + A^3$

$\qquad m = 4, \quad Q_1(A) = 2 + A^3 + A^4$

$\qquad m = 5, \quad Q_1(A) = 1 + 2A^4 + A^5,$

$p = 5$

$\qquad m = 2, \quad Q_1(A) = 3 + 3A + A^2$

$\qquad m = 3, \quad Q_1(A) = 3 + 4A^2 + A^3$

$\qquad m = 4, \quad Q_1(A) = 3 + 3A^2 + A^3 + A^4$

$\qquad m = 5, \quad Q_1(A) = 3 + 2A^2 + A^5,$

$p = 7$

$\qquad m = 2, \quad Q_1(A) = 5 + 5A + A^2$

$\qquad m = 3, \quad Q_1(A) = 4 + 5A^2 + A^3$

$\qquad m = 4, \quad Q_1(A) = 3 + 3A^2 + 2A^3 + A^4$

$\qquad m = 5, \quad Q_1(A) = 2 + 2A + A^5.$

3.8 Convolution Algorithms Over $Z(M)$ and NTTs

In this section, we develop a simple Legrange interpolation (LI) algorithm for computing acyclic convolution of two sequences defined over $Z(M)$. Secondly, the NTTs are demonstrated to be a special case of LI if all the interpolating points are in a special geometric order. This approach will be generalized in Chapter 4. To begin with, we review briefly the concept of LI.

Let $X(u)$ be a polynomial of degree $n-1$ and be represented as $X(u) = \sum_{i=0}^{n-1} X_i u^i$. LI deals with the problem of interpolating $X(u)$ through n points. This suggests a method for determining the coefficients of $X(u)$ based on the values of the polynomial $X(u)$ evaluated at n points A_i, $i = 1, 2, \ldots, n$; namely, $X(A_1)$, $X(A_2)$, ..., $X(A_n)$. First, we assume that all the coefficients of $X(u)$ are defined over a field. Then the LI algorithm is the same as the CRT-P when the modulo polynomial $M(u)$ is set equal to the product of n degree-one polynomials $(u - A_i)$. In other words, $m_i(u) = (u - A_i)$, $i = 1, 2, \ldots, n$. According to (3.4.2), $C(u)$ can be reconstructed using CRT-P or LI as follows,

$$X(u) = \sum_{i=1}^{n} X(A_i) \prod_{\substack{j=1 \\ j \neq i}}^{n} \frac{(u - A_j)}{(A_i - A_j)}. \tag{3.8.1}$$

Recall that the CRT-P requires that all factor polynomials $m_i(u) = (u - A_i)$ be pairwise relatively prime. When A_i, $i = 1, 2, \ldots, n$ are elements of a field, the only constraint for a unique solution is $A_i \neq A_j, i \neq j$.

Alternatively, the LI algorithm can be represented as a Vandermonde system of linear equations

$$\begin{bmatrix} 1 & A_1 & A_1^2 & \cdots & A_1^{n-1} \\ 1 & A_2 & A_2^2 & \cdots & A_2^{n-1} \\ \vdots & \vdots & \vdots & \ddots & \vdots \\ 1 & A_n & A_n^2 & \cdots & A_n^{n-1} \end{bmatrix} \begin{bmatrix} X_0 \\ X_1 \\ \vdots \\ X_{n-1} \end{bmatrix} = \begin{bmatrix} X(A_1) \\ X(A_2) \\ \vdots \\ X(A_n) \end{bmatrix}. \tag{3.8.2}$$

Let the Vandermonde matrix be denoted by $\mathbf{V}$. The determinant of $\mathbf{V}$ is given by

$$det(\mathbf{V}) = \prod_{i=1}^{n-1} \prod_{j=i+1}^{n} (A_i - A_j). \tag{3.8.3}$$

The condition for the existence of a unique solution to the Vandermonde system shown in (3.8.2) is $det(\mathbf{V}) \neq 0$ or $A_i \neq A_j, i \neq j$. *Consequently, we may interpret LI as a special case of the CRT-P when all the polynomial factors are of degree one.*

Next, we turn our attention to the finite ring of integers $Z(M)$ and demonstrate the existence of a LI algorithm for interpolating $X(u)$ defined over $Z(M)$.

Consider the system of linear congruences in (3.8.2) defined mod M. Using the Theorem 2.1 and (3.8.3), we see that a unique solution to the Vandermonde system exists *iff*

$$(A_i - A_j, M) = 1, \ 1 \le i, j \le n, i \ne j. \tag{3.8.4}$$

Recall from Section 3.5 that A_1, A_2, ..., A_n that satisfy (3.8.4) are said to be pairwise differentially coprime to M. Thus we conclude that

THEOREM 3.12 *The n coefficients of a degree $n-1$ polynomial $X(u)$ can be uniquely determined over $Z(M)$ from the values of $X(u)$ evaluated at $u = A_i, i = 1, 2, \ldots, n$, iff these n integers are pairwise differentially coprime to M.*

The Vandermonde system of linear congruences can be solved by either using the LI in (3.8.1) or the order-recursive algorithm for the Vandermonde systems. The order-recursive algorithm is applicable in the present situation as well since all the principal submatrices of $\mathbf{V}$ are also Vandermonde matrices having determinants that are coprime to M. The largest number of integers that are pairwise differentially coprime to M is p, $p = min\{p_1, p_2, \ldots, p_t\}$. Thus the largest value of n for a uniquely solvable Vandermonde system of linear congruences is $n = p$. Also, the condition in (3.8.4) for the existence of a unique solution simplifies to

$$(A_i - A_j, p_\theta) = 1, \ \theta = 1, 2, \ldots, t. \tag{3.8.5}$$

This approach can also be used to compute the acyclic convolution of two sequences over $Z(M)$ in a computationally efficient manner. Such an algorithm to compute $C(u) = A(u) \cdot B(u), deg(C(u)) < n$, consists of the following three steps:

1. Given n integers $D_1, D_2, \ldots, D_n$ pairwise differentially coprime to $M, n \le p$, compute
   $E_i = A(D_i) \bmod M$
   $F_i = B(D_i) \bmod M, \ i = 1, 2, \ldots, n.$

2. Evaluate
   $G_i = C(D_i) = E_i \cdot F_i \bmod M, \ i = 1, 2, \ldots, n.$

3. Use LI in (3.8.1) or solve the Vandermonde system of linear congruences, $\mathbf{V}\underline{C} \equiv \underline{G} \bmod M$ to get $C_0, C_1, \ldots, C_{n-1}$.

In the above algorithm, it is generally expected that the computations involved in steps 1 and 3 can be performed efficiently, since the scalars $D_1, D_2, \ldots, D_n$ are known a priori. Also, since the choice of $D_1, D_2, \ldots, D_n$ is non-unique, they may be chosen in a way so as to possess additional mathematical properties which, in turn, can be exploited to realize further computational savings. In polynomial notation, we can equivalently express the above LI based approach as a direct sum over a ring of polynomials defined over $Z(M)$

$$< X(u) >_{M(u)} \;\approx\; < X(u) >_{m_1(u)} \oplus < X(u) >_{m_2(u)}$$
$$\oplus \cdots \oplus < X(u) >_{m_n(u)} \tag{3.8.6}$$

$$m_i(u) = u - A_i \tag{3.8.7}$$

and

$$M(u) = \prod_{i=1}^{n} m_i(u). \tag{3.8.8}$$

This is analogous to the direct sum in (3.4.9) valid for the ring of polynomials defined over a field.

We finally observe that the above algorithm for computing acyclic convolution may be available only for small values of n as $n \leq p$. This algorithm will be generalized in Chapter 4. In Chapter 4, we establish the direct sum property and derive fast algorithms for acyclic and cyclic convolution of length n for sequences defined over $Z(M)$ for all values of n.

In many instances, there are computational benefits in expressing the interpolating points as powers of a single value. Let us say that the scalars $A_i \in Z(M)$ are to be chosen such that

$$A_i = G^i \bmod M, \; 1 \leq i \leq n. \tag{3.8.9}$$

Once again, $n \leq min\{p_1, p_2, \ldots, p_t\}$. Applying (3.8.4) to the above case, we see that a unique solution will exist *iff*

$$(G, M) = 1, \tag{3.8.10}$$

and

$$(G^i - 1, M) = 1, \; i = 1, 2, \ldots, n - 1. \tag{3.8.11}$$

Finding an integer G over $Z(M)$, $M = \prod_{j=1}^{t} p_j^{\alpha_j}$ is equivalent to finding an integer G_j over $Z(p_j^{\alpha_j})$ that satisfies

$$(G_j, p_j^{\alpha_j}) = 1 \tag{3.8.12}$$

and
$$(G_j^{i-1}, p_j^{\alpha_j}) = 1, \; i = 1, \ldots, n-1; \; j = 1, \ldots, t, \tag{3.8.13}$$

and then using the CRT-I to get G over $Z(M)$. The above conditions further simplify to
$$(G_j, p_j) = 1, \tag{3.8.14}$$

and
$$(G_j^i - 1, p_j) = 1. \tag{3.8.15}$$

Therefore, one has to compute an integer G over $Z(p^\alpha)$ such that
$$(G, p) = 1, \tag{3.8.16}$$

and
$$(G^i - 1, p) = 1, \; i = 1, 2, \ldots, n-1, \; n \leq p. \tag{3.8.17}$$

Consider an integer $A \in Z(p^\alpha)$ such that $(A, p) = 1$. We can write $\alpha A + \beta p = 1$. Also, dividing A by p, we have $A = Qp + A'$, $A' \in GF(p)$. Combining the two expressions leads to $\alpha' A' + \beta' p = 1$, that is, $(A', p) = 1$. An important consequence is that finding an integer G over $Z(p^\alpha)$ that satisfies (3.8.16) and (3.8.17) is the same as finding an integer over $GF(p)$. Expressed over $GF(p)$ (3.8.16) and (3.8.17) become,
$$(G, p) = 1, \quad G \in GF(p) \tag{3.8.18}$$

$$(G^i - 1, p) = 1, \; i = 1, 2, \ldots, n-1; \; n \leq p. \tag{3.8.19}$$

Since $G < p$, (3.8.18) is trivially satisfied for $G \neq 0$. Also, $G^i - 1 \bmod p < p$ and therefore (3.8.19) is satisfied as well except when $G^i - 1 \equiv 0 \bmod p$ or $G^i \equiv 1 \bmod p$, $i = 1, 2, \ldots, n-1$. Therefore, G can take only those values that have order $\geq n$ over $GF(p)$. In $GF(p)$, there exist integers of order up to $p - 1$. Consequently, a value can always be assigned to G provided that $n \leq p - 1$. This discussion leads to an important theorem.

THEOREM 3.13 *All the n interpolating points can be chosen as powers of an integer G in $Z(M)$, $M = \prod_{j=1}^t p_j^{\alpha_j}$ provided that G has order at least n in each of $GF(p_j)$, $n \leq p_j - 1$.*

The interpolating polynomials can now be written as
$$m_i(u) = u - G^i \tag{3.8.20}$$

$$M(u) = \prod_{i=1}^n m_i(u). \tag{3.8.21}$$

Thus acyclic convolution of length up to $p - 1$ can be computed by using interpolating points that are powers of a single integer.

A special case of the above approach to computing the acyclic convolution gives rise to number-theoretic-transforms (NTTs). In addition to the constraints in (3.8.12) and (3.8.13) for a unique solution to the Vandermonde system of linear congruences, NTTs further require that

$$M(u) = u^n - 1. \tag{3.8.22}$$

The resulting convolution is the *cyclic convolution* over $Z(M)$. Since G is a root of $M(u)$, $G^n - 1 = 0 \bmod M$, or G has order n over $Z(M)$. Equivalently, G has order n over $Z(p_j^{\alpha_j})$, $j = 1, 2, \ldots, t$. Also, according to (3.8.13), order of G cannot be less than n or else if order of $G = a$, $a < n$, then $G^a - 1 \equiv 0 \bmod p_j^{\alpha_j}$ and (3.8.13) is violated. Thus, the integer G is an element of order exactly equal to n over $Z(p_j^{\alpha_j})$, $j = 1, 2, \ldots, t$. An element of order n over $Z(p^\alpha)$ also implies that it has order n over $GF(p)$. Since the order of every element in $GF(p)$ divides $p - 1$ (Theorem 3.8), the value of n gets further constrained to $n | (p_j - 1)$, $j = 1, 2, \ldots, t$.

Given a primitive element G_1 over $GF(p)$, an element of order n over $Z(p^\alpha)$ can be obtained as

$$G \equiv G_1^{p^{\alpha-1}(p-1)/n} \bmod p^\alpha. \tag{3.8.23}$$

This statement can be proven as follows. Since $(G_1, p) = 1$, we have $(G_1, p^\alpha) = 1$ and $G^n = G_1^{p^{\alpha-1}(p-1)} \equiv 1 \bmod p^\alpha$ (Corollary 3.1). Also, if the order of G is a, $G^a \equiv 1 \bmod p^\alpha$, $a < n$, then

$$G_1^{ap^{\alpha-1}(p-1)/n} \equiv 1 \bmod p^\alpha. \tag{3.8.24}$$

Taking mod p of both sides leads to

$$G_1^{p^{\alpha-1}(p-1)a/n} \equiv 1 \bmod p.$$

Over $GF(p)$, $G_1^{p-1} \equiv 1 \bmod p$ or $G_1^p \equiv G_1 \bmod p$. Applying this argument to the above equation, we get

$$G_1^{(p-1)a/n} \equiv 1 \bmod p.$$

Since G_1 is a primitive element, we must have $a = n$ as $G_1^i \neq 1 \bmod p$ for $i < p - 1$. This discussion leads to the definition of NTTs which is summarized in the following.

THEOREM 3.14 *An NTT of length n exists over Z(M), $M = \prod_{i=1}^{t} p_i^{\alpha_i}$ provided that*

$$n|(p_i - 1), i = 1, 2, \ldots, t. \tag{3.8.25}$$

The value of the generator G over Z(M) can be computed using the CRT-I as

$$G = G_1 \oplus G_2 \oplus \ldots \oplus G_t,$$
$$G_i = A_i^{p_i^{\alpha_i - 1}(p_i - 1)/n}, \tag{3.8.26}$$

A_i *being a primitive element of $GF(p_i)$.*

Finally, it can be verified that in this case the inverse of the Vandermonde matrix is given by

$$\mathbf{V}^{-1} = \mathbf{W}^t, \tag{3.8.27}$$

where $\mathbf{W}$ is also a Vandermonde matrix. The (ij)th element of $\mathbf{W}$ is given by $G^{-i(j-1)}$.

The constraint on the length of the NTTs in (3.8.25) is a very severe one. For example, for even values of M, the maximum value of $n = 1$. This has restricted the usefulness of NTTs for computing the cyclic and acyclic convolution. This constraint is also known as the word sequence length constraint (WSLC) problem. There are two approaches to relaxing this limitation. The first is based on an extension of the CRT-P to polynomials over the ring $Z(M)$. It will be presented in Chapter 4. The second one is based on the application of the above described algorithm for computing the acyclic convolution.

A cyclic convolution of length n can be computed by first computing the ordinary polynomial product (acyclic convolution) of length $n' \leq 2n-1$ and then reducing the resulting polynomial mod $u^n - 1$. Acyclic convolution algorithms over $Z(M)$ exist for value of n' up to p, $p = min\{p_1, p_2, \ldots, p_t\}$, $M = \prod_{i=1}^{t} p_i^{\alpha_i}$. Therefore, cyclic convolution of any length $n \leq (p+1)/2$ can be computed using this approach.

3.8.1   Mersenne and Fermat Number Transforms

NTTs defined on certain integer rings have attracted more attention than others due to ease in the implementation of the associated arithmetic. Two such integer rings are the rings defined mod M, $M = 2^p - 1$ and $M = 2^{2^t} + 1$. Integers of the type $2^p - 1$ are called the Mersenne numbers and the corresponding NTT is called the Mersenne-Number-Transform (MNT). Integers of the type $M = 2^{2^t} + 1$ are called Fermat numbers and the corresponding NTT is called the Fermat-Number-Transform (FNT). However, like all

NTTs, the MNT and the FNT also suffer from the WSLC problem and will not be pursued further in this work.

3.9   The Problem of Error Control

With the advent of VLSI design methodology and concepts of parallel processing, there has been tremendous thrust toward high performance computing systems that possess fault tolerance capability. Such systems provide correct results in the presence of one or more than one faulty processors. Algorithm based fault tolerance is one of the most important and widely studied methods. It has found numerous applications in the design of fault-tolerant multiple processor systems for signal processing and related tasks.

Given a data vector $\underline{A}$ of length k defined over $Z(M)$, $\underline{A} = [A_1 A_2 \cdots A_k]$, $A_i \in Z(M)$, $i = 1, 2, \ldots, k$, algorithm based fault tolerance technique consists in the design of an (n, k) code over $Z(M)$ that is used to compute the corresponding codevector $\underline{V}$ of length n as a function of $\underline{A}$, that is, $\underline{V} = F(\underline{A})$. Let $\underline{V}$ be changed to a vector $\underline{R}$ due to presence of faults during computation. In this case, we write $\underline{R} = \underline{V} + \underline{E}$, or

$$R_i = V_i + E_i \bmod M, \ i = 1, 2, \ldots, n,$$

where $\underline{E}$ is the fault or error vector, $\underline{E} = [E_1 E_2 \ldots E_n]$. If $E_i \neq 0$, then an error is said to have taken place in the ith position.

In Chapter 6, we develop a mathematical framework for building algorithm based fault tolerance capability in vectors defined over $Z(M)$. Given this framework, we also focus on algorithms for processing $\underline{R}$ for the purpose of error detection and correction.

NOTES
This chapter is a potpourri of old and new results relevant to computing the cyclic and acyclic convolution in a computationally efficient manner. Polynomial algebra over a field is well established and we have selected those results that are useful in the further development and understanding of new algorithms described in Chapter 4. We have used the books by Mc-Clellan and Rader [1979], Nussbaumer [1981] and Blahut [1984] to enhance and select the topics in number theory that are relevant to digital signal processing. Specifically, the proof of CRT-P follows the proof in Blahut [1984] quite closely.

Polynomial algebra over an integer ring does not seem to be well studied from a digital signal processing point of view. We hope to rectify this situation in this and the next two chapters. In this regard, portions of Section 3.5 and the entire Section 3.8 are entirely new and based on our research

[Lin et al. 1994, Krishna et al. 1994]. The order recursive algorithm for solving a Vandermonde system of linear equations mentioned in Section 3.8 can be found in Golub and Van Loan [1983].

Factorization of $u^n - 1$ over $GF(p)$ is crucial to digital signal processing algorithms for sequences defined over $GF(p)$ as well as $Z(M)$. Section 3.7 is a collection of all those results that pertain to this factorization. The book by Peterson and Weldon [1972] has influenced our understanding of the algebras of finite fields and polynomials. The primitive polynomials listed in Section 3.7.2 are taken from Lidl and Niederreiter [1983] which contains a more extensive list of the same.

We have chosen to limit our attention to only those results in number theory that are of relevance in the design of digital signal processing algorithms and error control techniques. In our approach, the Euler's totient function *does not* play an important role. Consequently, we do not elaborate it much. Finally, we have proven every result in order to make this chapter completely self contained with perhaps the exception of the unique factorization property of polynomials over a field. This has been our philosophy throughout this book.

3.10   Bibliography

[3.1] R.E. Blahut, *Fast Algorithms for Digital Signal Processing*, Addison Wesley Publishing Co., 1984.

[3.2] J.H. McClellan and C.M. Rader, *Number Theory in Digital Signal Processing*, Prentice Hall, Inc., 1979.

[3.3] H.J. Nussbaumer, *Fast Fourier Transforms and Convolution Algorithms*, Springer Verlag, 1981.

[3.4] K.-Y. Lin, B. Krishna and H. Krishna, "Rings, Fields, the Chinese Remainder Theorem and an Extension, Part I: Theory," *IEEE Transactions on Circuits and Systems*, to appear in 1994.

[3.5] H. Krishna, K.-Y. Lin and B. Krishna, "Rings, Fields, the Chinese Remainder Theorem and an Extension, Part II: Application to Digital Signal Processing," *IEEE Transactions on Circuits and Systems*, to appear in 1994.

[3.6] W.W. Peterson and E.J. Weldon, Jr., *Error Correcting Codes*, IInd Edition, MIT Press, 1972.

[3.7] G.H. Golub and C.F. Van Loan, *Matrix Computations*, The John Hopkins University Press, 1983.

[3.8] R. Lidl and H. Niederreiter, *Finite Fields*, Addison-Wesley Publishing Co., 1983.

PART II

DIGITAL SIGNAL PROCESSING

Techniques for Convolution and Error Control

Chapter 4

New Algorithms Over Integer Rings

The CRT has been widely employed in designing computationally efficient algorithms in digital signal processing. It has two versions. One is over a ring of integers and the second is over a ring of polynomials with coefficients defined over a field. In this chapter, we extend the CRT to the case of a ring of polynomials with coefficients defined over a finite ring of integers. The entire work is correlated to the already established results on finite fields. This extension serves as a keystone in the design of number theoretic algorithms for performing some of the most computationally intensive tasks. This approach is far superior to the NTTs in the sense that the limitations on both the word length and the sequence length are completely removed. In fact, the NTTs may be considered as a very special case of the general approach. Furthermore, all the computations required in this approach, which inherits the merits of the CRT, can be performed in parallel.

Basically, NTTs have a transform structure identical to discrete Fourier transforms (DFTs), but with complex exponential roots of unity replaced by integer roots and the entire arithmetic defined over $Z(M)$. A major difference between NTTs and the DFTs is that NTTs give rise to *exact* arithmetic while DFTs introduce roundoff errors due to the complex arithmetic operating on the sine and cosine basis functions. In Theorem 3.14 of the previous chapter, we established the necessary and sufficient conditions for the existence of NTTs, with transform length n, in a finite ring of integers $Z(M)$. If M is expressed as $M = p_1^{\alpha_1} p_2^{\alpha_2} \cdots p_t^{\alpha_t}$, then the necessary and sufficient condition for an NTT of length n to exist requires that $n|(p_i - 1)$, $i = 1, 2, \ldots, t$. The above requirement on modulus (M)

and sequence length (n) gives rise to the well-known word sequence length constraint (WSLC) problem. In this case, the length of sequence, n, is severely constrained for a given value of the modulus M, if NTTs are to be employed to compute the cyclic convolution. Even though NTTs require only simple arithmetic like additions and bit shifts, and compute the exact value, their applications have been limited due to the WSLC problem.

There exists another trend in research activity wherein the algebraic structure of the ring of polynomials with coefficients defined over a field is exploited to compute the cyclic convolution in an efficient manner. Many such algorithms employ the CRT-P. In polynomial notation, the cyclic convolution of two length n discrete sequences can be represented as a polynomial product modulo $u^n - 1$. Winograd has derived optimal short-length convolution algorithms based on the CRT-P. Winograd's theorem states that the minimum number of MULT required to carry out a length n convolution is $2n - k$, where k is the number of irreducible polynomial factors of the modulo polynomial.

The DFT may also be interpreted as a special case of the CRT-P with coefficients defined over the field of complex numbers, where the computations are performed modulo $u^n - 1$. It is trivial to see that $u^n - 1$ factors into n degree-one polynomials of the type $u - w^i$, where w is the nth primitive root of unity in the field of complex numbers. This interpretation is also the major impetus for this work which focuses on finding the CRT-P for polynomials with coefficients defined over $Z(M)$, thereby generalizing the NTTs.

The goal of this work is to complete the theories of rings and fields and provide what we feel is a missing link between them. Specifically, we extend the CRT-P to the case of a ring of polynomials with coefficients from a finite ring of integers. We term this extension as the American-Indian-Chinese extension of the CRT (AICE-CRT). This extension serves a key role in the design of fast algorithms for computing the cyclic as well as acyclic convolution. In this regard, the WSLC problem caused by NTTs is completely solved. Also, parallel processing techniques inherent in the CRT may be employed in order to perform the necessary computations.

One of the major elements of the CRT-P is the factorization of the modulo polynomial $P(u)$ over $Z(M)$. We begin this chapter by developing techniques for monic polynomial factorization over $Z(M)$.

4.1   Monic Polynomial Factorization

In this section, we focus on factorization of *monic* polynomials over $Z(p^\alpha)$. Later, the CRT-I and the direct sum in (3.5.2) are used to obtain a non-unique factorization over $Z(M)$ based on the factors over the subrings. Results of this section will be utilized in developing subsequent sections.

The following three theorems establish relationships between polynomials defined over $Z(p^\alpha)$ and the corresponding polynomials defined over $GF(p)$. Theorem 4.1 is one of the versions of the Hensel's theorem. This modified version plays a pivotal role in our analysis. Theorems 4.2 and 4.3 can be used in determining the irreducibility and relatively prime properties for polynomials over $Z(p^\alpha)$. The definitions of factor polynomial, irreducible polynomial and relatively coprime polynomials are given in definitions 4.1, 4.2 and 4.3 respectively. The definition 4.1 is the same as the definition 3.5 in Section 3.5. Interestingly, these determinations are made in the finite field $GF(p)$. In this regard, many fruitful results that have been developed in the context of a field are applied to an integer ring.

DEFINITION 4.1 Let $f(u)$ be a polynomial over the ring $Z(M)$. A polynomial $h(u)$ defined over $Z(M)$ has $f(u)$ as its factor whenever $h(u)$ and $f(u)$ can be written as $h(u) = f(u)q(u)(\text{mod } M)$, where $q(u)$ is a polynomial defined over $Z(M)$.

One has to be careful in interpreting the above definition. Many results in polynomial algebra over a field are no longer valid over a ring. For example, $(3u^2 + 5u + 1)(6u^2 + 5u + 5) = u^2 + 3u + 5 = (u + 4)(u + 8)$ over $Z(9)$. This demonstrates that (i) the sum of the degrees of the factor polynomials may not be equal to the degree of the product, and (ii) the factorization of the polynomial over $Z(M)$ is non-unique. Furthermore, we can not even test if $3u^2 + 5u + 1$ divides $u^2 + 3u + 5$ over $Z(9)$ or not. To circumvent these problems, we will *limit our discussion to the factorization of monic polynomials over $Z(p^\alpha)$ into factor polynomials that are monic as well.* However, we choose to retain the term *monic* whenever needed for the sake of readability.

DEFINITION 4.2 A monic polynomial is termed *irreducible* if it has no other monic polynomial as its factor except itself.

DEFINITION 4.3 If two monic polynomials have no common monic factors over $Z(p^\alpha)$, then they are said to be *relatively prime* (or *coprime*).

It may be verified that for monic polynomials $f(u)$ and $h(u)$, if $h(u) = f(u)q(u) \ (\text{mod } M)$, then (i) $q(u)$ is monic, and (ii) $deg(h(u)) = deg(f(u)) + deg(q(u))$. Also, given $f(u)$ and $h(u)$, it is straightforward to verify if $f(u)$ is a factor of $h(u)$ or not.

THEOREM 4.1 (*Modified Hensel's theorem*): *Let $P(u)$ be a monic polynomial defined over $Z(p^\alpha)$, and $P(u) \equiv \prod_{i=1}^{\beta} P_{i,0}(u)(\mathrm{mod}\, p)$, where $P_{i,0}(u)$ and $P_{j,0}(u), i \neq j$ are coprime polynomials defined over $GF(p)$. Then, there exist β monic polynomials over $Z(p^\alpha)$, denoted by $P_i(u)$, such that $P_i(u) \equiv P_{i,0}(u)(\mathrm{mod}\, p)$ and $P(u) = \prod_{i=1}^{\beta} P_i(u)(\mathrm{mod}\, p^\alpha)$.*

PROOF. The original Hensel's theorem is based on the semi-infinite integer ring. On the other hand, the modified version is based on $Z(p^\alpha)$. As $\alpha \to \infty$, $Z(p^\alpha)$ becomes the semi-infinite integer ring. Hence, the Hensel's theorem presented here is a truncated version of the original Hensel's theorem. For the proof, the readers may refer to the detailed analysis, which is also the proof, described in Section 4.1.1.

THEOREM 4.2 *Two monic polynomials defined over $Z(p^\alpha)$, denoted by $P_1(u)$ and $P_2(u)$, have no monic factors in common* (mod p^α) *if they have no monic factors in common* (mod p).

PROOF. Proving the theorem is equivalent to proving the statement: *"If $P_1(u)$ and $P_2(u)$ have at least one monic factor in common* (mod p^α), *say $q(u)$, then they have at least one monic factor in common* (mod p)." Let $P_1(u)$ and $P_2(u)$ be expressed as

$$
\begin{aligned}
P_1(u) &= q(u)A(u) \quad (\mathrm{mod}\, p^\alpha), \\
P_2(u) &= q(u)B(u) \quad (\mathrm{mod}\, p^\alpha).
\end{aligned} \tag{4.1.1}
$$

After taking the modulo p operation in (4.1.1), the result is

$$
\begin{aligned}
P_1(u) &\equiv q_0(u)A_0(u) \quad (\mathrm{mod}\, p), \\
P_2(u) &\equiv q_0(u)B_0(u) \quad (\mathrm{mod}\, p),
\end{aligned} \tag{4.1.2}
$$

where $A(u), B(u)$ and $q(u)$ satisfy the following congruences

$$
\left\{
\begin{aligned}
A(u) &\equiv A_0(u) \quad (\mathrm{mod}\, p), \\
B(u) &\equiv B_0(u) \quad (\mathrm{mod}\, p), \\
q(u) &\equiv q_0(u) \quad (\mathrm{mod}\, p).
\end{aligned}
\right. \tag{4.1.3}
$$

Obviously, (4.1.2) tells us that $q_0(u)$ is the common monic factor between $P_1(u)$ and $P_2(u)$ mod p. This completes the proof.

We observe that Theorem 4.2 is only a sufficient condition. If $P_1(u)$ and $P_2(u)$ defined mod p^α have a monic factor $q_0(u)$ in common mod p, then we may conclude that $P_1(u)$ and $P_2(u)$ have monic factors $q_1(u)$ and $q_2(u)$ respectively mod p^α, such that $q_1(u) \equiv q_0(u)$ mod p and $q_2(u) \equiv$

$q_0(u)$ mod p. It is not necessary that $q_1(u) = q_2(u)$ mod p^α.

THEOREM 4.3 *A monic polynomial $P(u)$ defined over $Z(p^\alpha)$ is an irreducible polynomial iff $P_0(u) \equiv P(u)$ (mod p) is an irreducible polynomial over the finite field $GF(p)$.*

PROOF. We prove the sufficiency first. Assume that $P(u)$ is a reducible polynomial over $Z(p^\alpha)$ and has two monic factors; namely, $A(u)$ and $B(u)$. Therefore, $P(u) = A(u)B(u)$. Applying the mod p operation we have $P(u) \equiv A_0(u)B_0(u)$ mod p. Obviously, $P(u)$ has $A_0(u)$ and $B_0(u)$ as its factors mod p, and therefore is not an irreducible polynomial over $GF(p)$. Secondly, we prove the necessity. Let $P_0(u) \equiv P(u)$ mod p. Assume that $P_0(u)$ is a reducible monic polynomial defined over $GF(p)$ and has at least two relatively prime monic factors; namely, $A_0(u)$ and $B_0(u)$. By applying Theorem 4.1, $P(u) = A(u)B(u)(\text{mod } p^\alpha)$, where $A(u)$ and $B(u)$, both defined over $Z(p^\alpha)$, satisfy $A_0(u) \equiv A(u)$ mod p and $B_0(u) \equiv B(u)$ mod p respectively. Hence, $P(u)$ is a reducible monic polynomial over $Z(p^\alpha)$. This proves the theorem.

Theorems 4.1, 4.2 and 4.3 serve as a bridge connecting $GF(p)$ and $Z(p^\alpha)$. Indeed, we plan to present the procedure for polynomial factorization over an integer ring based on factorization over a finite field because researchers have already studied in great detail the algebraic structure of finite fields. Methods are well known for selecting relatively prime factors of a polynomial in a way that the multiplicative complexity of the CRT-P based algorithms for cyclic and acyclic convolution is as low as possible. The detailed analysis of polynomial factorization over $Z(p^\alpha)$ is presented in the following.

4.1.1 Monic Polynomial Factorization Over $Z(p^\alpha)$

In polynomial notation, a cyclic convolution can be viewed as a product of two polynomials modulo $(u^n - 1)$; therefore, research work relating to computation modulo $(u^n - 1)$ over a field has attracted a lot of attention. It was shown in Section 3.7 that, by using a primitive root of the extended finite field $GF(p^m)$ with an appropriate value of the integer m, $u^n - 1$ can be factorized systematically into the product of irreducible polynomials over $GF(p)$. In this regard, we may assume that the polynomial factorization of any polynomial over $GF(p)$ into a product of its irreducible or powers of irreducible factors is always known. Based on factorization over $GF(p)$, we present a procedure for factorizing a monic polynomial over the ring $Z(p^\alpha)$.

Let the monic polynomial be $P(u)$. According to Definition 3.4, $P(u)$ can be expressed in its p-adic form as shown in (3.5.3) and (3.5.4). Define a series of polynomials as

$$A_k(u) = P(u) \bmod p^{k+1}, \; 0 \le k \le \alpha - 1. \tag{4.1.4}$$

Clearly, $A_0(u) = P_0(u)$ and $A_{\alpha-1}(u) = P(u)$. Now, we introduce a recursive procedure that utilizes Theorem 4.1 and all the $A_k(u)$ polynomials to factorize $P(u)$. We start with $A_0(u)$ and $A_1(u)$. Assume that a set of relatively prime factors of $P_0(u)$ is known over $GF(p)$; namely, $\{B_{01}(u), B_{02}(u), \ldots, B_{0\beta}(u)\}$. In the most general case, these polynomials are either irreducible or power of an irreducible polynomial over $GF(p)$. Also, they are unique and pairwise relatively prime. However, in the present situation, we only require that the factor polynomials be relatively prime over $GF(p)$. Hence,

$$A_0(u) = \prod_{i=1}^{\beta} B_{0i}(u) \quad (\bmod \; p). \tag{4.1.5}$$

From (4.1.4), $A_1(u)$ equals

$$A_1(u) = P_0(u) + pP_1(u). \tag{4.1.6}$$

On the other hand, by applying Theorem 4.1, $A_1(u)$ may be expressed as

$$A_1(u) = \prod_{i=1}^{\beta} [B_{0i}(u) + pQ_{1i}(u)] \bmod p^2. \tag{4.1.7}$$

Equating (4.1.6) and (4.1.7) and rearranging terms, we have

$$P_0(u) + pP_1(u) = [A_0(u) + pA_0'(u)] + p\Big[\sum_{i=1}^{\beta} Q_{1i}(u) \prod_{\substack{j=1 \\ j \ne i}}^{\beta} B_{0j}(u)\Big] \bmod p^2, \tag{4.1.8}$$

where $A_0'(u)$ is determined by

$$p\, A_0'(u) = [(\prod_{i=1}^{\beta} B_{0i}(u)) - A_0(u)] \bmod p^2. \tag{4.1.9}$$

Cancelling out the prime integer p on both sides and taking mod $B_{0l}(u)$, $l = 1, 2, \ldots, \beta$, in (4.1.8), we get the following β congruences

$$[P_1(u) - A_0'(u)] \equiv Q_{1l}(u) \prod_{\substack{j=1 \\ j \ne l}}^{\beta} B_{0j}(u) \quad (\bmod \; B_{0l}(u), p), \tag{4.1.10}$$

where $A'_0(u)$ is given in (4.1.9), and all $B_{0j}(u)$ and $P_1(u)$ are known. Thus, the lth congruence shown in (4.1.10) will give a closed-form solution for $Q_{1l}(u)$. Note that the solution to congruences in (4.1.10) is unique. This is due to the reason that, first, the term $\prod_{\substack{j=1 \\ j \neq l}}^{\beta} B_{0j}(u)$ and $B_{0l}(u)$ are coprime (mod p); second, the degree of $Q_{1l}(u)$ is less than the degree of $B_{0l}(u)$. Once, we have solved for all $Q_{1l}(u)$, (4.1.7) reveals that $A_1(u)$, defined over $Z(p^2)$, has β factors; namely, $[B_{0i}(u) + pQ_{1i}(u)], i = 1, 2, \ldots, \beta$.

Let $B_{1i}(u) = B_{0i}(u) + pQ_{1i}(u)$ be the ith factor of $A_1(u)$. Also, $B_{1i}(u) \equiv B_{0i}(u)$ (mod p). Because $B_{0i}(u)$ and $B_{0j}(u)$ for $i \neq j$, are coprime, we know from Theorem 4.2 that $B_{1i}(u)$ and $B_{1j}(u)$, defined over $Z(p^2)$, have no common factors. This concludes that $A_1(u)$ have β pairwise relatively prime factors $B_{1i}(u)$. Hence,

$$A_1(u) = \prod_{i=1}^{\beta} B_{1i}(u) \bmod p^2. \tag{4.1.11}$$

Repeating the same analysis for $A_1(u)$ and $A_2(u)$, we have the following β congruences

$$[P_2(u) - A'_1(u)] \equiv Q_{2l}(u) \prod_{\substack{j=1 \\ j \neq l}}^{\beta} B_{1j}(u) \quad (\bmod \ B_{1l}(u), p), \tag{4.1.12}$$

where $A'_1(u)$ can be computed from

$$p^2 \, A'_1(u) = [(\prod_{i=1}^{\beta} B_{1i}(u)) - A_1(u)] \ \bmod p^3. \tag{4.1.13}$$

Once again, (4.1.12) and (4.1.13) give the polynomial factorization of $A_2(u)$ over $Z(p^3)$. Undoubtedly, all these factors $B_{2i}(u) = [B_{1i}(u) + p^2 Q_{2i}(u)], i = 1, 2, \ldots, \beta$ are relatively prime and can be used to construct the factors of $A_3(u)$. The same analysis repeats until $A_{\alpha-1}(u) = P(u)$ is factorized. This completes the systematic procedure.

The following theorem is very important as it states the unique factorization property of the above proposed procedure.

THEOREM 4.4 (*Unique factorization theorem*): *A monic polynomial defined over $Z(p^\alpha)$ has a unique factorization into a product of relatively prime monic polynomials that are either irreducible or power of an irreducible polynomial over $GF(p)$.*

PROOF. Let $P(u)$ be a monic polynomial defined over $Z(p^\alpha)$ and $P_0(u) \equiv P(u)(\bmod\ p)$. It is well known that the prime polynomial factorization of a polynomial with coefficients from a field satisfies the uniqueness property.

Based on this statement, $P_0(u)$ has a unique factorization into pairwise relatively prime factors

$$P_0(u) = \prod_{i=1}^{\beta} B_{i,0}(u) = B_{1,0}(u)B_0'(u), \qquad (4.1.14)$$

where $B_0'(u) = \prod_{i=2}^{\beta} B_{i,0}(u)$. Suppose that Theorem 4.4 fails, which means, the polynomial $P(u)$ has two distinct factorizations over $Z(p^\alpha)$. Thus

$$P(u) = B_1(u)B'(u) = C_1(u)C'(u) \quad (\bmod\ p^\alpha). \qquad (4.1.15)$$

Using the p-adic form and taking mod p^2 on (4.1.15), we have

$$[B_{1,0}(u) + pB_{1,1}(u)][B_0'(u) + pB_1'(u)]$$
$$= [C_{1,0}(u) + pC_{1,1}(u)][C_0'(u) + pC_1'(u)] \quad (\bmod\ p^2). \qquad (4.1.16)$$

Expanding (4.1.16), substituting $B_{1,0}(u) = C_{1,0}(u)$, $B_0'(u) = C_0'(u)$ (recall that $P_0(u)$ has unique factorization over $GF(p)$), and canceling common terms, we get

$$B_{1,0}(u)B_1'(u) + B_{1,1}(u)B_0'(u) = B_{1,0}(u)C_1'(u) + C_{1,1}(u)C_0'(u) \quad (\bmod\ p)$$

or

$$B_{1,0}(u)[B_1'(u) - C_1'(u)] = B_0'(u)[C_{1,1}(u) - B_{1,1}(u)] \quad (\bmod\ p) \qquad (4.1.17)$$

First, $B_{1,0}(u)$ and $B_0'(u)$ are coprime; second, degree of $B_{1,0}(u)$ is greater than degree of $C_{1,1}(u)$ and $B_{1,1}(u)$. Therefore, $B_{1,0}(u)$ does not divide $B_0'(u)[C_{1,1}(u) - B_{1,1}(u)]$. Similarly, $B_0'(u)$ does not divide $B_{1,0}(u)[B_1'(u) - C_1'(u)]$. Hence, the only solution to (4.1.17) is that $B_1'(u) = C_1'(u)$ and $B_{1,1}(u) = C_{1,1}(u)$.

Repeating the same analysis up to mod p^α in (4.1.15), we can conclude that the pairs $[B_1(u), C_1(u)]$ and $[B'(u), C'(u)]$ have the same p-adic form; therefore, they are equal. This proves the theorem.

It is worthwhile noting that each of the relatively prime monic factors in the unique factorization over $Z(p^\alpha)$ is either irreducible over $GF(p)$ or is a power of an irreducible polynomial over $GF(p)$. Interestingly, those monic factor polynomials over $Z(p^\alpha)$ that are powers of an irreducible polynomial over $GF(p)$, may further factor over $Z(p^\alpha)$. This further factorization of such polynomials may be non-unique and is *not* of interest to us in this work as it does not contribute to the AICE-CRT to be developed in Section 4.2. Consider the following example.

EXAMPLE 4.1 Let $P(u) = u^7 + 2u^4 + 2u^3$ (mod 4). Then the prime field is $GF(2)$ and $P_0(u) \equiv P(u)$ (mod 2) $= u^7$. Since $P_0(u)$ is a power of irreducible polynomial over $GF(2)$, factorization of $P(u)$ can not continue further on the basis of the factorization algorithm outlined above. However,

$$
\begin{aligned}
P(u) &= u^7 + 2u^4 + 2u^3 \quad \text{(mod 4)}, \\
&= u^3(u^4 + 2u + 2) \quad \text{(mod 4)}, \\
&= (u^3 + 2)(u^4 + 2) \quad \text{(mod 4)}.
\end{aligned}
$$

In the next subsection, we summarize the procedure by outlining steps involving closed-form equations of all intermediate polynomials. Later, we give two examples to illustrate the concepts.

4.1.2 A Systematic Procedure

Based on the concepts developed in the previous subsection, we summarize the procedure to be followed in order to factorize a given monic polynomial, $P(u)$, defined over $Z(p^\alpha)$.

1. According to the p-adic form of $P(u)$, we obtain $P_i(u)$, $i = 0, 1, \ldots,$ $\alpha - 1$. Note that all the coefficients of $P_i(u)$ are defined over $GF(p)$, even though $P(u)$ is defined over $Z(p^\alpha)$.

2. We form the following polynomials based on $P_i(u)$ obtained in the previous step.

$$
A_k(u) = \sum_{i=0}^{k} p^i P_i(u), \quad k = 0, 1, \ldots, \alpha - 1. \tag{4.1.18}
$$

Clearly, $A_0(u) = P_0(u)$ and $A_{\alpha-1}(u) = P(u)$. In order to factorize $P(u)$, we restate that $P_0(u) \equiv P(u)$ (mod p) has β relatively prime factors over $GF(p)$, denoted by $B_{0i}(u), i = 1, 2, \ldots, \beta$. Hence, the factorization of $A_0(u)$, which equals to $P_0(u)$, is shown in (4.1.5). In the following, we present a recursive methodology such that factorization of $A_k(u)$ can be computed from the factorization of $A_{k-1}(u)$.

3. Let $k = 1$.

4. Compute $A'_{k-1}(u)$ based on $A_{k-1}(u)$ and β factors $B_{k-1,j}(u), j = 1, 2, \ldots, \beta$.

$$p^k \, A'_{k-1}(u) = [(\prod_{i=1}^{\beta} B_{k-1,i}(u)) - A_{k-1}(u)] \mod p^{k+1}. \qquad (4.1.19)$$

5. Solve for $Q_{k,i}(u), i = 1, 2, \ldots, \beta$. The lth one is based on the congruence shown below.

$$[P_k(u) - A'_{k-1}(u)] \equiv Q_{k,l}(u) \prod_{\substack{j=1 \\ j \neq l}}^{\beta} B_{k-1,j}(u) \pmod{B_{k-1,l}(u), p}.$$

$$(4.1.20)$$

6. Then factorization of $A_k(u)$ is available and can be written as

$$A_k(u) = \prod_{i=1}^{\beta} B_{k,i}(u) \bmod p^{k+1},$$

where its ith factor equals to

$$B_{k,i}(u) = [B_{k-1,i}(u) + p^k Q_{k,i}(u)] \mod p^{k+1}. \qquad (4.1.21)$$

7. Increase k by one. If $k = \alpha$, then stop. Otherwise, go to step 4.

The following three examples are given to illustrate how the procedure works. Especially, we focus on the polynomials having the form $u^n \pm 1$ for their obvious importance in digital signal processing. It is seen that, even though two polynomials may have the same factorization over $GF(p)$, their individual factorization over $Z(p^\alpha)$ may be distinct. This reinforces the uniqueness property established in Theorem 4.4.

EXAMPLE 4.2 Factorize $P(u) = u^7 - 1$ over $Z(2^3)$. Obviously, $p = 2, \alpha = 3$. In step 1, we get the p-adic form of $P(u)$: $P_0(u) = u^7 + 1$, $P_1(u) = 1$ and $P_2(u) = 1$. In step 2, based on (4.1.18), we construct $A_1(u) = u^7 + 3$ and $A_2(u) = P(u) = u^7 + 7$. Also, factorization of $P(u)$ over $GF(2)$ is assumed to be known a priori:

$$P_0(u) = (u + 1)(u^3 + u + 1)(u^3 + u^2 + 1).$$

Namely, $B_{0,1}(u) = u + 1$, $B_{0,2}(u) = u^3 + u + 1$, $B_{0,3}(u) = u^3 + u^2 + 1$. We start the recursive procedure, outlined from step 4 to step 6, at $k =$

1. According to (4.1.19), $A'_0(u) = u^6 + u^5 + u^2 + u$. Next, we solve the congruences as shown in (4.1.20) for $Q_{1,i}(u)$, $i = 1, 2, 3$. They are

$$
\begin{cases}
Q_{1,1}(u) &= \quad\quad 1, & (\mathrm{mod}\ B_{0,1}(u)), \\
Q_{1,2}(u) &= \quad u^2 + 1, & (\mathrm{mod}\ B_{0,2}(u)), \\
Q_{1,3}(u) &= u^2 + u + 1, & (\mathrm{mod}\ B_{0,3}(u)).
\end{cases}
\tag{4.1.22}
$$

In step 6, based on (4.1.21), the factors of $A_1(u)$ (defined mod 4) are

$$B_{1,1}(u) = u + 3,\ B_{1,2}(u) = u^3 + 2u^2 + u + 3,\ B_{1,3}(u) = u^3 + 3u^2 + 2u + 3.$$

Repeating step 4 through step 6 for $k = 2$, we obtain

$$
\begin{cases}
Q_{2,1}(u) &= \quad\quad 1, & (\mathrm{mod}\ B_{0,1}(u)), \\
Q_{2,2}(u) &= \quad u^2 + u + 1, & (\mathrm{mod}\ B_{0,2}(u)), \\
Q_{2,3}(u) &= \quad\quad 1, & (\mathrm{mod}\ B_{0,3}(u)).
\end{cases}
\tag{4.1.23}
$$

and $B_{2,1}(u) = u + 7$, $B_{2,2}(u) = u^3 + 6u^2 + 5u + 7$, $B_{1,3}(u) = u^3 + 3u^2 + 2u + 7$. Since $k + 1 = \alpha$, the recursive procedure stops. We write down the polynomial factorization of $P(u)$ over $Z(8)$ as,

$$P(u) = (u + 7)(u^3 + 6u^2 + 5u + 7)(u^3 + 3u^2 + 2u + 7)\ \mathrm{mod}\ 8.$$

EXAMPLE 4.3 Factorize $P(u) = u^7 + 1$ over $Z(2^3)$. The p-adic form gives $P_0(u) = u^7 + 1$, $P_1(u) = 0$ and $P_2(u) = 0$. Note that $P_0(u) = u^7 + 1 = u^7 - 1$ over $GF(2)$, which is same as the polynomial in Example 4.2; therefore, $P_0(u)$ has the same factors: $B_{0,1}(u), B_{0,2}(u), B_{0,3}(u)$. Similarly, based on (4.1.18), we construct $A_1(u) = u^7 + 1$ and $A_2(u) = P(u) = u^7 + 1$. The intermediate polynomials are

$$
\begin{cases}
Q_{1,1}(u) &= \quad 0, & (\mathrm{mod}\ B_{0,1}(u)), \\
Q_{1,2}(u) &= \quad u^2, & (\mathrm{mod}\ B_{0,2}(u)), \\
Q_{1,3}(u) &= \quad u, & (\mathrm{mod}\ B_{0,3}(u)),
\end{cases}
\tag{4.1.24}
$$

$$B_{1,1}(u) = u + 1,\ B_{1,2}(u) = u^3 + 2u^2 + u + 1,\ B_{1,3}(u) = u^3 + u^2 + 2u + 1,$$

$$
\begin{cases}
Q_{2,1}(u) &= \quad 0, & (\mathrm{mod}\ B_{0,1}(u)), \\
Q_{2,2}(u) &= \quad u, & (\mathrm{mod}\ B_{0,2}(u)), \\
Q_{2,3}(u) &= \quad u^2, & (\mathrm{mod}\ B_{0,3}(u)),
\end{cases}
\tag{4.1.25}
$$

$$B_{2,1}(u) = u + 1, B_{2,2}(u) = u^3 + 2u^2 + 5u + 1, B_{2,3}(u) = u^3 + 5u^2 + 2u + 1.$$
Hence, factorization of $P(u)$ over $Z(8)$ is

$$u^7 + 1 = (u + 1)(u^3 + 2u^2 + 5u + 1)(u^3 + 5u^2 + 2u + 1)\ \mathrm{mod}\ 8.$$

EXAMPLE 4.4 Factorize $P(u) = u^8 - 1$ over $Z(3^2)$. Obviously, $p = 3$ and $\alpha = 2$. The p-adic form gives $P_0(u) = u^8 + 2$, $P_1(u) = 2$. Since $\alpha = 2$, in step 2 we obtain $A_0(u) = P_0(u), A_1(u) = P(u)$. The factorization of $P(u)$ over $GF(3)$ is:

$$P_0(u) = (u+1)(u+2)(u^2+1)(u^2+2u+2)(u^2+u+2).$$

Namely, $B_{0,1}(u) = u + 1$, $B_{0,2}(u) = u + 2$, $B_{0,3}(u) = u^2 + 1$, $B_{0,4}(u) = u^2 + 2u + 2$, $B_{0,5}(u) = u^2 + u + 2$, and $\beta = 5$. We start the recursive procedure, outlined from step 4 to step 6, at $k = 1$. According to (4.1.19), $A_0'(u) = 2u^7 + 2u^4 + 2u^2 + 2u + 2$. Next, we solve the congruences as shown in (4.1.20) for $Q_{1,i}(u)$, $i = 1, 2, \ldots, 5$. They are

$$\begin{cases} Q_{1,1}(u) &= \quad 0, & (\bmod\ B_{0,1}(u)), \\ Q_{1,2}(u) &= \quad 2, & (\bmod\ B_{0,2}(u)), \\ Q_{1,3}(u) &= \quad 0, & (\bmod\ B_{0,1}(u)), \\ Q_{1,4}(u) &= u+2, & (\bmod\ B_{0,2}(u)), \\ Q_{1,5}(u) &= u+2, & (\bmod\ B_{0,3}(u)). \end{cases} \qquad (4.1.26)$$

Finally, based on (4.1.21), the monic factors of $P(u)$ over $Z(9)$ are $B_{1,1}(u) = u + 1$, $B_{1,2}(u) = u + 8$, $B_{1,3}(u) = u^2 + 1, B_{1,4}(u) = u^2 + 5u + 8, B_{1,5}(u) = u^2 + 4u + 8$.

Once the monic polynomial factorizations over the rings of integers $Z(p^\alpha)$ are done, we extend the results further by employing CRT-I to obtain factorization over $Z(M)$. This general factorization over $Z(M)$ is non-unique in many respects, and is studied in the following subsection.

4.1.3   Monic Polynomial Factorization over $Z(M)$

The factorization of a monic polynomial $P(u)$ of degree n over $Z(M)$ is considered. Recall that since M is a composite integer, the standard factorization can be applied to M such that $M = p_1^{\alpha_1} p_2^{\alpha_2} \cdots p_t^{\alpha_t}$. Further, we assume that factorizations over the individual subrings $Z(p^\alpha)$ have been done. Namely, they are

$$\begin{cases} P(u) &\equiv P_1^{(1)}(u) \cdot P_2^{(1)}(u) \cdots P_{\gamma_1}^{(1)}(u) & (\bmod\ p_1^{\alpha_1}), \\ P(u) &\equiv P_1^{(2)}(u) \cdot P_2^{(2)}(u) \cdots P_{\gamma_2}^{(2)}(u) & (\bmod\ p_2^{\alpha_2}), \\ &\vdots \\ P(u) &\equiv P_1^{(t)}(u) \cdot P_2^{(t)}(u) \cdots P_{\gamma_t}^{(t)}(u) & (\bmod\ p_t^{\alpha_t}). \end{cases} \qquad (4.1.27)$$

The main idea is to combine the coefficients of the factor polynomials shown in (4.1.27) using the CRT-I to form factorization over $Z(M)$. Assume that

$P(u)$ has γ factors in $Z(M)$; the goal is to find these γ factors. We will soon see that γ can take more than one values; therefore, by setting $\gamma = 2$ tentatively, the main idea can be conceptually illustrated. Later, γ can be generalized to any arbitrary integer values. Let $P_1(u)$ and $P_2(u)$ be the factors to be found. The first step is to divide γ_i factors in $Z(p_i^{\alpha_i})$ into two groups. The products of polynomial factors of each group give rise to two factors in $Z(p_i^{\alpha_i})$, denoted by $P_1'^{(i)}(u)$ and $P_2'^{(i)}(u)$. Then (4.1.27) becomes:

$$\begin{cases} P(u) \equiv P_1'^{(1)}(u) \cdot P_2'^{(1)}(u) & (\text{mod } p_1^{\alpha_1}), \\ P(u) \equiv P_1'^{(2)}(u) \cdot P_2'^{(2)}(u) & (\text{mod } p_2^{\alpha_2}), \\ \quad \vdots \\ P(u) \equiv P_1'^{(t)}(u) \cdot P_2'^{(t)}(u) & (\text{mod } p_t^{\alpha_t}). \end{cases} \quad (4.1.28)$$

By claiming that $P_k(u) \equiv P_k'^{(i)}(u)$, mod $p_i^{\alpha_i}$, $k = 1,2$; $i = 1,2, \ldots, t$, the coefficients of the factors $P_1(u)$ and $P_2(u)$ can be computed using the CRT-I. The same analysis can be extended easily to γ by dividing γ_i factors in $Z(p_i^{\alpha_i})$ into γ groups. In most cases, $\gamma_1, \gamma_2, \ldots, \gamma_t$ are non-equal. If $\gamma > \gamma_i$, then additional 1 (degree zero polynomials) can be attached. It is clear from this description that the factorization over $Z(M)$ is not unique. One reason is that γ can be chosen arbitrarily; the other reason is that there are many ways to divide groups in the first step. If we remove the freedom in selecting γ by restricting $\gamma = max\{\gamma_1, \gamma_2, \cdots, \gamma_t\}$ then the total number of different ways of combination (or factorization) is $(\gamma_1! \gamma_2! \cdots \gamma_t!)$. A very special case arises when $\gamma_i, i = 1,2,\ldots,t$ are all equal to γ and all $P_k'^{(i)}(u), i = 1,2,\ldots,t; k = 1,2,\ldots,\gamma$ are degree-one factors. In this special case, there are a total of $(\gamma!)^t$ different factorizations possible over $Z(M)$. Example 4.5 is given to illustrate the procedure.

EXAMPLE 4.5 Factorize $P(u) = u^7 - 1$ defined over $Z(2^3 3^2) = Z(72)$. Clearly, $p_1 = 2, \alpha_1 = 3$ and $p_2 = 3, \alpha_2 = 2$. In Example 4.2, we have done the polynomial factorization over $Z(2^3)$:

$$P(u) \equiv (u + 7)(u^3 + 6u^2 + 5u + 7)(u^3 + 3u^2 + 2u + 7) \text{ (mod 8)}.$$

Following the systematic procedure outlined in Section 4.1.2, we obtain factorization over $Z(3^2)$ as,

$$P(u) \equiv (u + 8)(u^6 + u^5 + u^4 + u^3 + u^2 + u + 1) \text{ (mod 9)}.$$

Next, we propose two factorizations over $Z(72)$ by different combinations of the factors in $Z(2^3)$ and $Z(3^2)$. Because the CRT-I is used to compute the coefficients of the factors in $Z(72)$, the following information about the scalars T_i in (2.4.2) is necessary: $T_1 = 1, T_2 = 8$. If we combine $(u + 7) \in Z(8)$ and $(u + 8) \in Z(9)$ together, a factor in $Z(72)$ is generated as $(u+71)$. Further, we combine $(u^3 + 6u^2 + 5u + 7)(u^3 + 3u^2 + 2u + 7) \in Z(8)$ and $(u^6 + u^5 + u^4 + u^3 + u^2 + u + 1) \in Z(9)$ to form the second factor $(u^6 + u^5 + u^4 + u^3 + u^2 + u + 1) \in Z(72)$. Hence, one factorization of $P(u)$ that has two factors is

$$P(u) = (u + 71)(u^6 + u^5 + u^4 + u^3 + u^2 + u + 1) \quad (\text{mod } 72).$$

Another factorization which has three factors can be found by employing the following combination: $((u + 7) \in Z(8)$ and $(u + 8) \in Z(9))$, $((u^3 + 6u^2 + 5u + 7) \in Z(8)$ and $(u^6 + u^5 + u^4 + u^3 + u^2 + u + 1) \in Z(9))$, $((u^3 + 3u^2 + 2u + 7) \in Z(8)$ and $1 \in Z(9))$. It leads to $P(u) = (u + 71)(64u^6 + 64u^5 + 64u^4 + u^3 + 46u^2 + 37u + 55)(9u^3 + 27u^2 + 18u + 55)$ (mod 72).

4.2   The AIC Extension of the CRT

In the previous section, we introduced a systematic procedure to factorize a monic polynomial over the ring $Z(p^\alpha)$. This methodology is primarily based on the known factorization of the monic polynomial over the field $GF(p)$. In this section, we present an extension of the CRT-P when coefficients are defined over the integer ring $Z(p^\alpha)$. This extension is termed the AICE-CRT. We demonstrate that the results obtained in the previous sections play a fundamental role in deriving the AICE-CRT. As its predecessor, we expect that the proposed AICE-CRT will find a large number of applications in designing efficient algorithms for many computationally intensive tasks involving arithmetic over integer rings. Later sections of this chapter will be oriented towards the application of AICE-CRT to deriving fast algorithms for cyclic and acyclic convolution of sequences defined over finite integer rings.

Let $X(u)$ and $P(u)$ be two polynomials defined over $Z(p^\alpha)$. The polynomial $X(u)$ is an arbitrary polynomial over $Z(p^\alpha)$ having degree at most $n-1$. *It is necessary and sufficient that $P(u)$ be monic and of degree greater than $n - 1$.* Assume that factorization of $P(u)$ over $Z(p^\alpha)$ has been done by following the steps listed in the previous section; that is,

$$P(u) = \prod_{i=1}^{\beta} P_i(u) \bmod p^\alpha. \tag{4.2.1}$$

Recall that the monic factors $P_i(u)$ and $P_j(u)$, for $i \neq j, 1 \leq i, j \leq \beta$, are relatively prime. This important property will lead to the proposed

AICE-CRT. Define a vector of polynomials, $[X_1(u)\ X_2(u) \cdots X_\beta(u)]$ with the component $X_i(u)$ computed using

$$X_i(u) \equiv X(u) \pmod{P_i(u), p^\alpha}, \quad i = 1, 2, \ldots, \beta. \qquad (4.2.2)$$

The AICE-CRT delivers a unique relationship between the polynomial $X(u)$ and its vector representation. Theorem 4.5 plays a preliminary role in proving the one-to-one correspondence. The AICE-CRT is described in Theorem 4.6.

THEOREM 4.5 *Consider the polynomials $X(u)$, $P_1(u)$ and $P_2(u)$, all defined over $Z(p^\alpha)$. Let $P_1(u)$ and $P_2(u)$ be monic and coprime (mod p). If $P_1(u)$ is a factor of $X(u)$ and $P_2(u)$ is a factor of $X(u)$ then the statement "$P_1(u)P_2(u)$ divides $X(u)$" is true.*

PROOF. Since $P_1(u)$ is a factor of $X(u)$, it follows

$$X(u) = P_1(u)Q(u) \pmod{p^\alpha}. \qquad (4.2.3)$$

Dividing $Q(u)$ by $P_2(u)$ (recall that such a division is possible as $P_2(u)$ is monic), we may write

$$Q(u) = P_2(u)Q'(u) + R(u) \pmod{p^\alpha}, \qquad (4.2.4)$$

where $R(u)$ is the remainder whose degree is less than the degree of $P_2(u)$. Substituting (4.2.4) into (4.2.3), we obtain

$$X(u) = P_1(u)P_2(u)Q'(u) + P_1(u)R(u) \pmod{p^\alpha}. \qquad (4.2.5)$$

Since we have assumed that $P_2(u)$ divides $X(u)$, (4.2.5) reveals that

$$R(u)P_1(u) \equiv 0 \mod{(P_2(u), p^\alpha)}. \qquad (4.2.6)$$

Obviously, (4.2.6) holds true mod p; therefore,

$$R(u)P_1(u) \equiv 0 \mod{(P_2(u), p)}. \qquad (4.2.7)$$

Because $P_1(u)$ and $P_2(u)$ are coprime mod p, in order for (4.2.7) to be true, we must have $R(u) \equiv 0 \mod p$ or

$$R(u) = pR_1(u). \qquad (4.2.8)$$

Substituting (4.2.8) into (4.2.6), and cancelling out common terms, we have

$$R_1(u)P_1(u) \equiv 0 \mod{(P_2(u), p^{\alpha-1})}. \qquad (4.2.9)$$

This analysis is repeated until the arithmetic operations reduce to modulo p. Finally, we have

$$R_{\alpha-1}(u)P_1(u) \equiv 0 \mod (P_2(u), p). \qquad (4.2.10)$$

where $R_{\alpha-1}(u)$ satisfies $R(u) = p^{\alpha-1}R_{\alpha-1}(u) \mod p^\alpha$. Without any doubt, $deg(R_{\alpha-1}(u)) = deg(R(u)) < deg(P_2(u))$. Based on the above statements, along with the fact that $P_1(u)$ and $P_2(u)$ are coprime, the only solution to (4.2.10) is $R_{\alpha-1}(u) = 0$, or $R(u) = 0$. From (4.2.5), if $R(u) = 0$, then $P_1(u)P_2(u)$ is a factor of $X(u)$. This proves the theorem.

THEOREM 4.6 *There exists an one-to-one correspondence between $X(u)$ and the vector representation $[X_1(u)\ X_2(u) \cdots X_\beta(u)]$, as long as $deg(X(u)) < deg(P(u))$ and the monic factors of $P(u)$, denoted by $P_1(u)$, $P_2(u)$, ..., $P_\beta(u)$, are pairwise relatively prime. Here the entire theorem is defined over the ring $Z(p^\alpha)$.*

PROOF. Assume that the theorem is false. This implies that there exists another polynomial of degree less than n defined over $Z(p^\alpha)$ such that it has the same representation vector as the one for $X(u)$. Let this polynomial be $X'(u)$ and $X''(u) = X(u) - X'(u)$. Then, $X''(u)$ has all zeros in its vector representation. In other words, the following equation holds for all the monic factors of $P(u)$

$$P_i(u)|X''(u) \pmod{p^\alpha}, \ i = 1, 2, \ldots, \beta. \qquad (4.2.11)$$

According to Theorem 4.5, (4.2.11) indicates that $[\prod_{i=1}^{\beta} P_i(u)]|X''(u)$. This is not possible as $P(u)$ is a monic polynomial of degree n while $deg(X''(u)) < n$. This proves the theorem.

On the basis of the AICE-CRT, one can form an alternative system for representing polynomials with coefficients from an integer ring. In fact, we can now express the direct sum property of the polynomials over $Z(p^\alpha)$ as,

$$< X(u) >_{P(u)} \ \approx \ < X(u) >_{P_1(u)} \oplus < X(u) >_{P_2(u)}$$
$$\oplus \cdots \oplus < X(u) >_{P_\beta(u)} . \qquad (4.2.12)$$

in a manner analogous to (3.4.9). As its two counterparts (the CRT-I and CRT-P), we expect the AICE-CRT to find extensive applications as a tool to partition a large size computation task into a number of smaller but independent subtasks which may be performed in parallel. We will exploit this belief in succeeding work.

Strictly speaking, the AICE-CRT has two phases. In Theorem 4.6, we have proven that the solution to the set of congruences shown in (4.2.2) is unique. Next, we prove the existence of a solution by giving a procedure to find it.

4.2.1 The AICE-CRT Reconstruction

Assume a monic polynomial $P(u)$ defined over $Z(p^\alpha)$ has β relatively prime factors, $P_1(u)$, $P_2(u)$, ..., $P_\beta(u)$. These factors are obtained based on the systematic procedure introduced in Section 4.1. Every polynomial $X(u)$ with degree less than the degree of $P(u)$ can be represented by an unique vector $\underline{X}(u)$ having β components $X_1(u), X_2(u), \ldots, X_\beta(u)$,

$$X(u) \leftrightarrow \underline{X}(u) = [X_1(u)\, X_2(u) \cdots X_\beta(u)],$$

where $X_i(u)$, the ith residue polynomial corresponding to $P_i(u)$, is given in (4.2.2). According to the AICE-CRT, if the vector $\underline{X}(u)$ is given, then the corresponding polynomial $X(u)$ can be uniquely determined by solving the β congruences in (4.2.2). A closed-form expression for such a solution for $X(u)$ is given by

$$X(u) \equiv \sum_{i=1}^{\beta} X_i(u)T_i(u)P_i'(u) \pmod{P(u), p^\alpha}, \qquad (4.2.13)$$

where $P_i'(u) = \frac{P(u)}{P_i(u)}$. The polynomials $T_i(u), i = 1, 2, \ldots, \beta$ satisfy the congruences,

$$T_i(u)P_i'(u) \equiv 1 \pmod{P_i(u), p^\alpha}. \qquad (4.2.14)$$

Our belief is that everything that can be done in the field $GF(p)$ can be extended to the ring $Z(p^\alpha)$. We now propose a procedure for reconstructing the polynomial $X(u)$ over $Z(p^\alpha)$ from its corresponding residue vector. The goal is to find the polynomials $T_i(u)$, $i = 1, 2, \ldots, \beta$ such that (4.2.14) is satisfied. First, we represent $T_i(u)$ and $P_i'(u)$ in the p-adic form and then take mod p on both sides. Then (4.2.14) becomes

$$T_{i,0}(u)P_{i,0}'(u) \equiv 1 \pmod{P_i(u), p}. \qquad (4.2.15)$$

or $T_{i,0}(u)$ is the inverse of $P_{i,0}'(u)$. The solution to (4.2.15) exists because $P_{i,0}'(u)$ and $P_i(u)$ are coprime. If we perform the mod p^2 operation on the p-adic form, it becomes

$$[T_{i,0}(u) + pT_{i,1}(u)][P_{i,0}'(u) + pP_{i,1}'(u)] \equiv 1 \pmod{P_i(u), p^2}. \qquad (4.2.16)$$

Carrying out the product and substituting

$$T_{i,0}(u)P_{i,0}'(u) \equiv 1 + pQ_{i,1}(u) \pmod{P_i(u), p^2}, \qquad (4.2.17)$$

we get

$$T_{i,0}(u)P_{i,1}'(u) \equiv -[T_{i,1}(u)P_{i,0}'(u) + Q_{i,1}(u)] \pmod{P_i(u), p}. \qquad (4.2.18)$$

Hence, $T_{i,1}(u)$ can be solved as

$$T_{i,1}(u) = -T_{i,0}(u)[T_{i,0}(u)P'_{i,1}(u) + Q_{i,1}(u)] \quad (\text{mod } P_i(u), p). \quad (4.2.19)$$

The same analysis is repeated until $T_{i,2}(u), \cdots, T_{i,\alpha-1}(u)$ are computed. In summary, we may express the closed-form solutions for finding $T_{i,k}(u)$, $k = 1, 2, \ldots, \alpha - 1$ as

$$T_{i,k}(u) = -T_{i,0}(u)[T_{i,0}(u)P'_{i,k}(u) + Q_{i,k}(u)] \quad (\text{mod } P_i(u), p). \quad (4.2.20)$$

Here, $Q_{i,k}(u)$ satisfies the congruence

$$(\sum_{j=0}^{k-1} p^j T_{i,j}(u))(\sum_{j=0}^{k-1} p^j P'_{i,j}(u)) \equiv 1 + p^k Q_{i,k}(u) \quad (\text{mod } P_i(u), p^{k+1}). \quad (4.2.21)$$

Once $T_{i,k}(u)$ are computed, they can be used to construct $T_i(u)$ as

$$T_i(u) = \sum_{k=0}^{\alpha-1} p^k T_{i,k}(u) \quad (\text{mod } p^\alpha), \ i = 1, 2, \ldots, \beta. \quad (4.2.22)$$

The polynomials $T_i(u), i = 1, 2, \ldots, \beta$ satisfy the important property that they are *unique* over $Z(p^\alpha)$. The following example is given to illustrate the reconstruction procedure.

EXAMPLE 4.6 Let $P(u) = u^7 + 1 \quad (\text{mod } 2^3)$. Based on factorization shown in Example 4.3, its monic factors over $Z(8)$ are $P_1(u) = u+1, P_2(u) = u^3 + 2u^2 + 5u + 1$ and $P_3(u) = u^3 + 5u^2 + 2u + 1$. Consequently, $P'_1(u) = u^6 + 7u^5 + u^4 + 7u^3 + u^2 + 7u + 1$, $P'_2(u) = u^4 + 6u^3 + 7u^2 + 3u + 1$ and $P'_3(u) = u^4 + 3u^3 + 7u^2 + 6u + 1$. Based on their p-adic forms, we have $P'_{1,0}(u) = u^6 + u^5 + u^4 + u^3 + u^2 + u + 1$, $P'_{1,1}(u) = P'_{1,2}(u) = u^5 + u^3 + u$; $P'_{2,0}(u) = u^4 + u^2 + u + 1$, $P'_{2,1}(u) = u^3 + u^2 + u$, $P'_{2,2}(u) = u^3 + u^2$; $P'_{3,0}(u) = u^4 + u^3 + u^2 + u$, $P'_{3,1}(u) = P'_{2,1}(u)$, $P'_{3,2}(u) = u^2 + u$.

First, we solve for $T_{1,0}(u), T_{2,0}(u)$ and $T_{3,0}(u)$ according to (4.2.15). They are $T_{1,0}(u) = T_{2,0}(u) = 1$, $T_{3,0}(u) = u^2 + 1$. Then, we compute $Q_{i,1}(u)$, for $i = 1, 2, 3$, from (4.2.21) to get $Q_{1,1}(u) = 0$, $Q_{2,1}(u) = u + 1$, and $Q_{3,1}(u) = u^2$. Substituting into (4.2.20), we solve for $T_{i,1}(u), i = 1, 2, 3$: $T_{1,1}(u) = u$, $T_{2,1}(u) = u^2 + u$, $T_{3,1}(u) = u^2$. Finally, repeating the same procedure for $k = 2$, we get $Q_{1,2}(u) = 1$, $Q_{2,2}(u) = 0$, $Q_{3,2}(u) = 0$; and $T_{1,2}(u) = 0$, $T_{2,2}(u) = u^2 + u + 1$, $T_{3,2}(u) = u + 1$. By (4.2.22), $T_1(u) = 7$, $T_2(u) = 6u^2 + 6u + 5$, $T_3(u) = 3u^2 + 4u + 5$.

Consider the polynomial $X(u) = u^6 + 3u^5 + 7u^3 + u^2 + 6u + 3$ defined over $Z(8)$. Its residue polynomials corresponding to $P_1(u), P_2(u), P_3(u)$ are $X_1(u) = 5$, $X_2(u) = 5u^2 + 2u + 4$, $X_3(u) = u^2 + 2u + 1$, respectively. If we recompute $X(u)$ from (4.2.13), the result is $X(u) = u^6 + 3u^5 + 7u^3 + u^2 + 6u + 3$ which verifies the AICE-CRT and the reconstruction procedure.

4.2.2 AICE-CRT in the Matrix Form

Conceptually, data sequences are converted to polynomials for deriving computationally efficient algorithms. One example is the Winograd's algorithm that employs the CRT-P for convolving two sequences with data defined over finite or infinite fields. In real-time computations, the matrix-vector forms may attract more attention. The matrix form of the algorithm for computing the cyclic convolution of two sequences that employs the Winograd's algorithm is known. We emphasize that the Winograd's algorithm (or the CRT-P) is applied to data sequences that are defined over a field. In this subsection, we complete the analysis on the AICE-CRT by expressing the AICE-CRT in matrix form.

Assume that a monic polynomial $P(u)$, defined over the ring $Z(p^\alpha)$, has β relatively prime factors, $P_1(u)$, $P_2(u)$, ..., $P_\beta(u)$. Every polynomial $X(u)$ with degree less than the degree of $P(u)$ can be represented by an unique vector $\underline{X}(u)$ having β components $X_1(u), X_2(u), \ldots, X_\beta(u)$, where $X_i(u)$, the ith residue polynomial corresponding to $P_i(u)$, is given in (4.2.2). Let the coefficient vectors of $X(u)$ and $X_i(u)$ be $\underline{C}$ and $\underline{C}_i$, respectively. If the polynomial degrees are $deg(X(u)) = n - 1$, $deg(P(u)) = d$, $d \geq n$, $deg(P_i(u)) = d_i$ and $d = d_1 + d_2 + \cdots + d_\beta$, then vector $\underline{C}$ has length n and $\underline{C}_i$ has length d_i. Based on (4.2.2), there exists a matrix $\mathbf{A}_i$ of dimensions $d_i \times n$ such that $\underline{C}$, $\underline{C}_i$ and $\mathbf{A}_i$ are related as

$$\underline{C}_i = \mathbf{A}_i \, \underline{C} \quad (\bmod \; p^\alpha). \tag{4.2.23}$$

The expression in (4.2.23) states the individual relation between $\underline{C}$ and $\underline{C}_i$ for $i = 1, 2, \ldots, \beta$. Combining all β equations as shown in (4.2.23) together, we obtain the following matrix equation:

$$\begin{bmatrix} \underline{C}_1 \\ \underline{C}_2 \\ \vdots \\ \underline{C}_\beta \end{bmatrix}_{d \times 1} = \begin{bmatrix} \mathbf{A}_1 \\ \mathbf{A}_2 \\ \vdots \\ \mathbf{A}_\beta \end{bmatrix}_{d \times n} \underline{C} \quad (\bmod \; p^\alpha). \tag{4.2.24}$$

Note that matrix $\mathbf{A}_i$ depends only on the monic factor polynomial $P_i(u)$, $i = 1, 2, \ldots, \beta$. Thus, they are computed a priori.

On the other hand, the AICE-CRT reconstruction of the polynomial $X(u)$ based on its residue polynomials has been shown in (4.2.13). The matrix form of the AICE-CRT reconstruction is

$$\underline{C} = \sum_{i=1}^{\beta} \mathbf{B}_i \, \underline{C}_i \quad (\bmod \; p^\alpha). \tag{4.2.25}$$

Each matrix $\mathbf{B}_i$ has dimensions $n \times d_i$. Substituting (4.2.23) into (4.2.25), we obtain

$$\underline{C} = [\mathbf{B}_1 \mathbf{B}_2 \cdots \mathbf{B}_\beta] \begin{bmatrix} \mathbf{A}_1 \\ \mathbf{A}_2 \\ \vdots \\ \mathbf{A}_\beta \end{bmatrix} \underline{C} \pmod{p^\alpha}. \qquad (4.2.26)$$

Let $\mathbf{A}^t = [\mathbf{A}_1^t \mathbf{A}_2^t \cdots \mathbf{A}_\beta^t]$, $\mathbf{B} = [\mathbf{B}_1 \mathbf{B}_2 \cdots \mathbf{B}_\beta]$ and $\mathbf{I}_{n \times n}$ be the identity matrix. The superscript t denotes matrix transposition. Then (4.2.26) implies that

$$\mathbf{B}\,\mathbf{A} = \mathbf{I}_{n \times n} \pmod{p^\alpha}. \qquad (4.2.27)$$

Example 4.7 is given to verify the above matrix equality based on the results shown in the previous example.

EXAMPLE 4.7 Let $P(u) = u^7 + 1 \pmod{2^3}$. Factors of $P(u)$ over $Z(8)$ are $P_1(u) = u+1$, $P_2(u) = u^3 + 2u^2 + 5u + 1$, and $P_3(u) = u^3 + 5u^2 + 2u + 1$. Consequently,

$$P_1'(u) = u^6 + 7u^5 + u^4 + 7u^3 + u^2 + 7u + 1,$$

$$P_2'(u) = u^4 + 6u^3 + 7u^2 + 3u + 1,$$

$$P_3'(u) = u^4 + 3u^3 + 7u^2 + 6u + 1.$$

By following the material presented in Section 4.2.1, unique polynomials $T_1(u) = 7$, $T_2(u) = 6u^2 + 6u + 5$, and $T_3(u) = 3u^2 + 4u + 5$ are obtained.

Based on the factors $P_1(u)$, $P_2(u)$ and $P_3(u)$, the matrices $\mathbf{A}_i$, $i = 1, 2, 3$, can be computed. The joint matrix $\mathbf{A}$ is seen to be

$$\mathbf{A} = \begin{bmatrix} \mathbf{A}_1 \\ \mathbf{A}_2 \\ \mathbf{A}_3 \end{bmatrix} = \left[\begin{array}{ccccccc} 1 & 7 & 1 & 7 & 1 & 7 & 1 \\ \hline 1 & 0 & 0 & 7 & 2 & 1 & 5 \\ 0 & 1 & 0 & 3 & 1 & 7 & 2 \\ 0 & 0 & 1 & 6 & 7 & 3 & 1 \\ \hline 1 & 0 & 0 & 7 & 5 & 1 & 2 \\ 0 & 1 & 0 & 6 & 1 & 7 & 5 \\ 0 & 0 & 1 & 3 & 7 & 6 & 1 \end{array} \right].$$

Similarly, based on the polynomials $T_i(u)$ and $P_i'(u)$, matrices $\mathbf{B}_i$, $i =$

1, 2, 3, can be computed. The joint matrix **B** is shown as

$$
\mathbf{B} = [\mathbf{B}_1 \mathbf{B}_2 \mathbf{B}_3] = \begin{bmatrix}
7 & | & 5 & 2 & 6 & | & 5 & 5 & 3 \\
1 & | & 5 & 5 & 2 & | & 2 & 5 & 5 \\
7 & | & 3 & 5 & 5 & | & 6 & 2 & 5 \\
1 & | & 2 & 3 & 5 & | & 5 & 6 & 2 \\
7 & | & 3 & 2 & 3 & | & 6 & 5 & 6 \\
1 & | & 2 & 3 & 2 & | & 5 & 6 & 5 \\
7 & | & 6 & 2 & 3 & | & 3 & 5 & 6
\end{bmatrix}.
$$

It can be verified that the product of matrices **A** and **B** over $Z(8)$ gives the identity matrix having dimension 7. The above matrix representation over $Z(p^\alpha)$ can be extended to matrix representation over $Z(M)$ using the CRT-I and the direct sum approach. This will be further analyzed in later sections.

In Section 3.8, we established the LI based algorithm for computing the cyclic and acyclic convolutions over $Z(M)$ and expressed it as a direct sum in (3.8.6). The AICE-CRT may also be interpreted as a generalization of the LI over $Z(M)$ in the same manner as the CRT-P may be interpreted as a generalization of the LI over a field.

4.3 Convolution Algorithms Over $Z(M)$

In this section, we focus on the application of the AICE-CRT to deriving computationally efficient algorithms for computing cyclic and acyclic convolution of data sequences over $Z(M)$. Several examples are presented to illustrate the approach.

Consider a system which computes the output sequence $\underline{Z}$ as a function of sequences $\underline{X}_1$, $\underline{X}_2$, ... defined over $Z(M)$. The computation is termed *multilinear* if the functional relationship between Z_j, the jth component of $\underline{Z}$, and $\underline{X}_1$, $\underline{X}_2$, ... has the form,

$$
Z_j = \sum_{\forall i_1} \sum_{\forall i_2} \cdots \sum_{\forall i_k} \beta_{j,i_1,i_2,\cdots,i_k} X_{1,i_1} X_{2,i_2} \cdots X_{k,i_k} \mod M. \tag{4.3.1}
$$

Here $\beta_{j,i_1,\cdots,i_k}$ is a constant and X_{θ,i_θ}, the i_θth component of $\underline{X}_\theta$, $\theta = 1, 2, \ldots, k$ is an indeterminate variable over $Z(M)$. The following theorem is fundamental in deriving algorithms for computing multilinear forms over $Z(M)$.

THEOREM 4.7 *An algorithm for computing a multilinear form over $Z(M)$ can equivalently be represented as a direct sum of t algorithms for* *computing the multilinear form over $Z(p_i^{\alpha_i})$, $i = 1, 2, \ldots, t$, $M = \prod_{i=1}^{t} p_i^{\alpha_i}$.*

PROOF. The statement is a result of the direct sum property of the CRT-I in (2.4.9) and the linearity of the modulo operation. Applying the direct sum property to (4.3.1), we get

$$Z_j \bmod M \approx \sum_{\substack{\oplus \\ i=1}}^{t} Z_j \bmod p_i^{\alpha_i}$$

$$= \sum_{\substack{\oplus \\ i=1}}^{t} (\sum_{\forall i_1} \cdots \sum_{\forall i_k} \beta_{j,i_1,i_2,\cdots,i_k} X_{1,i_1} X_{2,i_2} \cdots X_{k,i_k}) \bmod p_i^{\alpha_i}$$

$$= \sum_{\substack{\oplus \\ i=1}}^{t} (\sum_{\forall i_1} \cdots \sum_{\forall i_k} (\beta_{j,i_1,i_2,\cdots,i_k} \bmod p_i^{\alpha_i})(X_{1,i_1} \bmod p_i^{\alpha_i}) \cdots$$

$$\cdots (X_{k,i_k} \bmod p_i^{\alpha_i})) \bmod p_i^{\alpha_i}$$

$$= \sum_{\substack{\oplus \\ i=1}}^{t} \{\sum_{\forall i_1} \cdots \sum_{\forall i_k} \beta_{j,i_1,i_2,\cdots,i_k}^{(i)} X_{1,i_1}^{(i)} X_{2,i_2}^{(i)} \cdots X_{k,i_k}^{(i)}\} \bmod p_i^{\alpha_i}$$

The statement of the theorem follows from the above expression.

A system of bilinear forms over $Z(M)$ in the indeterminates X_i, $i = 1, 2, \ldots, a$ and Y_j, $j = 1, 2, \ldots, b$ can be expressed as

$$Z_k = \sum_{i=1}^{a} \sum_{j=1}^{b} \beta_{k,i,j} X_i Y_j \bmod M, \quad k = 1, 2, \ldots, n. \tag{4.3.2}$$

A bilinear form is a special case of a multilinear form and, therefore, Theorem 4.7 is valid in this case as well. It is well known that without division, an algorithm for computing a system of bilinear forms consists in computing linear combinations of products of pairs of linear forms in X_i, $i = 1, 2, \ldots, a$ and Y_k, $k = 1, 2, \ldots, b$. Thus, these algorithms take the form,

$$\underline{Z} = \mathbf{C} \, \{\mathbf{A} \, \underline{X} \bigodot \mathbf{B} \, \underline{Y}\} \bmod M, \tag{4.3.3}$$

A direct application of Theorem 4.7 implies that

$$\underline{Z} \approx \sum_{\substack{\oplus \\ i=1}}^{t} \underline{Z}^{(i)}, \tag{4.3.4}$$

$$\underline{Z}^{(i)} = \mathbf{C}^{(i)} \{\mathbf{A}^{(i)} \underline{X}^{(i)} \bigodot \mathbf{B}^{(i)} \underline{Y}^{(i)}\}, \tag{4.3.5}$$

$$\mathbf{A}^{(i)} \equiv \mathbf{A} \bmod p_i^{\alpha_i}, \tag{4.3.6}$$

$$\mathbf{B}^{(i)} \equiv \mathbf{B} \bmod p_i^{\alpha_i}, \tag{4.3.7}$$

$$\mathbf{C}^{(i)} \equiv \mathbf{C} \bmod p_i^{\alpha_i}, i = 1, 2, \ldots, t, \tag{4.3.8}$$

and

$$\mathbf{A} \approx \sum_{\substack{\oplus \\ i=1}}^{t} \mathbf{A}^{(i)}, \tag{4.3.9}$$

$$\mathbf{B} \approx \sum_{\substack{\oplus \\ i=1}}^{t} \mathbf{B}^{(i)}, \tag{4.3.10}$$

$$\mathbf{C} \approx \sum_{\substack{\oplus \\ i=1}}^{t} \mathbf{C}^{(i)}. \tag{4.3.11}$$

As a result, the problem of deriving an algorithm for computing the bilinear form over $Z(M)$ (which consists in determining the matrices $\mathbf{A}$, $\mathbf{B}$ and $\mathbf{C}$ over $Z(M)$) is reduced to finding the algorithm for computing the same bilinear form over $Z(p_i^{\alpha_i})$. The matrices $\mathbf{A}$, $\mathbf{B}$ and $\mathbf{C}$ over $Z(M)$ are a direct sum of the corresponding matrices over $Z(p_i^{\alpha_i})$. We will therefore focus on constructing algorithms for computing a bilinear form over $Z(p^\alpha)$.

The acyclic convolution of two sequences $X_i, i = 0, 1, \ldots, d_x - 1$ and $Y_j, j = 0, 1, \ldots, d_y - 1$ over $Z(M)$ can be expressed as the bilinear form,

$$Z_k = \sum_{i=0}^{k} X_i Y_{k-i} \bmod M, \ k = 0, 1, \ldots, n - 1, \tag{4.3.12}$$

where

$$n = d_x + d_y - 1. \tag{4.3.13}$$

Expressed as a polynomial product, it takes the form

$$Z(u) = X(u)Y(u) \pmod{P(u), M}, \tag{4.3.14}$$

where $deg(X(u)) = d_x - 1$, $deg(Y(u)) = d_y - 1$, and $P(u)$ is a monic polynomial that satisfies $deg(P(u)) \geq n$.

The cyclic convolution of two sequences $\underline{X}$ and $\underline{Y}$, $d_x = d_y = n$, can be expressed as the bilinear form

$$Z_k = \sum_{i=0}^{n-1} X_i Y_{<k-i>_n} \bmod M, \ k = 0, 1, \ldots, n - 1, \tag{4.3.15}$$

where the subscript $< k - i >_n$ means $(k - i)$ mod n. Expressed as a polynomial product, it takes the form

$$Z(u) = X(u)Y(u) \quad (\text{mod } u^n - 1, M). \qquad (4.3.16)$$

The fundamental difference between the cyclic and acyclic convolutions is in the choice of the modulo polynomial $P(u)$. For cyclic convolution, $P(u) = u^n - 1$, while for acyclic convolution, $P(u)$ is any arbitrary monic polynomial. This provides considerable flexibility in the design of algorithms for acyclic convolution. In that case, $P(u)$ is selected to reduce the multiplicative complexity of the overall algorithm. We now present the design of computationally efficient algorithms for computing cyclic and acyclic convolutions.

4.3.1   AICE-CRT Based Algorithms for Cyclic Convolution

Consider the cyclic convolution over $Z(p^\alpha)$ expressed as the polynomial product

$$Z(u) = X(u)Y(u) \quad (\text{mod } u^n - 1, p^\alpha). \qquad (4.3.17)$$

Let $P(u) = u^n - 1$. A procedure to obtain the factorization of $u^n - 1$ over $Z(p^\alpha)$ in terms of its irreducible factors begins with the factorization of $u^n - 1$ over $GF(p)$ and employs the Hensel's theorem to obtain recursively the factorization over $Z(p^i)$, $i = 2, 3, \ldots, \alpha$. The factorization of $u^n - 1$ over $GF(p)$ is established in Section 3.7 and the pertinent details of this factorization are summarized in tables 3.1 and 3.2.

Given the factorization of $u^n - 1$ over $Z(p^\alpha)$ as $u^n - 1 = \prod_{j=1}^{s} P_j(u)$, the AICE-CRT based fast algorithm for computing the cyclic convolution over $Z(p^\alpha)$ can be described in three steps as follows,

1. AICE-CRT reduction. Reduce $X(u)$ and $Y(u)$ mod $P_j(u)$ to get

$$
\begin{aligned}
X_j(u) &\equiv X(u) \text{ mod } P_j(u), \\
Y_j(u) &\equiv Y(u) \text{ mod } P_j(u), \quad j = 1, 2, \ldots, s, \qquad (4.3.18)
\end{aligned}
$$

2. Polynomial multiplication.

$$Z_j(u) \equiv X_j(u)Y_j(u) \text{ mod } P_j(u), \quad j = 1, 2, \ldots, s. \qquad (4.3.19)$$

3. AICE-CRT reconstruction.

$$Z(u) \equiv \sum_{j=1}^{s} Z_j(u)T_j(u)P_j'(u) \quad (\text{mod } u^n - 1). \qquad (4.3.20)$$

Since, in general, $deg(P_j(u)) \ll deg(P(u))$, the computation in (4.3.19) can be performed using the algorithms for small degree polynomial multiplication. Let the coefficient vectors of $X(u)$, $Y(u)$, $Z(u)$, $X_j(u)$, $Y_j(u)$ and $Z_j(u)$ be $\underline{X}$, $\underline{Y}$, $\underline{Z}$, $\underline{X}_j$, $\underline{Y}_j$ and $\underline{Z}_j$ respectively. The computation of $\underline{Z}_j$ in (4.3.19) has the form,

$$\underline{Z}_j = \mathbf{D}_j \left\{ \mathbf{E}_j \, \underline{X}_j \bigodot \mathbf{F}_j \, \underline{Y}_j \right\}. \tag{4.3.21}$$

The various polynomial degrees are $deg(X(u)) = deg(Y(u)) = n - 1$, $deg(P_j(u)) = n_j$ and $n = n_1 + n_2 + \cdots + n_s$. According to (4.3.18), there exists a matrix $\mathbf{G}_j$ of dimensions $n_j \times n$ such that $\underline{X}$, $\underline{Y}$, $\underline{X}_j$, $\underline{Y}_j$ and $\mathbf{G}_j$ are related as,

$$\begin{aligned} \underline{X}_j &= \mathbf{G}_j \, \underline{X}, & (4.3.22) \\ \underline{Y}_j &= \mathbf{G}_j \, \underline{Y}, \; j = 1, 2, \ldots, s. & (4.3.23) \end{aligned}$$

The matrix $\mathbf{G}_j$ depends only on the monic factor polynomial $P_j(u), j = 1, 2, \ldots, s$ and is known a priori. The AICE-CRT reconstruction in (4.3.20) can be represented in the matrix form as,

$$\underline{Z} = \sum_{j=1}^{s} \mathbf{H}_j \, \underline{Z}_j. \tag{4.3.24}$$

Substituting (4.3.22) and (4.3.23) in (4.3.21) and replacing $\underline{Z}_j$ in (4.3.24) by the resulting expression, we get,

$$\underline{Z} = \mathbf{C} \left\{ \mathbf{A} \, \underline{X} \bigodot \mathbf{B} \, \underline{Y} \right\}, \tag{4.3.25}$$

where

$$\begin{aligned} \mathbf{A} &= \begin{bmatrix} \mathbf{E}_1 & & \cdots & \\ & \mathbf{E}_2 & \cdots & \\ & & \ddots & \\ & & \cdots & \mathbf{E}_s \end{bmatrix} \begin{bmatrix} \mathbf{G}_1 \\ \mathbf{G}_2 \\ \vdots \\ \mathbf{G}_s \end{bmatrix} \\ &= \mathbf{E} \, \mathbf{G}, & (4.3.26) \end{aligned}$$

$$\begin{aligned} \mathbf{B} &= \begin{bmatrix} \mathbf{F}_1 & & \cdots & \\ & \mathbf{F}_2 & \cdots & \\ & & \ddots & \\ & & \cdots & \mathbf{F}_s \end{bmatrix} \begin{bmatrix} \mathbf{G}_1 \\ \mathbf{G}_2 \\ \vdots \\ \mathbf{G}_s \end{bmatrix} \\ &= \mathbf{F} \, \mathbf{G}, & (4.3.27) \end{aligned}$$

and

$$
\mathbf{C} = [\mathbf{H}_1\ \mathbf{H}_2 \cdots \mathbf{H}_s]
\begin{bmatrix}
\mathbf{D}_1 & & \cdots & \\
 & \mathbf{D}_2 & \cdots & \\
 & & \ddots & \\
 & & \cdots & \mathbf{D}_s
\end{bmatrix}
$$

$$
= \ \mathbf{H}\,\mathbf{D}. \tag{4.3.28}
$$

The matrices $\mathbf{G}$ and $\mathbf{H}$ will be referred to as the AICE-CRT reduction and reconstruction matrices respectively. The number of MULT required for the cyclic convolution over $Z(p^\alpha)$ is the same as sum of the MULT required for the computation in (4.3.19). In other words,

$$
MULT_{Z(p^\alpha)}(n) = \sum_{j=1}^{s} MULT_{Z(p^\alpha)}(n_j). \tag{4.3.29}
$$

Here, $n_j = deg(P_j(u))$. Finally, the overall algorithm for cyclic convolution over $Z(M)$ is a direct sum of the algorithms for cyclic convolution over $Z(p_i^{\alpha_i})$, $M = \prod_{i=1}^{t} p_i^{\alpha_i}$. Therefore, the multiplicative complexity of the cyclic convolution algorithm over $Z(M)$ is the maximum of the multiplicative complexities over $Z(p_i^{\alpha_i})$, $i = 1, 2, \ldots, t$. Interestingly, the computational complexity of the cyclic convolution algorithm over $Z(p^\alpha)$ is the same as the complexity over $GF(p)$. This leads to,

$$
MULT_{Z(M)}(n) = max\{MULT_{GF(p_i)}(n), i = 1, 2, \ldots, t\}. \tag{4.3.30}
$$

We now turn to specific examples to illustrate this approach. The superscript (i) refers to the particular quantity over $Z(p_i^{\alpha_i})$, $i = 1, 2, \ldots, t$.

EXAMPLE 4.8 Let $M = 15$ and $n = 8$. Thus $t = 2$, $p_1 = 3$, $p_2 = 5$, $\alpha_1 = \alpha_2 = 1$. This example demonstrates a simple algorithm obtained by using the CRT-P and the direct sum property. The polynomial factorizations of $P(u) = u^8 - 1$ over $GF(3)$ and $GF(5)$ are given by

$$
u^8 - 1 = (u+1)(u+2)(u^2+1)(u^2+2u+2)(u^2+u+2) \bmod 3,
$$
$$
u^8 - 1 = (u+1)(u+4)(u+2)(u+3)(u^2+2)(u^2+3) \bmod 5.
$$

Hence, the number of irreducible factor polynomials over $GF(3)$ and $GF(5)$ is $s^{(1)} = 5$ and $s^{(2)} = 6$, respectively. Based on these factor polynomials, $T_j^{(i)}(u)$ and $P_j^{'(i)}(u)$ are computed as

$$
P_1^{'(1)}(u) = u^7 + 2u^6 + u^5 + 2u^4 + u^3 + 2u^2 + u + 2,
$$

$$T_1^{(1)}(u) = 1 \mod (u+1),$$
$$P_2'^{(1)}(u) = u^7 + u^6 + u^5 + u^4 + u^3 + u^2 + u + 1,$$
$$T_2^{(1)}(u) = 2 \mod (u+2),$$
$$P_3'^{(1)}(u) = u^6 + 2u^4 + u^2 + 2,$$
$$T_3^{(1)}(u) = 2 \mod (u^2+1),$$
$$P_4'^{(1)}(u) = u^6 + u^5 + 2u^4 + 2u^2 + 2u + 1,$$
$$T_4^{(1)}(u) = 2u+1 \mod (u^2+2u+2),$$
$$P_5'^{(1)}(u) = u^6 + 2u^5 + 2u^4 + 2u^2 + u + 1,$$
$$T_5^{(1)}(u) = u+1 \mod (u^2+u+2),$$
$$P_1'^{(2)}(u) = u^7 + 4u^6 + u^5 + 4u^4 + u^3 + 4u^2 + u + 4,$$
$$T_1^{(2)}(u) = 3 \mod (u+1,)$$
$$P_2'^{(2)}(u) = u^7 + u^6 + u^5 + u^4 + u^3 + u^2 + u + 1,$$
$$T_2^{(2)}(u) = 2 \mod (u+4),$$
$$P_3'^{(2)}(u) = u^7 + 3u^6 + 4u^5 + 2u^4 + u^3 + 3u^2 + 4u + 2,$$
$$T_3^{(2)}(u) = 1 \mod (u+2),$$
$$P_4'^{(2)}(u) = u^7 + 2u^6 + 4u^5 + 3u^4 + u^3 + 2u^2 + 4u + 3,$$
$$T_4^{(2)}(u) = 4 \mod (u+3),$$
$$P_5'^{(2)}(u) = u^6 + 3u^4 + 4u^2 + 2,$$
$$T_5^{(2)}(u) = 2 \mod (u^2+2),$$
$$P_6'^{(2)}(u) = u^6 + 2u^4 + 4u^2 + 3,$$
$$T_5^{(2)}(u) = 3 \mod (u^2+3).$$

Consequently, the AICE-CRT reconstruction matrices $\mathbf{H}^{(1)}$ and $\mathbf{H}^{(2)}$ are generated as,

$$\mathbf{H}^{(1)} = \begin{bmatrix} 2 & | & 2 & | & 1 & 0 & | & 1 & 2 & | & 1 & 1 \\ 1 & | & 2 & | & 0 & 1 & | & 1 & 1 & | & 2 & 1 \\ 2 & | & 2 & | & 2 & 0 & | & 0 & 1 & | & 0 & 2 \\ 1 & | & 2 & | & 0 & 2 & | & 1 & 0 & | & 2 & 0 \\ 2 & | & 2 & | & 1 & 0 & | & 2 & 1 & | & 2 & 2 \\ 1 & | & 2 & | & 0 & 1 & | & 2 & 2 & | & 1 & 2 \\ 2 & | & 2 & | & 2 & 0 & | & 0 & 2 & | & 0 & 1 \\ 1 & | & 2 & | & 0 & 2 & | & 2 & 0 & | & 1 & 0 \end{bmatrix},$$

and

$$\mathbf{H}^{(2)} = \left[\begin{array}{c|c|c|c|cc|cc}
2 & 2 & 2 & 2 & 4 & 0 & 4 & 0 \\
3 & 2 & 4 & 1 & 0 & 4 & 0 & 4 \\
2 & 2 & 3 & 3 & 3 & 0 & 2 & 0 \\
3 & 2 & 1 & 4 & 0 & 3 & 0 & 2 \\
2 & 2 & 2 & 2 & 1 & 0 & 1 & 0 \\
3 & 2 & 4 & 1 & 0 & 1 & 0 & 1 \\
2 & 2 & 3 & 3 & 2 & 0 & 3 & 0 \\
3 & 2 & 1 & 4 & 0 & 2 & 0 & 3
\end{array}\right].$$

At step 1, we proceed with the AICE-CRT reduction to generate the matrices $\mathbf{G}^{(1)}$ and $\mathbf{G}^{(2)}$. They are given by

$$\mathbf{G}^{(1)} = \left[\begin{array}{cccccccc}
1 & 2 & 1 & 2 & 1 & 2 & 1 & 2 \\ \hline
1 & 1 & 1 & 1 & 1 & 1 & 1 & 1 \\ \hline
1 & 0 & 2 & 0 & 1 & 0 & 2 & 0 \\
0 & 1 & 0 & 2 & 0 & 1 & 0 & 2 \\ \hline
1 & 0 & 1 & 1 & 2 & 0 & 2 & 2 \\
0 & 1 & 1 & 2 & 0 & 2 & 2 & 1 \\ \hline
1 & 0 & 1 & 2 & 2 & 0 & 2 & 1 \\
0 & 1 & 2 & 2 & 0 & 2 & 1 & 1
\end{array}\right],$$

and

$$\mathbf{G}^{(2)} = \left[\begin{array}{cccccccc}
1 & 4 & 1 & 4 & 1 & 4 & 1 & 4 \\ \hline
1 & 1 & 1 & 1 & 1 & 1 & 1 & 1 \\ \hline
1 & 3 & 4 & 2 & 1 & 3 & 4 & 2 \\ \hline
1 & 2 & 4 & 3 & 1 & 2 & 4 & 3 \\ \hline
1 & 0 & 3 & 0 & 4 & 0 & 2 & 0 \\
0 & 1 & 0 & 3 & 0 & 4 & 0 & 2 \\ \hline
1 & 0 & 2 & 0 & 4 & 0 & 3 & 0 \\
0 & 1 & 0 & 2 & 0 & 4 & 0 & 3
\end{array}\right].$$

The matrices $\mathbf{D}^{(i)}$, $\mathbf{E}^{(i)}$ and $\mathbf{F}^{(i)}, i = 1, 2$ are given by

$$
\mathbf{D}^{(1)} =
\begin{bmatrix}
1 & & & & & & & & \\
 & 1 & & & & & & & \\
 & & 0 & 1 & 2 & & & & \\
 & & 1 & 2 & 2 & & & & \\
 & & & & & 0 & 1 & 1 & \\
 & & & & & 1 & 2 & 0 & \\
 & & & & & & & 0 & 1 & 1 \\
 & & & & & & & 1 & 2 & 1
\end{bmatrix},
$$

$$
\mathbf{D}^{(2)} =
\begin{bmatrix}
1 & & & & & & & \\
 & 1 & & & & & & \\
 & & 1 & & & & & \\
 & & & 1 & & & & \\
 & & & & 0 & 1 & 3 & \\
 & & & & 1 & 4 & 4 & \\
 & & & & & & 0 & 1 & 2 \\
 & & & & & & 1 & 4 & 4
\end{bmatrix},
$$

$$
\mathbf{E}^{(1)} = \mathbf{F}^{(1)} =
\begin{bmatrix}
1 & & & & & & \\
 & 1 & & & & & \\
 & & 1 & 1 & & & \\
 & & 1 & 0 & & & \\
 & & 0 & 1 & & & \\
 & & & & 1 & 1 & \\
 & & & & 1 & 0 & \\
 & & & & 0 & 1 & \\
 & & & & & & 1 & 1 \\
 & & & & & & 1 & 0 \\
 & & & & & & 0 & 1
\end{bmatrix},
$$

and

$$
\mathbf{E}^{(2)} = \mathbf{F}^{(2)} =
\begin{bmatrix}
1 & & & & & & \\
 & 1 & & & & & \\
 & & 1 & & & & \\
 & & & 1 & & & \\
 & & & & 1 & 1 & \\
 & & & & 1 & 0 & \\
 & & & & 0 & 1 & \\
 & & & & & & 1 & 1 \\
 & & & & & & 1 & 0 \\
 & & & & & & 0 & 1
\end{bmatrix}.
$$

The matrices $\mathbf{A}^{(i)}$, $\mathbf{B}^{(i)}$ and $\mathbf{C}^{(i)} = \mathbf{H}^{(i)}\mathbf{D}^{(i)}, i = 1, 2$ can now be computed as

$$\mathbf{A}^{(1)} = \mathbf{B}^{(1)} = \mathbf{E}^{(1)}\mathbf{G}^{(1)} = \left[\begin{array}{cccccccc}
1 & 2 & 1 & 2 & 1 & 2 & 1 & 2 \\
\hline
1 & 1 & 1 & 1 & 1 & 1 & 1 & 1 \\
\hline
1 & 1 & 2 & 2 & 1 & 1 & 2 & 2 \\
1 & 0 & 2 & 0 & 1 & 0 & 2 & 0 \\
0 & 1 & 0 & 2 & 0 & 1 & 0 & 2 \\
\hline
1 & 1 & 2 & 0 & 2 & 2 & 1 & 0 \\
1 & 0 & 1 & 1 & 2 & 0 & 2 & 2 \\
0 & 1 & 1 & 2 & 0 & 2 & 2 & 1 \\
\hline
1 & 1 & 0 & 1 & 2 & 2 & 0 & 2 \\
1 & 0 & 1 & 2 & 2 & 0 & 2 & 1 \\
0 & 1 & 2 & 2 & 0 & 2 & 1 & 1
\end{array}\right] ,$$

$$\mathbf{A}^{(2)} = \mathbf{B}^{(2)} = \mathbf{E}^{(2)}\mathbf{G}^{(2)} = \left[\begin{array}{cccccccc}
1 & 4 & 1 & 4 & 1 & 4 & 1 & 4 \\
\hline
1 & 1 & 1 & 1 & 1 & 1 & 1 & 1 \\
\hline
1 & 3 & 4 & 2 & 1 & 3 & 4 & 2 \\
\hline
1 & 2 & 3 & 4 & 1 & 2 & 3 & 4 \\
\hline
1 & 1 & 3 & 3 & 4 & 4 & 2 & 2 \\
1 & 0 & 3 & 0 & 4 & 0 & 2 & 0 \\
0 & 1 & 0 & 3 & 0 & 4 & 0 & 2 \\
\hline
1 & 1 & 2 & 2 & 4 & 4 & 3 & 3 \\
1 & 0 & 2 & 0 & 4 & 0 & 3 & 0 \\
0 & 1 & 0 & 2 & 0 & 4 & 0 & 3
\end{array}\right] ,$$

$$\mathbf{C}^{(1)} = \mathbf{H}^{(1)}\mathbf{D}^{(1)} = \left[\begin{array}{c|c|ccc|ccc|ccc}
2 & 2 & 0 & 1 & 2 & 2 & 2 & 1 & 1 & 0 & 2 \\
1 & 2 & 1 & 2 & 2 & 1 & 0 & 1 & 1 & 1 & 0 \\
2 & 2 & 0 & 2 & 1 & 1 & 2 & 0 & 2 & 1 & 2 \\
1 & 2 & 2 & 1 & 1 & 0 & 1 & 1 & 0 & 2 & 2 \\
2 & 2 & 0 & 1 & 2 & 1 & 1 & 2 & 2 & 0 & 1 \\
1 & 2 & 1 & 2 & 2 & 2 & 0 & 2 & 2 & 2 & 0 \\
2 & 2 & 0 & 2 & 1 & 2 & 1 & 0 & 1 & 2 & 1 \\
1 & 2 & 2 & 1 & 1 & 0 & 2 & 2 & 0 & 1 & 1
\end{array}\right],$$

and

$$\mathbf{C}^{(2)} = \mathbf{H}^{(2)}\mathbf{D}^{(2)} = \left[\begin{array}{c|c|c|c|ccc|ccc}
2 & 2 & 2 & 2 & 0 & 4 & 2 & 0 & 4 & 3 \\
3 & 2 & 4 & 1 & 4 & 1 & 1 & 4 & 1 & 1 \\
2 & 2 & 3 & 3 & 0 & 3 & 4 & 0 & 2 & 4 \\
3 & 2 & 1 & 4 & 3 & 2 & 2 & 2 & 3 & 3 \\
2 & 2 & 2 & 2 & 0 & 1 & 3 & 0 & 1 & 2 \\
3 & 2 & 4 & 1 & 1 & 4 & 4 & 1 & 4 & 4 \\
2 & 2 & 3 & 3 & 0 & 2 & 1 & 0 & 3 & 1 \\
3 & 2 & 1 & 4 & 2 & 3 & 3 & 3 & 2 & 2
\end{array}\right].$$

Finally, we combine the matrices $\mathbf{C}^{(1)}$ and $\mathbf{C}^{(2)}$, $\mathbf{A}^{(1)}$ and $\mathbf{A}^{(2)}$, and $\mathbf{B}^{(1)}$ and $\mathbf{B}^{(2)}$ component-by-component using the direct sum to get the algorithm for cyclic convolution over $Z(15)$. We note that proper zero-appending may have to be done to match the dimensions of all of $\mathbf{A}^{(i)}$, $\mathbf{B}^{(i)}$ and $\mathbf{C}^{(i)}$ before direct sum is performed as stated in (4.3.9)–(4.3.11). From (4.3.30), we see that the multiplicative complexity of computing a cyclic convolution of length 8 over $Z(15)$ is 11 MULT.

EXAMPLE 4.9 Let $M = 3^2 \cdot 5^3$ and $n = 8$. Thus $t = 2$, $p_1 = 3$, $p_2 = 5$, $\alpha_1 = 2$, $\alpha_2 = 3$. The polynomial factorizations of $P(u) = u^8 - 1$ over $Z(9)$ and $Z(125)$ are obtained by using the factorizations over $GF(3)$ and $GF(5)$ respectively and then applying Hensel's theorem to them. These factorizations are given by,

$$u^8 - 1 = (u+1)(u+8)(u^2+1)(u^2+5u+8)(u^2+4u+8) \bmod 9,$$
$$u^8 - 1 = (u+1)(u+124)(u+57)(u+68)(u^2+57)(u^2+68) \bmod 125.$$

Clearly $s^{(1)} = 5$ and $s^{(2)} = 6$. Based on these factors polynomials, polynomials $P_j^{'(i)}(u)$ and $T_j^{(i)}(u)$ are computed as,

$$P_1^{'(1)}(u) = u^7 + 8u^6 + u^5 + 8u^4 + u^3 + 8u^2 + u + 8,$$

$$T_1^{(1)}(u) = 1 \bmod (u+1),$$
$$P_2'^{(1)}(u) = u^7 + u^6 + u^5 + u^4 + u^3 + u^2 + u + 1,$$
$$T_2^{(1)}(u) = 8 \bmod (u+8),$$
$$P_3'^{(1)}(u) = u^6 + 8u^4 + u^2 + 8,$$
$$T_3^{(1)}(u) = 2 \bmod (u^2+1),$$
$$P_4'^{(1)}(u) = u^6 + 4u^5 + 8u^4 + 8u^2 + 5u + 1,$$
$$T_4^{(1)}(u) = 5u + 7 \bmod (u^2+5u+8),$$
$$P_5'^{(1)}(u) = u^6 + 5u^5 + 8u^4 + 8u^2 + 4u + 1,$$
$$T_5^{(1)}(u) = 4u + 7 \bmod (u^2+4u+8),$$
$$P_1'^{(2)}(u) = u^7 + 124u^6 + u^5 + 124u^4 + u^3 + 124u^2 + u + 124,$$
$$T_1^{(2)}(u) = 78 \bmod (u+1),$$
$$P_2'^{(2)}(u) = u^7 + u^6 + u^5 + u^4 + u^3 + u^2 + u + 1,$$
$$T_2^{(2)}(u) = 47 \bmod (u+124),$$
$$P_3'^{(2)}(u) = u^7 + 68u^6 + 124u^5 + 57u^4 + u^3 + 68u^2 + 124u + 57,$$
$$T_3^{(2)}(u) = 71 \bmod (u+57),$$
$$P_4'^{(2)}(u) = u^7 + 57u^6 + 124u^5 + 68u^4 + u^3 + 57u^2 + 124u + 68,$$
$$T_4^{(2)}(u) = 54 \bmod (u+68),$$
$$P_5'^{(2)}(u) = u^6 + 68u^4 + 124u^2 + 57,$$
$$T_5^{(2)}(u) = 17 \bmod (u^2+57),$$
$$P_6'^{(2)}(u) = u^6 + 57u^4 + 124u^2 + 68,$$
$$T_5^{(2)}(u) = 108 \bmod (u^2+68).$$

Consequently, the AICE-CRT reconstruction matrices $\mathbf{H}^{(1)}$ and $\mathbf{H}^{(2)}$ are generated as,

$$\mathbf{H}^{(1)} = \begin{bmatrix} 8 & 8 & 7 & 0 & 7 & 5 & 7 & 4 \\ 1 & 8 & 0 & 7 & 4 & 7 & 5 & 7 \\ 8 & 8 & 2 & 0 & 0 & 4 & 0 & 5 \\ 1 & 8 & 0 & 2 & 4 & 0 & 5 & 0 \\ 8 & 8 & 7 & 0 & 2 & 4 & 2 & 5 \\ 1 & 8 & 0 & 7 & 5 & 2 & 4 & 2 \\ 8 & 8 & 2 & 0 & 0 & 5 & 0 & 4 \\ 1 & 8 & 0 & 2 & 5 & 0 & 4 & 0 \end{bmatrix},$$

and

$$\mathbf{H}^{(2)} = \begin{bmatrix} 47 & 47 & 47 & 47 & 94 & 0 & 94 & 0 \\ 78 & 47 & 54 & 71 & 0 & 94 & 0 & 94 \\ 47 & 47 & 78 & 78 & 108 & 0 & 17 & 0 \\ 78 & 47 & 71 & 54 & 0 & 108 & 0 & 17 \\ 47 & 47 & 47 & 47 & 31 & 0 & 31 & 0 \\ 78 & 47 & 54 & 71 & 0 & 31 & 0 & 31 \\ 47 & 47 & 78 & 78 & 17 & 0 & 108 & 0 \\ 78 & 47 & 71 & 54 & 0 & 17 & 0 & 108 \end{bmatrix}.$$

At step 1, we proceed with the AICE-CRT reduction to generate the matrices $\mathbf{G}^{(1)}$ and $\mathbf{G}^{(2)}$. They are given by

$$\mathbf{G}^{(1)} = \begin{bmatrix} 1 & 8 & 1 & 8 & 1 & 8 & 1 & 8 \\ \hline 1 & 1 & 1 & 1 & 1 & 1 & 1 & 1 \\ \hline 1 & 0 & 8 & 0 & 1 & 0 & 8 & 0 \\ 0 & 1 & 0 & 8 & 0 & 1 & 0 & 8 \\ \hline 1 & 0 & 1 & 4 & 8 & 0 & 8 & 5 \\ 0 & 1 & 4 & 8 & 0 & 8 & 5 & 1 \\ \hline 1 & 0 & 1 & 5 & 8 & 0 & 8 & 4 \\ 0 & 1 & 5 & 8 & 0 & 8 & 4 & 1 \end{bmatrix},$$

$$\mathbf{G}^{(2)} = \begin{bmatrix} 1 & 124 & 1 & 124 & 1 & 124 & 1 & 124 \\ \hline 1 & 1 & 1 & 1 & 1 & 1 & 1 & 1 \\ \hline 1 & 68 & 124 & 57 & 1 & 68 & 124 & 57 \\ \hline 1 & 57 & 124 & 68 & 1 & 57 & 124 & 68 \\ \hline 1 & 0 & 68 & 0 & 124 & 0 & 57 & 0 \\ 0 & 1 & 0 & 68 & 0 & 124 & 0 & 57 \\ \hline 1 & 0 & 57 & 0 & 124 & 0 & 68 & 0 \\ 0 & 1 & 0 & 57 & 0 & 124 & 0 & 68 \end{bmatrix}.$$

The matrices $\mathbf{D}^{(1)}$ and $\mathbf{D}^{(2)}$ are given by,

$$
\mathbf{D}^{(1)} =
\begin{bmatrix}
1 & & & & & & & & \\
& 1 & & & & & & & \\
& & 0 & 1 & 8 & & & & \\
& & 1 & 8 & 8 & & & & \\
& & & & & 0 & 1 & 1 & \\
& & & & & 1 & 8 & 3 & \\
& & & & & & & & 0 & 1 & 1 \\
& & & & & & & & 1 & 8 & 4 \\
\end{bmatrix},
$$

and

$$
\mathbf{D}^{(2)} =
\begin{bmatrix}
1 & & & & & & & \\
& 1 & & & & & & \\
& & 1 & & & & & \\
& & & 1 & & & & \\
& & & & 0 & 1 & 68 & \\
& & & & 1 & 124 & 124 & \\
& & & & & & & 0 & 1 & 57 \\
& & & & & & & 1 & 124 & 124 \\
\end{bmatrix}.
$$

The matrices $\mathbf{E}^{(i)}$ and $\mathbf{F}^{(i)}, i = 1, 2$ given in (4.3.26) and (4.3.27) can be constructed based on their individual partitioned matrices shown below.

$$
\mathbf{E}_1^{(1)} = \mathbf{E}_2^{(1)} = 1,
$$

$$
\mathbf{E}_3^{(1)} = \mathbf{E}_4^{(1)} = \mathbf{E}_5^{(1)} =
\begin{bmatrix}
1 & 1 \\
1 & 0 \\
0 & 1
\end{bmatrix},
$$

and

$$
\mathbf{E}_1^{(2)} = \mathbf{E}_2^{(2)} = \mathbf{E}_3^{(2)} = \mathbf{E}_4^{(2)} = 1,
$$

$$
\mathbf{E}_5^{(2)} = \mathbf{E}_6^{(2)} =
\begin{bmatrix}
1 & 1 \\
1 & 0 \\
0 & 1
\end{bmatrix}.
$$

The matrices $\mathbf{A}^{(i)}$, $\mathbf{B}^{(i)}$ and $\mathbf{C}^{(i)}, i = 1, 2$ can now be computed as

$$
\mathbf{A}^{(1)} = \mathbf{B}^{(1)} =
\left[
\begin{array}{cccccccc}
1 & 8 & 1 & 8 & 1 & 8 & 1 & 8 \\
\hline
1 & 1 & 1 & 1 & 1 & 1 & 1 & 1 \\
\hline
1 & 1 & 8 & 8 & 1 & 1 & 8 & 8 \\
1 & 0 & 8 & 0 & 1 & 0 & 8 & 0 \\
0 & 1 & 0 & 8 & 0 & 1 & 0 & 8 \\
\hline
1 & 1 & 5 & 3 & 8 & 8 & 4 & 6 \\
1 & 0 & 1 & 4 & 8 & 0 & 8 & 5 \\
0 & 1 & 4 & 8 & 0 & 8 & 5 & 1 \\
\hline
1 & 1 & 6 & 4 & 8 & 8 & 3 & 5 \\
1 & 0 & 1 & 5 & 8 & 0 & 8 & 4 \\
0 & 1 & 5 & 8 & 0 & 8 & 4 & 1 \\
\end{array}
\right],
$$

$$
\mathbf{A}^{(2)} = \mathbf{B}^{(2)} =
\left[
\begin{array}{cccccccc}
1 & 124 & 1 & 124 & 1 & 124 & 1 & 124 \\
\hline
1 & 1 & 1 & 1 & 1 & 1 & 1 & 1 \\
\hline
1 & 68 & 124 & 57 & 1 & 68 & 124 & 57 \\
\hline
1 & 57 & 124 & 68 & 1 & 57 & 124 & 68 \\
\hline
1 & 1 & 68 & 68 & 124 & 124 & 57 & 57 \\
1 & 0 & 68 & 0 & 124 & 0 & 57 & 0 \\
0 & 1 & 0 & 68 & 0 & 124 & 0 & 57 \\
\hline
1 & 1 & 57 & 57 & 124 & 124 & 68 & 68 \\
1 & 0 & 57 & 0 & 124 & 0 & 68 & 0 \\
0 & 1 & 0 & 57 & 0 & 124 & 0 & 68 \\
\end{array}
\right],
$$

$$
\mathbf{C}^{(1)} =
\left[
\begin{array}{c|c|ccc|ccc|ccc}
8 & 8 & 0 & 7 & 2 & 5 & 2 & 4 & 4 & 3 & 5 \\
1 & 8 & 7 & 2 & 2 & 7 & 6 & 7 & 7 & 7 & 6 \\
8 & 8 & 0 & 2 & 7 & 4 & 5 & 3 & 5 & 4 & 2 \\
1 & 8 & 2 & 7 & 7 & 0 & 4 & 4 & 0 & 5 & 5 \\
8 & 8 & 0 & 7 & 2 & 4 & 7 & 5 & 5 & 6 & 4 \\
1 & 8 & 7 & 2 & 2 & 2 & 3 & 2 & 2 & 2 & 3 \\
8 & 8 & 0 & 2 & 7 & 5 & 4 & 6 & 4 & 5 & 7 \\
1 & 8 & 2 & 7 & 7 & 0 & 5 & 5 & 0 & 4 & 4 \\
\end{array}
\right],
$$

and

$$
\mathbf{C}^{(2)} =
\begin{bmatrix}
47\,|\,47\,|\,47\,|\,47\,| & 0 & 94 & 17 & | & 0 & 94 & 108 \\
78\,|\,47\,|\,54\,|\,71\,| & 94 & 31 & 31 & | & 94 & 31 & 31 \\
47\,|\,47\,|\,78\,|\,78\,| & 0 & 108 & 94 & | & 0 & 108 & 94 \\
78\,|\,47\,|\,71\,|\,54\,| & 108 & 17 & 17 & | & 17 & 17 & 108 \\
47\,|\,47\,|\,47\,|\,47\,| & 0 & 31 & 108 & | & 0 & 31 & 17 \\
78\,|\,47\,|\,54\,|\,71\,| & 31 & 94 & 94 & | & 31 & 94 & 94 \\
47\,|\,47\,|\,78\,|\,78\,| & 0 & 17 & 31 & | & 0 & 17 & 31 \\
78\,|\,47\,|\,71\,|\,54\,| & 17 & 108 & 108 & | & 108 & 108 & 17
\end{bmatrix}.
$$

Finally, we combine the matrices $\mathbf{C}^{(1)}$ and $\mathbf{C}^{(2)}$, $\mathbf{A}^{(1)}$ and $\mathbf{A}^{(2)}$, and $\mathbf{B}^{(1)}$ and $\mathbf{B}^{(2)}$ component-by-component using the direct sum to get the algorithm for cyclic convolution over $Z(1,125)$. Again, we note that proper zero-appending may have to be done to match the dimensions of all of $\mathbf{A}^{(i)}$, $\mathbf{B}^{(i)}$ and $\mathbf{C}^{(i)}$ before direct sum is performed. From (4.3.30), we see that the multiplicative complexity of computing a cyclic convolution of length 8 over $Z(1,125)$ is 11 MULT.

4.3.2 AICE-CRT Based Algorithms for Acyclic Convolution

Once again, we represent the computation of the acyclic convolution of length n over $Z(M)$ in (4.3.14) as a direct sum of the computation over $Z(p_i^{\alpha_i})$, $i = 1, 2, \ldots, t$.

In this approach, the polynomial $P(u)$ for computing the acyclic convolution over $Z(p^\alpha)$ is selected in terms of its factor polynomials over $GF(p)$. The criterion used is the minimum number of MULT. The algorithm consists in selecting a degree d polynomial $P(u)$ over $Z(p^\alpha)$ in terms of its factor polynomials and performing the computation

$$Z(u) \equiv X(u)Y(u) \bmod P(u). \tag{4.3.31}$$

This results in a wraparound of the last $w = n - d$ coefficients in the ordinary polynomial product. This is termed as intentional wraparound. The last w coefficients of the product $X(u)Y(u)$ are computed separately and them combined with the computation in (4.3.31) to get the complete algorithm for acyclic convolution over $Z(p^\alpha)$. The remaining analysis is very similar to the analysis for the case of cyclic convolution. We illustrate the approach with an example.

EXAMPLE 4.10 Let $n = d_x + d_y - 1 = 10$ and $M = 3^2 \cdot 5^3 = 1,125$. The irreducible factor polynomials of $P(u)$ over $GF(3)$ are chosen to be

$P_1(u) = u$, $P_2(u) = u+1$, $P_3(u) = u+2$, $P_4(u) = u^2+1$, $P_5(u) = u^2+u+2$ and $P_6(u) = u^2 + 2u + 2$. Clearly, $deg(P(u)) = 9$. Note that $\prod_{i=1}^{6} P_i(u) = u^9 - u$. In this case, we have a wraparound of 1 coefficient as $d = 9$. In the same way, the irreducible factor polynomials of $P(u)$ over $GF(5)$ are chosen to be $P_1(u) = u$, $P_2(u) = u + 1$, $P_3(u) = u + 2$, $P_4(u) = u + 3$, $P_5(u) = u + 4$, $P_6(u) = u^2 + 2$ and $P_7(u) = u^2 + 3$.

Let $d_x = 6$ and $d_y = 5$. The various $\mathbf{A}^{(i)}$, $\mathbf{B}^{(i)}$ and $\mathbf{C}^{(i)}$ matrices for computing the acyclic convolutions over $Z(p_i^{\alpha_i})$, $i = 1, 2$ are given by

$$
\mathbf{A}^{(1)} =
\left[
\begin{array}{cccccc}
1 & 0 & 0 & 0 & 0 & 0 \\
\hline
1 & 8 & 1 & 8 & 1 & 8 \\
\hline
1 & 1 & 1 & 1 & 1 & 1 \\
\hline
1 & 1 & 8 & 8 & 1 & 1 \\
1 & 0 & 8 & 0 & 1 & 0 \\
0 & 1 & 0 & 8 & 0 & 1 \\
\hline
1 & 1 & 5 & 3 & 8 & 8 \\
1 & 0 & 1 & 4 & 8 & 0 \\
0 & 1 & 4 & 8 & 0 & 8 \\
\hline
1 & 1 & 6 & 4 & 8 & 8 \\
1 & 0 & 1 & 5 & 8 & 0 \\
0 & 1 & 5 & 8 & 0 & 8 \\
\end{array}
\right] ,
$$

$$
\mathbf{A}^{(2)} =
\left[
\begin{array}{cccccc}
1 & 0 & 0 & 0 & 0 & 0 \\
\hline
1 & 124 & 1 & 124 & 1 & 124 \\
\hline
1 & 1 & 1 & 1 & 1 & 1 \\
\hline
1 & 68 & 124 & 57 & 1 & 68 \\
\hline
1 & 57 & 124 & 68 & 1 & 57 \\
\hline
1 & 1 & 68 & 68 & 124 & 124 \\
1 & 0 & 68 & 0 & 124 & 0 \\
0 & 1 & 0 & 68 & 0 & 124 \\
\hline
1 & 1 & 57 & 57 & 124 & 124 \\
1 & 0 & 57 & 0 & 124 & 0 \\
0 & 1 & 0 & 57 & 0 & 124
\end{array}
\right] ,
$$

$$
\mathbf{B}^{(1)} =
\left[
\begin{array}{ccccc}
1 & 0 & 0 & 0 & 0 \\
\hline
1 & 8 & 1 & 8 & 1 \\
\hline
1 & 1 & 1 & 1 & 1 \\
\hline
1 & 1 & 8 & 8 & 1 \\
1 & 0 & 8 & 0 & 1 \\
0 & 1 & 0 & 8 & 0 \\
\hline
1 & 1 & 5 & 3 & 8 \\
1 & 0 & 1 & 4 & 8 \\
0 & 1 & 4 & 8 & 0 \\
\hline
1 & 1 & 6 & 4 & 8 \\
1 & 0 & 1 & 5 & 8 \\
0 & 1 & 5 & 8 & 0
\end{array}
\right] ,
$$

$$\mathbf{B}^{(2)} = \begin{bmatrix}
1 & 0 & 0 & 0 & 0 \\
\hline
1 & 124 & 1 & 124 & 1 \\
\hline
1 & 1 & 1 & 1 & 1 \\
\hline
1 & 68 & 124 & 57 & 1 \\
\hline
1 & 57 & 124 & 68 & 1 \\
\hline
1 & 1 & 68 & 68 & 124 \\
1 & 0 & 68 & 0 & 124 \\
0 & 1 & 0 & 68 & 0 \\
\hline
1 & 1 & 57 & 57 & 124 \\
1 & 0 & 57 & 0 & 124 \\
0 & 1 & 0 & 57 & 0
\end{bmatrix},$$

$$\mathbf{C}^{(1)} = \begin{bmatrix}
1 & | & 0 & | & 0 & | & 0 & 0 & 0 & | & 0 & 0 & 0 & | & 0 & 0 & 0 \\
0 & | & 1 & | & 8 & | & 7 & 2 & 2 & | & 7 & 7 & 7 & | & 7 & 7 & 6 \\
0 & | & 8 & | & 8 & | & 0 & 2 & 7 & | & 4 & 5 & 3 & | & 5 & 4 & 2 \\
0 & | & 1 & | & 8 & | & 2 & 7 & 7 & | & 0 & 5 & 4 & | & 0 & 5 & 5 \\
0 & | & 8 & | & 8 & | & 0 & 7 & 2 & | & 4 & 3 & 5 & | & 5 & 6 & 4 \\
0 & | & 1 & | & 8 & | & 7 & 2 & 2 & | & 2 & 2 & 2 & | & 2 & 2 & 3 \\
0 & | & 8 & | & 8 & | & 0 & 2 & 7 & | & 5 & 4 & 6 & | & 4 & 5 & 7 \\
0 & | & 1 & | & 8 & | & 2 & 7 & 7 & | & 0 & 4 & 5 & | & 0 & 4 & 4 \\
8 & | & 8 & | & 8 & | & 0 & 7 & 2 & | & 5 & 6 & 4 & | & 4 & 3 & 5
\end{bmatrix},$$

and

$$\mathbf{C}^{(2)} = \begin{bmatrix}
1 & | & 0 & | & 0 & | & 0 & | & 0 & | & 0 & 0 & 0 & | & 0 & 0 & 0 \\
0 & | & 78 & | & 47 & | & 54 & | & 71 & | & 94 & 31 & 31 & | & 94 & 31 & 31 \\
0 & | & 47 & | & 47 & | & 78 & | & 78 & | & 0 & 108 & 94 & | & 0 & 17 & 94 \\
0 & | & 78 & | & 47 & | & 71 & | & 54 & | & 108 & 17 & 17 & | & 17 & 108 & 108 \\
0 & | & 47 & | & 47 & | & 47 & | & 47 & | & 0 & 31 & 108 & | & 0 & 31 & 17 \\
0 & | & 78 & | & 47 & | & 54 & | & 71 & | & 31 & 94 & 94 & | & 31 & 94 & 94 \\
0 & | & 47 & | & 47 & | & 78 & | & 78 & | & 0 & 17 & 31 & | & 0 & 108 & 31 \\
0 & | & 78 & | & 47 & | & 71 & | & 54 & | & 17 & 108 & 108 & | & 108 & 17 & 17 \\
124 & | & 47 & | & 47 & | & 47 & | & 47 & | & 0 & 94 & 17 & | & 0 & 94 & 108
\end{bmatrix}.$$

4.4   Computational Complexity Analysis

Expressions (4.3.29) and (4.3.30) form the basis of the analysis for multiplicative complexity associated with the algorithms for cyclic and acyclic convolutions. Table 4.1 summarizes the multiplicative complexities of the AICE-CRT based algorithms that we have derived during the course of this work. In most instances, the NTT based approach is *not* valid due to the WSLC problem.

One of the most promising methods known for convolving sequences having large lengths is to convert an one-dimensional cyclic convolution into a multidimensional convolution. If $n = n_1 n_2 \cdots n_k$, $(n_i, n_j) = 1, i \neq j$, then a length-n cyclic convolution can be converted into a k-dimensional cyclic convolution using the CRT-I. Once again, the AICE-CRT based approach can be employed to design the cyclic convolution algorithms of lengths $n_i, i = 1, 2, \ldots, k$. The multiplicative complexity of the multidimensional convolution algorithm is given by

$$MULT(n) = \prod_{i=1}^{k} MULT(n_i). \qquad (4.4.1)$$

An important question is "Should we employ the AICE-CRT to directly design a length-n cyclic convolution or should we use the AICE-CRT to design length-$n_i, i = 1, 2, \ldots, k$ cyclic convolution algorithms and employ the multidimensional approach?" This was explored in the context of cyclic convolution over $GF(p)$ in Morgera and Krishna [1989]. Based on [ch. 4, Morgera and Krishna, 1989], (4.3.29), (4.3.30), (4.4.1) and the discussion preceding (4.3.30), we conclude that the direct design of a length-n cyclic convolution using the AICE-CRT is computationally superior to the multidimensional approach in terms of multiplicative complexity. This comparison is also illustrated in Table 4.1.

NOTES

The CRT has been widely employed in designing fast computationally efficient algorithms in digital signal processing. It is developed in two versions. One is in a ring of integers (CRT-I) and the other is in a ring of polynomials defined over a finite or infinite field (CRT-P). In many computing systems, one is faced with the task of processing data which is inherently defined on a ring of integers associated with modulo arithmetic. In this regard, researchers have introduced and defined NTTs for processing integer valued sequences. However, the necessary and sufficient conditions on moduli, word length and sequence length for NTTs to exist reveal the WSLC problem that severely limits the application of the NTTs. In this chapter, we

Table 4.1: The Multiplicative Complexity of Cyclic and Acyclic Convolution Algorithms Over Integer Rings

	1-D Cyclic #MULT			1-D Acyclic #MULT			M-D Cyclic #MULT		
n	$Z(2^\alpha)$	$Z(3^\alpha)$	$Z(5^\alpha)$	$Z(2^\alpha)$	$Z(3^\alpha)$	$Z(5^\alpha)$	$Z(2^\alpha)$	$Z(3^\alpha)$	$Z(5^\alpha)$
2	3	2	2	2	2	2	-	-	-
3	4	6	4	3	3	3	-	-	-
4	8	5	4	5	4	4	-	-	-
5	11	10	10	6	6	5	-	-	-
8	24	11	10	12	10	9	-	-	-
11		25	23	18	15	14	-	-	-
13	55	25	22	22	19	17	-	-	-
15	31			26	23	20	44	60	40
16		29	26	28	25	22	-		-
22		50	46	39	37	31		50	46
24		57	34	43	41	34	96	66	40
26		50	44	48	45	37	165	50	44
31	85		51	60	55	46	-	-	-
32		75	64	63	57	48	-	-	-
104		239	202				1320	275	220

1. All blanks can be filled under the AICE-CRT.

2. NTT exists over $Z(p^\alpha)$ *iff* $n|(p-1)$.

3. Multidimensional approach is valid if $n = n_1 n_2, (n_1, n_2) = 1$.

presented the AICE-CRT that extends the CRT-P to the case of a ring of polynomials with coefficients defined over a finite ring of integers. This theoretical work provides us with what we believe is a missing link between fields and rings. It removes the limitations on both word length and sequence length. In fact, the NTTs are a special case of LI which in turn is a special case of our general approach when all factor polynomials have degree equal to one. The AICE-CRT plays a fundamental role in designing computationally efficient number theoretic algorithms for performing cyclic and acyclic convolution of data sequences.

This chapter is based on the research work reported in Lin et al. [1994] and Krishna et al. [1994].

4.5   Bibliography

[4.1] J.W. Cooley and J.W. Tukey, "An Algorithm for the Machine Calculation of Complex Fourier Series," *Math. Comput.*, vol. 19, pp. 297–301, April 1965.

[4.2] C.M. Rader, "Discrete Convolutions via Mersenne Transforms,"

IEEE Transactions on Computers, vol. C-21, pp. 1269–1273, December 1972.

[4.3] R.C. Agarwal and C.S. Burrus, "Fast Convolution Using Fermat Number Transforms with Applications to Digital Filtering," *IEEE Transactions on Acoustics, Speech and Signal Processing,* vol. ASSP-22, no. 2, pp. 87–97, April 1974.

[4.4] R.C. Agarwal and C.S. Burrus, "Number Theoretic Transforms to Implement Fast Digital Convolution," *Proceeding IEEE,* vol. 63, pp. 550–560, April 1975.

[4.5] J.H. McClellan, "Hardware Realization of a Fermat Number Transform," *IEEE Transactions on Acoustics, Speech and Signal Processing,* vol. ASSP-24, no. 3, pp. 216–225, June, 1976.

[4.6] H. Lu and S.C. Lee, "A New Approach to Solve the Sequence Length Constraint Problem in Circular Convolution Using Number Theoretic Transform," *IEEE Transactions on Signal Processing,* vol. 39, no. 6, pp. 1314–1321, June 1991.

[4.7] I.S. Reed and T.K. Truong, "The Use of Finite Fields to Compute Convolutions," *IEEE Transactions on Information Theory,* vol. IT-21, no. 2, pp. 208–213, March 1975.

[4.8] H.J. Nussbaumer, *Fast Fourier Transform and Convolution Algorithms,* Second Edition, Springer-Verlag, 1982.

[4.9] D.G. Myers, *Digital Signal Processing - Efficient Convolution and Fourier Transform Techniques,* Prentice-Hall, Inc., 1990.

[4.10] J.H. McClellan and C.M. Rader, *Number Theory in Digital Signal Processing,* Prentice Hall, Inc., 1979.

[4.11] R.E. Blahut, *Fast Algorithms for Digital Signal Processing,* Addison-Wesley, 1985.

[4.12] L.K. Hua, *Introduction to Number Theory,* Springer-Verlag, 1982.

[4.13] R.E. Blahut, *Theory and Practice of Error Control Codes,* Addison-Wesley, 1983.

[4.14] R.C. Agarwal and J.W. Cooley, "New Algorithms for Digital Convolution," *IEEE Transactions on Acoustics, Speech and Signal Processing,* vol. ASSP-25, pp. 392–410, October 1977.

[4.15] D.P. Kolba and T.W. Parks, "A Prime Factor FFT Algorithm Using High-Speed Convolution," *IEEE Transactions on Acoustics, Speech and Signal Processing*, vol. ASSP-25, pp. 281–294, August 1977.

[4.16] S. Winograd, "On Computing the Discrete Fourier Transform," *Mathematics of Computation*, vol. 32, no. 41, pp. 175–199, January 1978.

[4.17] S. Winograd, "Some Bilinear Forms Whose Multiplicative Complexity Depends on the Field of Constants," *Math. Systems Theory*, vol. 10, pp. 169–180, 1977.

[4.18] M.D. Wagh and S. D. Morgera, "A New Structured Design Method for Convolutions Over Finite Fields, Part I," *IEEE Transactions on Information Theory*, vol. IT-29, no. 4, pp. 583–595, July 1983.

[4.19] S.D. Morgera and H. Krishna, *Digital Signal Processing - Applications to Communications and Algebraic Coding Theories*, Academic Press, 1989.

[4.20] A. Skavantzos and F.J. Taylor, "On the Polynomial Residue Number System," *IEEE Transactions on Signal Processing*, vol. 39, no.2, pp. 376–382, February 1992.

[4.21] G.H. Golub and C.F. Van Loan, *Matrix Computations*, the Johns Hopkins University Press, 1983.

[4.22] E.R. Berlekamp, "Factoring Polynomials Over Large Finite Fields," *Math. Comp.*, vol. 24, no. 11, pp. 713–735, July 1970.

[4.23] A.K. Lenstra, "Factoring Multivariate Polynomials over Algebraic Number Fields," *SIAM J. on Computing*, vol. 16, no. 3, pp. 591–598, June 1987.

[4.24] D.R. Musser, "Multivariate Polynomial Factorization," *J. Asso. Comput. Mach.*, vol. 22, no. 2, pp. 291–308, April 1975.

[4.25] P.S. Wang and L.P. Rothschild, "Factoring Multivariate Polynomials over the Integers," *Math. Comp.*, vol. 29, no. 131, pp. 935–950, July 1975.

[4.26] D.E. Knuth, *The Art of Computer Programming, vol. II, Seminumerical Algorithm*, Addison-Wesley, 1969.

[4.27] H. Zassenhaus, "On Hensel Factorization," *Int. J. Nummer. Theory*, vol. 1, pp. 291–311, 1969.

[**4.28**] K.-Y.Lin, B. Krishna and H. Krishna, "Rings, Fields, the Chinese Remainder Theorem and an Extension, Part I: Theory," *IEEE Transactions on Circuits and Systems*, to appear in 1994.

[**4.29**] H. Krishna, K.-Y. Lin and B. Krishna, "Rings, Fields, the Chinese Remainder Theorem and an Extension, Part II: Application to Digital Signal Processing," *IEEE Transactions on Circuits and Systems*, to appear in 1994.

Chapter 5

AICE-CRT: The Complex Case

The CRT is a fundamental technique widely employed in designing computationally efficient convolution algorithms. In the previous chapter, we presented the AICE-CRT, an extension to the CRT for the case of a ring of polynomials with coefficients defined over a finite integer ring, and developed fast algorithms for computing cyclic and acyclic convolutions. The objective is to generalize NTTs which turn out to be a special case of this extension. This chapter focuses on the extension of the CRT for processing complex valued integer sequences. Once again, the present work generalizes the complex-number-theoretic-transforms (CNTTs).

The impetus for this work is provided by the occurrence of complex valued integer sequences in digital signal processing and the desire to process them using exact arithmetic. Based on the factorization properties of polynomials defined over complex integer rings, we develop a generalization of the CNTTs. This generalization is obtained by developing the complex version of the AICE-CRT. It was demonstrated in the previous chapter that the AICE-CRT can be used to compute the convolution of two real-valued sequences defined over an integer ring. The objective of this chapter is to establish a generalization of the CNTTs by deriving the complex version of the AICE-CRT. This also leads to a complete solution of the WSLC problem which has restricted the application of NTTs and CNTTs in digital signal processing.

The complex integer ring $CZ(M)$ is a generalization of the $Z(M)$. It consists of all complex integers of the type

$$W = U + jV, \tag{5.0.1}$$

111

where $U, V \in Z(M)$, and

$$j^2 \equiv -1 \bmod M. \tag{5.0.2}$$

The arithmetic operations of ADD and MULT in $CZ(M)$ are the same as those for complex numbers with the exception that both real and the imaginary parts are defined modulo M. The $CZ(M)$ can be expressed as the direct sum

$$CZ(M) \approx \sum_{\substack{\oplus \\ i=1}}^{t} CZ(q_i), \tag{5.0.3}$$

where $M = \prod_{i=1}^{t} q_i, q_i = p_i^{\alpha_i}$, the standard factorization of M. There is a one-to-one correspondence between the complex integer $W \in CZ(M)$ and a vector of residues $W = [W_1 W_2 \ldots W_t]$, where

$$W_i \equiv W \bmod q_i, \ i = 1, 2, \ldots, t. \tag{5.0.4}$$

A multilinear form over $CZ(M)$ can be expressed as

$$Z_\ell = \sum_{\forall i_1} \sum_{\forall i_2} \cdots \sum_{\forall i_k} \beta_{\ell,i_1,i_2,\cdots,i_k} X_{1,i_1} X_{2,i_2} \cdots X_{k,i_k} \bmod M, \tag{5.0.5}$$

where the scalars $\beta_{\ell,i_1,i_2,\cdots,i_k}$ are elements of $CZ(M)$ and X_{θ,i_θ}, an indeterminate over $CZ(M)$, is the i_θth component of the input sequence $\underline{X}_\theta$, $\theta = 1, 2, \ldots, k$. Here, we are interested in the computation of the cyclic and acyclic convolution of sequences over $CZ(M)$ which are special cases of bilinear forms. Bilinear forms, in turn, are special cases of the multilinear form in (5.0.5).

A useful interpretation of the description of $CZ(M)$ is that every element in $CZ(M)$ can be expressed as a polynomial of degree 1 over $Z(M)$. Here j is an indeterminate. The arithmetic operations of MULT and ADD between two elements of $CZ(M)$ can equivalently be represented as operations between two polynomials in j of degree 1 modulo $(j^2 + 1)$ and M, denoted by $(\bmod j^2 + 1, M)$. Similarly, the computation of a complex multilinear form over $CZ(M)$ can be represented as sum of products of degree 1 polynomials in j over $Z(M)$ $(\bmod j^2 + 1, M)$. Given the direct sum properties of $Z(M)$ and $CZ(M)$ in (2.4.9) and (5.0.3) respectively and the nature of the modulo operation, the following lemma can be established in a straightforward manner.

LEMMA 5.1 *An algorithm for computing a multilinear form over $CZ(M)$ can equivalently be expressed as a direct sum of t algorithms for computing the multilinear forms $(\bmod j^2 + 1, q_i)$, $i = 1, 2, \ldots, t$, $M = \prod_{i=1}^{t} q_i$.*

Given two sequences $X_i, i = 0, 1, \ldots, n-1$ and $Y_h, h = 0, 1, \ldots, n-1$ defined over $CZ(M)$, the cyclic convolution is the computation of the polynomial product,

$$Z(u) = X(u)Y(u) \quad (\text{mod } u^n - 1, j^2 + 1, M). \tag{5.0.6}$$

Similarly, the acyclic convolution of two sequences $X_i, i = 0, 1, \ldots, d_x - 1$ and $Y_h, h = 0, 1, \ldots, d_y - 1$ defined over $CZ(M)$ can be expressed as

$$Z(u) = X(u)Y(u). \tag{5.0.7}$$

It can also be expressed as

$$Z(u) = X(u)Y(u) \quad (\text{mod } P(u), j^2 + 1, M), \tag{5.0.8}$$

where $deg(X(u)) = d_x - 1$, $deg(Y(u)) = d_y - 1$ and $P(u)$ is a monic polynomial that satisfies $deg(P(u)) \geq n$. In general, $P(u)$ is selected to be a polynomial that results in an algorithm having the least possible multiplicative complexity.

Using lemma 5.1, we see that an algorithm for computing the convolution over $CZ(M)$ is a direct sum of corresponding algorithms over $CZ(q_i), i = 1, 2, \ldots, t$. Consequently, in the remainder of this chapter, we focus on developing algorithms for the following computations:

1. Cyclic convolution :

$$Z(u) = X(u)Y(u) \quad (\text{mod } u^n - 1, j^2 + 1, p^\alpha). \tag{5.0.9}$$

2. Acyclic convolution :

$$Z(u) = X(u)Y(u) \quad (\text{mod } P(u), j^2 + 1, p^\alpha). \tag{5.0.10}$$

We observe here that any algorithm designed for computing the convolution of sequences defined over $Z(M)$ may also be used for computing the convolution of complex sequences defined over $CZ(M)$. The main thrust of this chapter is in deriving algorithms for convolving complex sequences which exploit the number theoretic properties of $CZ(M)$ and thereby require fewer computations than the algorithms designed for convolving sequences over $Z(M)$.

5.1 Multilinear Forms Over Complex Integer Rings

In this section, we analyze the factorization of $j^2 + 1$ over $Z(q), q = p^\alpha$, and the resulting AICE-CRT based algorithms for computing the convolutions in (5.0.9) and (5.0.10).

Consider a multilinear form over $CZ(q)$ expressed as a sum of products of polynomials in j of degree 1 over $Z(q)$ (mod $j^2 + 1, q$). If $j^2 + 1$ factors into the product of two relatively prime polynomials over $Z(q)$, say $j^2+1 = (j - a)(j - b)$ mod q, then the computation of the multilinear form can be performed in the following three steps :

1. Reduce the multilinear form over $CZ(q)$ by evaluating the various quantities for $j = a$ and $j = b$. This results in two multilinear forms over $Z(q)$.

2. Evaluate the two resulting multilinear forms over $Z(q)$.

3. Reconstruct the multilinear form over $CZ(q)$ from the two multilinear forms over $Z(q)$.

The above three steps are based on a straightforward application of the AICE-CRT to evaluating multilinear forms   (mod $j^2 + 1, q$). The factorization of $j^2 + 1$ over $Z(q)$ begins with its factorization over $GF(p)$ and employs the Hensel's theorem in a recursive manner. Two degree one polynomials $j + a$ and $j + b$ are relatively prime over $Z(q)$ (or equivalently over $GF(p)$) *iff* a mod $p \neq b$ mod p, that is, the factors are distinct mod p.

5.1.1   Factorization of $j^2 + 1$ Over $GF(p)$

For $p = 2, j^2 + 1 = (j + 1)^2$. Therefore, $j^2 + 1$ does not factor into two distinct factors over $GF(2)$. However, it does factor which implies that it is not irreducible over $GF(2)$ and the complex integers over $GF(2)$ along with the arithmetic operations of MULT and ADD mod $(j^2 + 1)$ do not constitute a finite field.

Now consider the factorization of j^2+1 over $GF(p), p \neq 2$. The order of every non-zero element in $GF(p)$ divides $p - 1$. Since $p - 1$ is always even, factorization of $j^2 - 1$ always exists and is given by $j^2 - 1 = (j + 1)(j - 1)$. Therefore, the factorization of $j^2 + 1$ over $GF(p)$ can be determined from the factorization of $(j^2 + 1)(j^2 - 1) = j^4 - 1$. The polynomial $j^2 + 1$ factors over $GF(p)$ *iff* $j^4 - 1$ has four distinct factors, that is there is an element of order 4 in $GF(p)$. Recalling that the order of every non-zero element in $GF(p)$ divides $p - 1$, we get the following lemma.

LEMMA 5.2 *For $p \neq 2$, $j^2 + 1$ is irreducible over $GF(p)$ iff 4 is not a factor of $(p - 1)$ or $p = 4L + 3$. It factors into the product of two relatively prime polynomials over $GF(p)$ iff $4|(p - 1)$ or $p = 4L + 1$.*

For example, j^2+1 is irreducible over $GF(3)$ and therefore over $Z(3^\alpha)$ as well. Over $GF(5)$, $j^2+1 = (j+2)(j+3)$ which leads to $j^2+1 = (j+7)(j+18)$

over $Z(5^2)$ using Hensel's theorem. The factorization of $j^2 + 1$ over $GF(p)$, or equivalently over $Z(q)$, is summarized as follows.

$p = 2$	$j^2 + 1 = (j + 1)^2.$
$p = 4L + 3$	$j^2 + 1$ is irreducible over $GF(p)$.
$p = 4L + 1$	$j^2 + 1 = (j - a)(j - b), a, b \in GF(p)$.

We draw an important distinction between the irreducibility of $j^2 + 1$ over $GF(p)$, $p = 4L + 3$, and the repeated factors of $j^2 + 1$ over $GF(2)$. In the former case, the complex integers over $GF(p)$ constitute a finite field while in the latter case, no such property holds. This plays a crucial role in the factorization of polynomials over complex integer rings and in the design of fast algorithms for computing convolutions.

We now turn to the computation of cyclic and acyclic convolutions in (5.0.9) and (5.0.10) respectively.

5.1.2 Algorithms for Convolutions: Special Cases

Consider the cyclic and acyclic convolutions expressed as polynomial products $(\mod u^n - 1, j^2 + 1, q)$ and $(\mod P(u), j^2 + 1, q)$ in (5.0.9) and (5.0.10) respectively.

Case I. $p = 2$. For $p = 2, j^2 + 1 = (j + 1)^2$. In this case, the algorithm for computing the convolution of sequences defined over $CZ(2^\alpha)$ is the same as the algorithm for computing the convolution of sequences defined over $Z(2^\alpha)$. No further computational savings can be realized.

Case II. $p = 4L + 1$. In this case, $j^2 + 1 = (j - a)(j - b)$ over $Z(q)$, $a \mod p \neq b \mod p$. The algorithms for computing (5.0.9) and (5.0.10) can be outlined as follows:

Step 1. Evaluate $X(u)$ and $Y(u)$ at $j = a$ and $j = b$ to get polynomials $X_1(u)$, $X_2(u)$, $Y_1(u)$ and $Y_2(u)$ over $Z(q)$. Similarly, evaluate $P(u)$ to get $P_1(u)$ and $P_2(u)$ (required for (5.0.10) only). This is the AICE-CRT reduction with regard to the polynomial $j^2 + 1$.

Step 2. Evaluate $Z_1(u) = X_1(u)Y_1(u) \mod (u^n - 1, q)$ and $Z_2(u) = X_2(u) Y_2(u) \mod (u^n - 1, q)$ for (5.0.9); $Z_1(u) = X_1(u)Y_1(u) \mod (P_1(u), q)$ and $Z_2(u) = X_2(u)Y_2(u) \mod (P_2(u), q)$ for (5.0.10) using the AICE-CRT over $Z(q)$.

Step 3. Reconstruct $Z(u)$ from $Z_1(u)$ and $Z_2(u)$ using the AICE-CRT.

This special-case algorithm with $p = 4L + 1$ is summarized in Table 5.1. As an example, if the polynomials $X(u)$ and $Y(u)$ are defined over $CZ(5^2)$, then in step (1), they are evaluated at $j = 7$ and $j = 18$. In step (3), $Z(u)$ is constructed from $Z_1(u)$ and $Z_2(u)$ using the AICE-CRT reconstruction mod $(j^2 + 1, 5^2)$ as

$$
\begin{aligned}
Z(u) &= 16(j + 18)Z_1(u) + 9(j + 7)Z_2(u) \\
     &= 13(Z_1(u) + Z_2(u)) + j(16Z_1(u) + 9Z_2(u)).
\end{aligned}
$$

Factorization properties of polynomials over $CZ(q)$, $q = p^\alpha$, $p = 4L + 3$, the AICE-CRT and the corresponding algorithms for computing the convolution form the topic of the following sections.

5.2   Factorization Over a Complex Integer Ring

Throughout this section, we assume that the polynomials under discussion are defined over $CZ(q)$, $q = p^\alpha$, $p = 4L+3$. In this case, $j^2 + 1$ is irreducible (not primitive) over $GF(p)$, and therefore, $CZ(p)$ is same as the finite field $GF(p^2)$.

DEFINITION 5.1 Let $f(u) = \sum_{i=0}^{n} c_i u^i$, $c_i \in CZ(q), q = p^\alpha$, a polynomial of degree n. The p-adic representation of $f(u)$ is given by

$$
f(u) = \sum_{i=0}^{\alpha-1} f_i(u)p^i, \tag{5.2.1}
$$

where

$$
f_i(u) = \sum_{k=0}^{n} c_{i,k}\, u^k, \; c_{i,k} \in GF(p^2). \tag{5.2.2}
$$

This definition is based on the p-adic representation of a scalar $c \in CZ(q)$ as

$$
c = c_R + jc_I = \sum_{i=0}^{\alpha-1}(c_{i,R} + jc_{i,I})p^i. \tag{5.2.3}
$$

DEFINITION 5.2 Given two polynomials $f(u)$ and $h(u)$, $h(u)$ has $f(u)$ as its factor whenever $h(u)$ and $f(u)$ can be written as $h(u) = f(u)q(u)$.

DEFINITION 5.3 A monic polynomial is termed *irreducible* if it has no other monic polynomial as its factor except itself.

Table 5.1: Special-case Algorithm for Computing Convolution Over $CZ(p^\alpha)$

$p = 4L + 1$	$j^2 + 1 = (j - a)(j - b) \bmod q$
Step 1	Evaluate $X_1(u) = X(u, j = a)$ and $Y_1(u) = Y(u, j = a)$ $X_2(u) = X(u, j = b)$ and $Y_2(u) = Y(u, j = b)$. Similarly, evaluate $P_1(u)$ and $P_2(u)$ for acyclic convolution. This is the AICE-CRT reduction with regard to $j^2 + 1$. $X_1(u)$, $X_2(u)$, $Y_1(u)$, $Y_2(u)$, $P_1(u)$ and $P_2(u)$ are polynomials defined over $Z(q)$.
Step 2	For cyclic convolution, evaluate $Z_1(u) = X_1(u)Y_1(u) \bmod (u^n - 1, q)$ $Z_2(u) = X_2(u)Y_2(u) \bmod (u^n - 1, q)$ For acyclic convolution, evaluate $Z_1(u) = X_1(u)Y_1(u) \bmod (P_1(u), q)$ $Z_2(u) = X_2(u)Y_2(u) \bmod (P_2(u), q)$ Note that AICE-CRT based algorithms over $Z(q)$ from Chapter 4 are employed to compute $Z_1(u)$ and $Z_2(u)$ respectively.
Step 3	Reconstruct $Z(u)$ from $Z_1(u)$ and $Z_2(u)$ using the AICE-CRT $Z(u) = Z_1(u)T_1(j)(j - b) +$ $Z_2(u)T_2(j)(j - a)$.

DEFINITION 5.4 If two monic polynomials have no common monic factors, they are said to be *relatively prime* (or coprime).

THEOREM 5.1 *Let $P(u)$ be a monic polynomial defined over $CZ(q)$, and $P(u) \equiv \prod_{i=1}^{\beta} P_{i,0}(u)$ (mod p), where $P_{i,0}(u)$ and $P_{k,0}(u), i \neq k$ are coprime polynomials defined over $GF(p^2)$. Then, there exist β monic polynomials over $CZ(q)$, denoted by $P_i(u)$, such that $P_i(u) \equiv P_{i,0}(u)$ (mod p) and $P(u) = \prod_{i=1}^{\beta} P_i(u)$.*

This theorem may be treated as an extension of the Hensel's theorem as the original Hensel's theorem deals with polynomials defined over $Z(q)$.

For the proof, the readers are referred to the detailed analysis in Section 4.1 which also constitutes the proof. We observe that if a polynomial defined over $CZ(q)$ is reduced mod p, it results in a polynomial defined over $GF(p^2)$.

THEOREM 5.2 *Two monic polynomials defined over $CZ(q)$, denoted by $P_1(u)$ and $P_2(u)$, have no monic factors in common over $CZ(q)$, if they have no monic factors in common mod p.*

Theorem 5.2 describes only a sufficient condition. If $P_1(u)$ and $P_2(u)$ defined over $CZ(q)$ have a monic factor $q_0(u)$ in common mod p, then we may conclude that $P_1(u)$ and $P_2(u)$ have monic factors $q_1(u)$ and $q_2(u)$ over $CZ(q)$, such that $q_1(u) \equiv q_0(u) \bmod p$ and $q_2(u) \equiv q_0(u) \bmod p$. It is not necessary that $q_1(u) = q_2(u)$.

THEOREM 5.3 *A monic polynomial $P(u)$ defined over $CZ(q)$ is an irreducible polynomial iff $P_0(u) \equiv P(u) \,(\bmod\ p)$ is an irreducible polynomial over the finite field $GF(p^2)$.*

A detailed discussion of these theorems may be found in Chapter 4. We omit the proofs of these theorems as they are similar to the proofs for polynomials defined over $Z(q)$ which can also be found in Chapter 4.

Given a monic polynomial $P(u)$ and its p-adic form as in (5.2.1) and (5.2.2), define a series of polynomials as

$$A_k(u) = P(u) \bmod p^{k+1}, \ 0 \le k \le \alpha - 1. \tag{5.2.4}$$

Clearly, $A_0(u) = P_0(u)$, and $A_{\alpha-1}(u) = P(u)$. In the following, we describe a recursive procedure that utilizes Theorem 5.1 and all the $A_k(u)$ polynomials to factorize $P(u)$. Assume that a factorization of $P_0(u)$ in terms of pairwise relatively prime factors is known over $GF(p^2)$; namely, $B_{01}(u)$, $B_{02}(u)$, ..., $B_{0\beta}(u)$. In the most general case, these polynomials are irreducible or powers of an irreducible polynomial over $GF(p^2)$. In that case, they are unique and pairwise relatively prime (unique factorization property over a field). Hence,

$$A_0(u) = \prod_{i=1}^{\beta} B_{0i}(u) \quad (\bmod\ p). \tag{5.2.5}$$

We begin with $A_0(u)$ and $A_1(u)$. From (5.2.4), we have

$$A_1(u) = P_0(u) + pP_1(u). \tag{5.2.6}$$

Applying Theorem 5.1, we express $A_1(u)$ as

$$A_1(u) = \prod_{i=1}^{\beta} [B_{0i}(u) + pQ_{1i}(u)] \mod p^2. \qquad (5.2.7)$$

Equating (5.2.6) and (5.2.7), and rearranging terms, we get

$$P_0(u) + pP_1(u) = [A_0(u) + pA_0'(u)] + p[\sum_{i=1}^{\beta} Q_{1i}(u) \prod_{\substack{k=1 \\ k \neq i}}^{\beta} B_{0k}(u)] \mod p^2,$$

$$(5.2.8)$$

where $A_0'(u)$ is determined as

$$p \, A_0'(u) = [(\prod_{i=1}^{\beta} B_{0i}(u)) - A_0(u)] \mod p^2. \qquad (5.2.9)$$

Cancelling out p on both sides in (5.2.8) and taking modulo $B_{0l}(u)$, $l = 1, 2, \ldots, \beta$, in (5.2.8), we get the following k congruences

$$Q_{1l}(u) \prod_{\substack{k=1 \\ k \neq l}}^{\beta} B_{0k}(u) \equiv [P_1(u) - A_0'(u)] \pmod{B_{0l}(u), p}. \qquad (5.2.10)$$

All the quantities in the above polynomial congruence are known except $Q_{1l}(u)$. A unique and closed form solution to $Q_{1l}(u)$ can be obtained from (5.2.10). This is possible due to the reasons that (i) the polynomials $B_{01}(u), B_{02}(u), \cdots, B_{0\beta}(u)$ are relatively coprime, and (ii) $deg(Q_{1l}(u)) < deg(B_{0l}(u))$. Clearly, $A_1(u)$ also has β relatively coprime factors as in (5.2.7) and we can write

$$A_1(u) = \prod_{i=1}^{\beta} B_{1i}(u) \mod p^2. \qquad (5.2.11)$$

The same procedure may be employed in a recursive manner to obtain $A_2(u)$, $A_3(u)$ and so on till $A_{\alpha-1}(u) = P(u)$ is reached. All the factor polynomials are relatively coprime over the corresponding complex integer ring. We summarize this procedure to factorize a monic polynomial $P(u)$ over $CZ(q)$ in the following.

1. Express $P(u)$ in p-adic form to obtain $P_i(u)$, $i = 0, 1, 2, \ldots, \alpha-1$. Note that $P(u)$ is defined over $CZ(q)$ while $P_i(u)$ is defined over $GF(p^2)$.

2. Form the polynomials

$$A_k(u) = \sum_{i=0}^{k} p^i P_i(u), \ k = 0, 1, \ldots, \alpha - 1. \tag{5.2.12}$$

Clearly, $A_0(u) = P_0(u)$ and $A_{\alpha-1}(u) = P(u)$. We emphasize that the factorization of $P_0(u)$ over $GF(p^2)$ is assumed to be known.

3. Let $k = 1$.

4. Compute $A'_{k-1}(u)$ based on $A_{k-1}(u)$ and the factors $B_{k-1,i}(u)$, $i = 1, 2, \ldots, \beta$ as

$$p^k A'_{k-1}(u) = [(\prod_{i=1}^{\beta} B_{k-1,i}(u)) - A_{k-1}(u)] \ \text{mod} \ p^{k+1}. \tag{5.2.13}$$

5. Solve for $Q_{k,l}(u), l = 1, 2, \ldots, \beta$ from the congruence

$$Q_{k,l}(u) \prod_{\substack{a=1 \\ a \neq l}}^{\beta} B_{k-1,a}(u) \equiv [P_k(u) - A'_{k-1}(u)] \quad (\text{mod} \ B_{k-1,l}(u), p).$$

$$\tag{5.2.14}$$

6. Then factorization of $A_k(u)$ is given by

$$A_k(u) = \prod_{i=1}^{k} B_{k,i}(u) \ \text{mod} \ p^{k+1},$$

where

$$B_{k,i}(u) = [B_{k-1,i}(u) + p^k Q_{k,i}(u)] \ \text{mod} \ p^{k+1}. \tag{5.2.15}$$

7. Increase k by 1. If $k = \alpha$, stop; else go to step 4.

The most interesting property associated with the factorization of $P(u)$ over $CZ(q)$ is stated in the following theorem.

THEOREM 5.4 (*Unique factorization theorem*): *A monic polynomial defined over $CZ(q)$ has a unique factorization into a product of relatively prime and irreducible (or power of irreducible) monic polynomials.*

We now present an example to illustrate this factorization procedure.

EXAMPLE 5.1 Factorize $P(u) = u^5 - 1$ over $CZ(3^2)$. Clearly, $p = 3$ and $\alpha = 2$. In step 1, we get the p-adic form of $P(u)$: $P_0(u) = u^5 + 2$ and $P_1(u) = 2$. In step 2, we get $A_0(u) = P_0(u) = u^5 + 2$ and $A_1(u) = P(u) = u^5 + 8$. The factorization of $P(u)$ over $GF(3^2)$ is given by (assumed to be known a priori)

$$P_0(u) = (u+2)(u^2 + (2+j)u + 1)(u^2 + (2+2j)u + 1).$$

Thus, $B_{0,1}(u) = u + 2$, $B_{0,2}(u) = u^2 + (2+j)u + 1$, $B_{0,3}(u) = u^2 + (2+2j)u + 1$. For $k = 1$, we get $A'_0(u) = (2+j)u^4 + (1+j)u^3 + (1+2j)u^2 + 2ju$ in step 4. In step 5, the three congruences ($\beta = 3$) to be solved are

$$\begin{cases} Q_{1,1}(u)(1+j)(1+2j) & \equiv \quad 1 \quad (\text{mod } u+2, 3), \\ Q_{1,2}(u)(u+2j) & \equiv \quad (1+2j)u + 2 \quad (\text{mod } u^2 + (2+j)u + 1, 3), \\ Q_{1,3}(u)(u+j) & \equiv \quad ju + (2+j) \quad (\text{mod } u^2 + (2+2j)u + 1, 3). \end{cases}$$
$$(5.2.16)$$

The solutions are $Q_{1,1}(u) = 2$, $Q_{1,2}(u) = u$ and $Q_{1,3}(u) = (1+2j)u$. Finally, the factorization of $A_1(u) = P(u)$ over $CZ(9)$ is given by (step 6)

$$P(u) = (u+8)(u^2 + (5+j)u + 1)(u^2 + (5+8j)u + 1).$$

Given the factorization of a monic polynomial over $CZ(q)$, we now present our main result in the next section.

5.3 The AICE-CRT Over Complex Integer Rings

In this section, we present the complex version of the AICE-CRT. Once again, we assume that all the polynomials under discussion are defined over $CZ(q)$, $q = p^\alpha$, $p = 4L + 3$.

Let $X(u)$ and $P(u)$ be two polynomials such that $deg(X(u)) \leq n - 1$ and $deg(P(u)) \geq n$. $P(u)$ is a monic polynomial. Also, we are given the factorization of $P(u)$ as

$$P(u) = \prod_{i=1}^{\beta} P_i(u). \qquad (5.3.1)$$

We note that the monic factors $P_i(u)$ and $P_k(u)$, $i \neq k$, $1 \leq i, k \leq \beta$, are relatively prime. Given $X(u)$, the AICE-CRT over $CZ(q)$ establishes a one-to-one correspondence between $X(u)$ and the residue polynomials $X_1(u), X_2(u), \cdots, X_\beta(u)$, where

$$X_i(u) \equiv X(u) \mod P_i(u), \quad i = 1, 2, \ldots, \beta. \qquad (5.3.2)$$

Before proving the AICE-CRT over $CZ(q)$, we state the following fundamental result.

THEOREM 5.5 *Let $P_1(u)$ and $P_2(u)$ be two monic and relatively coprime polynomials. For a polynomial $X(u)$, if $P_1(u)|X(u)$ and $P_2(u)|X(u)$, then $[P_1(u)P_2(u)]|X(u)$.*

The proof of the above theorem may be found in Theorem 4.5 in the context of polynomials defined over $Z(q)$. The proof of the complex case is very similar to the proof of Theorem 4.5 and is omitted here. The following theorem is the key result of this chapter.

THEOREM 5.6 (*The AICE-CRT over $CZ(q)$*). *There exists a one-to-one correspondence between the polynomial $X(u)$, $deg(X(u)) \leq n - 1$, and the residue polynomials $X_1(u)$, $X_2(u)$, $\cdots$, $X_\beta(u)$ as defined in (5.3.2), provided that the modulo polynomials $P_1(u)$, $P_2(u)$, $\cdots$, $P_\beta(u)$ are relatively coprime and $deg(P(u)) \geq n$, $P(u) = \prod_{i=1}^{\beta} P_i(u)$.*

PROOF. If the theorem is false, then there exists at lease one set of residue polynomials, say $X_1(u)$, $X_2(u)$, $\cdots$, $X_\beta(u)$, that can be obtained from two distinct polynomials, say $X'(u)$ and $X''(u)$, of degree less than n. Define a polynomial $X(u)$ as $X(u) = X'(u) - X''(u)$. Clearly, $deg(X(u)) \leq n - 1$ and $X(u)$ has all zeros in its residue representation. Thus, we have $P_i(u)|X(u), i = 1, 2, \ldots, \beta$, which also implies that $\prod_{i=1}^{\beta} P_i(u)|X(u)$ or $P(u)|X(u)$ (Theorem 5.5). This is not possible as $P(u)$ is a monic polynomial and $deg(X(u)) < deg(P(u))$. This proves the theorem.

The above theorem leads to the direct sum property for polynomials defined over $CZ(q)$ which can be expressed as

$$< X(u) >_{P(u)} \approx \sum_{\substack{\oplus \\ i=1}}^{\beta} < X(u) >_{P_i(u)} . \qquad (5.3.3)$$

Given $X(u)$, (5.3.2) also provides a computational procedure to obtain the residue polynomials $X_i(u)$, $i = 1, 2, \ldots, \beta$. Finally, one must establish a reconstruction procedure to obtain $X(u)$ from its residue polynomials. The polynomial $X(u)$ can be obtained from the residue polynomials $X_i(u)$, $i = 1, 2, \ldots, \beta$ by using the expression

$$X(u) \equiv \sum_{i=1}^{k} X_i(u)T_i(u)P_i'(u) \mod P(u), \qquad (5.3.4)$$

where

$$P_i'(u) = \frac{P(u)}{P_i(u)}. \tag{5.3.5}$$

The polynomials $T_i(u), i = 1, 2, \ldots, \beta$ are determined a priori by solving the congruences,

$$T_i(u)P_i'(u) \equiv 1 \mod P_i(u). \tag{5.3.6}$$

Once again, $T_i(u)$ is determined in a recursive manner by first expressing it in its p-adic form as

$$T_i(u) = \sum_{l=0}^{\alpha-1} p^l T_{i,l}(u). \tag{5.3.7}$$

Also, let the p-adic representation of $P_i'(u)$ be

$$P_i'(u) = \sum_{l=0}^{\alpha-1} p^l P_{i,l}'(u). \tag{5.3.8}$$

Substituting the p-adic forms of $T_i(u)$ and $P_i'(u)$ in (5.3.6) and taking mod p on both sides, we get

$$T_{i,0}(u)P_{i,0}'(u) \equiv 1 \quad (\bmod\ P_i(u), p). \tag{5.3.9}$$

The polynomial congruence in (5.3.9) can be solved over $GF(p^2)$ by using the Euclid's algorithm as the polynomials $P_i(u)$ and $P_i'(u)$ are relatively coprime. Once $T_{i,0}(u)$ is known, we take mod p^2 on both sides in (5.3.6) to get,

$$[T_{i,0}(u) + pT_{i,1}(u)][P_{i,0}'(u) + pP_{i,1}'(u)] \equiv 1 \quad (\bmod\ P_i(u), p^2) \tag{5.3.10}$$

or

$$T_{i,1}(u)P_{i,0}'(u) \equiv -[T_{i,0}(u)P_{i,1}'(u) + Q_{i,1}(u)] \quad (\bmod\ P_i(u), p), \tag{5.3.11}$$

where $Q_{i,1}(u)$ is determined by solving,

$$T_{i,0}(u)P_{i,0}'(u) \equiv 1 + pQ_{i,1}(u) \quad (\bmod\ P_i(u), p^2). \tag{5.3.12}$$

The solution to (5.3.11) leads to

$$T_{i,1}(u) = -T_{i,0}(u)[T_{i,0}(u)P_{i,1}'(u) + Q_{i,1}(u)] \quad (\bmod\ P_i(u), p). \tag{5.3.13}$$

Similarly, $T_{i,2}(u), \ldots, T_{i,\alpha-1}(u)$ can be determined in an order-recursive manner. In summary,

$$T_{i,k}(u) = -T_{i,0}(u)[T_{i,0}(u)P_{i,k}'(u) + Q_{i,k}(u)] \quad (\bmod\ P_i(u), p), \tag{5.3.14}$$

where

$$p^k Q_{i,k}(u) \equiv [(\sum_{j=0}^{k-1} p^j T_{i,j}(u))(\sum_{j=0}^{k-1} p^j P'_{i,j}(u)) - 1] \pmod{P_i(u), p^{k+1}}.$$

$$(5.3.15)$$

We observe here that all the computations on $CZ(q)$ are reduced to equivalent computations on $GF(p^2)$ using the p-adic representation of the various quantities. Also, the reconstruction polynomials $T_i(u), i = 1, 2, \ldots,$ β are unique. In the following, we illustrate the above reconstruction procedure for the factorization of $u^5 - 1$ over $CZ(3^2)$.

EXAMPLE 5.2 The monic factors of $u^5 - 1$ over $CZ(3^2)$ are $P_1(u) = u - 1$, $P_2(u) = u^2 + (5 + j)u + 1$ and $P_3(u) = u^2 + (5 + 8j)u + 1$. Therefore, $P'_1(u) = u^4 + u^3 + u^2 + u + 1$, $P'_2(u) = u^3 + (4 + 8j)u^2 + (5 + j)u + 8$ and $P'_3(u) = u^3 + (4 + j)u^2 + (5 + 8j)u + 8$. Based on their p-adic forms, we have $P'_{1,0}(u) = u^4 + u^3 + u^2 + u + 1$, $P'_{1,1}(u) = 0$, $P'_{2,0}(u) = u^3 + (1 + 2j)u^2 + (2 + j)u + 2$, $P'_{2,1}(u) = (1 + 2j)u^2 + u + 2$, $P'_{3,0}(u) = u^3 + (1 + j)u^2 + (2 + 2j)u + 2$ and $P'_{3,1}(u) = u^2 + (1 + 2j)u + 2$. The equations (5.3.9), (5.3.12) and (5.3.13) are used to obtain $T_{i,0}(u)$, $Q_{i,1}(u)$ and $T_{i,1}(u)$ respectively, for $i = 1, 2, 3$. In this case, we get

$$i = 1 \begin{cases} T_{1,0}(u) &= 2 \\ Q_{1,1}(u) &= 0 \\ T_{1,1}(u) &= 0, \end{cases}$$

$$i = 2 \begin{cases} T_{2,0}(u) &= (2 + j)u + 2 \\ Q_{2,1}(u) &= (1 + j)u + (1 + j) \\ T_{2,1}(u) &= (2 + 2j)u + 1, \end{cases}$$

$$i = 3 \begin{cases} T_{3,0}(u) &= (2 + 2j)u + 2 \\ Q_{3,1}(u) &= ju + 1 \\ T_{3,1}(u) &= 2u + 1. \end{cases}$$

The reconstruction polynomials are obtained as

$$\begin{aligned} T_1(u) &= T_{1,0}(u) + 3T_{1,1}(u) = 2 \\ T_2(u) &= T_{2,0}(u) + 3T_{2,1}(u) = (8 + 7j)u + 5 \\ T_3(u) &= T_{3,0}(u) + 3T_{3,1}(u) = (8 + 2j)u + 5. \end{aligned}$$

Finally,

$$\begin{aligned} T_1(u)P'_1(u) &= 2u^4 + 2u^3 + 2u^2 + 2u + 2 \\ T_2(u)P'_2(u) &= (8 + 7j)u^4 + (8 + 2j)u^3 + (8 + 2j)u^2 + (8 + 7j)u + 4 \\ T_3(u)P'_3(u) &= (8 + 2j)u^4 + (8 + 7j)u^3 + (8 + 7j)u^2 + (8 + 2j)u + 4. \end{aligned}$$

It is easily verified that $T_i(u)P'_i(u) \equiv 1 \mod P_i(u), i = 1, 2, 3$.

5.4 Algorithms for Convolution

In this section, we describe the design of fast algorithms for computing the cyclic and acyclic convolutions as stated in (5.0.9) and (5.0.10) respectively for the case $p = 4L + 3$.

5.4.1 Algorithms for Cyclic Convolution

For cyclic convolution, we need the factorization of $u^n - 1$ over $CZ(q)$. Such a factorization begins with the factorization over $GF(p^2)$. After obtaining the factors of $u^n - 1$ over $GF(p^2)$, the recursive procedure described in Section 5.2 gives rise to the desired factors over $CZ(q)$. The techniques to factor $u^n - 1$ over a given finite field are well established. A summary of these techniques is presented in tables 5.2 and 5.3.

Given the factorization of $u^n - 1$ over $CZ(q)$ as

$$u^n - 1 = \prod_{i=1}^{s} P_i(u),$$

an algorithm for computing the cyclic convolution can be outlined as follows:

1. Step 1 (AICE-CRT reduction). Compute

$$\begin{aligned} X_i(u) &\equiv X(u) \bmod P_i(u), \\ Y_i(u) &\equiv Y(u) \bmod P_i(u), \quad i = 1, 2, \ldots, s. \end{aligned}$$

2. Step 2. Compute

$$Z_i(u) = X_i(u)Y_i(u) \bmod P_i(u), \quad i = 1, 2, \ldots, s.$$

3. Step 3 (AICE-CRT reconstruction). Obtain $Z(u)$ as

$$Z(u) \equiv \sum_{i=1}^{s} Z_i(u)T_i(u)P_i'(u) \pmod{u^n - 1}.$$

5.4.2 Algorithms for Acyclic Convolution

For acyclic convolution, the choice of $P(u)$ is arbitrary. In this case, $P(u)$ is selected in terms of its factor polynomials over $GF(p^2)$. In general, a large number of factors results in an algorithm having lower multiplicative complexity. For example, over $CZ(3^2)$, we may choose $P(u) = u(u+1)(u+$

Table 5.2: Factorization of $u^n - 1$ Over $GF(p^2)$

	Let A be a primitive element in $GF(p^2)$.
$n = p^2 - 1$	$u^n - 1 = \prod_{i=1}^{n}(u - A^i)$
$n\|(p^2 - 1)$	$u^n - 1 = \prod_{i=1}^{n}(u - B^i)$ $B = A^{\frac{p^2-1}{n}}$
	Note: A primitive element in $GF(p^2)$ always exists. All elements of $GF(p^2)$ are represented as complex integers $a + jb, a, b \in GF(p)$.

$2)(u+1+j)(u+1+2j)$ or $P(u) = (u+8)(u^2+(5+j)u+1)(u^2+(5+8j)u+1)$ for $n = 5$. For the former choice, the corresponding algorithm will require 5 MULT while for the latter choice the count is 7 MULT.

The notion of intentional wraparound is valid for these algorithms as well. This is done in order to reduce the computational complexity even further. The basic idea is to compute a length n acyclic convolution using a polynomial $P(u)$ of degree d. This leads to a wraparound of the last $w = n - d$ coefficients which are computed separately and then combined to get the correct result.

5.5   Discussion

All the algorithms described in this chapter can be put in a matrix-vector form. In general, the algorithms for computing the bilinear forms can be expressed as $\mathbf{C}\{\mathbf{A}\ \underline{X} \odot \mathbf{B}\ \underline{Y}\}$ where $\underline{X}$ and $\underline{Y}$ represent the two inputs, $\mathbf{A}, \mathbf{B}, \mathbf{C}$ are suitable matrices, and $\odot$ denotes the component-by-component multiplication of two vectors.

Let us assume that the product of two degree one polynomials is computed in 3 MULT as

$$(x_0 + x_1 u)(y_0 + y_1 u) = x_0 y_0 + ((x_0 + x_1)(y_0 + y_1) - x_0 y_0 - x_1 y_1)u + x_1 y_1 u^2.$$

Table 5.3: Factorization of $u^n - 1$ Over $GF(p^2)$ Using $GF(p^{2m})$

	Let A be a primitive element in $GF(p^{2m})$.
$n = (q-1)$ $q = p^{2m}$	$u^n - 1 = \prod_{k \in \mathcal{I}} Q_k(u)$ $Q_k(u) = \prod_{i=0}^{l-1}(u - A^{kp^{2i}}), l \le m$ $\mathcal{I}$ is the set of indices of the cyclotomic sets $\mho_k$, $\mho_k = \{k, kp^2, kp^4, \cdots\}$. Here, k is the smallest integer not appearing in $\mho_1, \cdots,$ $\mho_{k-1}$. The elements of $\mho_k$ are defined modulo $(p^{2m} - 1)$.
$n\|(q-1)$	Let $B = A^{(q-1)/n}$. $u^n - 1 = \prod_{k \in \mathcal{I}} Q_k(u)$ $Q_k(u) = \prod_{i=0}^{l-1}(u - B^{kp^{2i}}), l \le m$ $\mathcal{I}$ is the set of the indices of the cyclotomic set $\mho_k$. $\mho_k = \{k, kp^2, kp^4, \cdots\}$. Here, k is the smallest integer not appearing in $\mho_1, \cdots,$ $\mho_{k-1}$. The elements of $\mho_k$ are defined modulo n.
$(n, p) = 1$	Find the smallest integer m such that $n\|(p^{2m} - 1)$. Apply the previous case to factorize $u^n - 1$.
$n = cp$ $(c, p) = 1$	$u^n - 1 = (u^c - 1)^p$ Factorize $u^c - 1$ using the techniques described above.
	Note: A primitive element in $GF(p^{2m})$ always exists. The polynomials $Q_k(u)$ are irreducible.

Using the various polynomials given in examples 5.1 and 5.2, the various matrices for the length 5 cyclic convolution of integer sequences over $CZ(3^2)$ are found to be

$$
\mathbf{A} = \mathbf{B} =
\begin{bmatrix}
1 & 1 & 1 & 1 & 1 \\
\hline
1 & 0 & 8 & 5+j & 4+8j \\
0 & 1 & 4+8j & 5+j & 8 \\
1 & 1 & 3+8j & 1+2j & 3+8j \\
\hline
1 & 0 & 8 & 5+8j & 4+j \\
0 & 1 & 4+j & 5+8j & 8 \\
1 & 1 & 3+j & 1+7j & 3+j
\end{bmatrix},
$$

$$
\mathbf{C} =
\begin{bmatrix}
2 & | & 5+2j & 4j & 8+7j & | & 5+7j & 5j & 8+2j \\
2 & | & 4+7j & 4+7j & 4 & | & 4+2j & 4+2j & 4 \\
2 & | & 4j & 5+2j & 8+7j & | & 5j & 5+7j & 8+2j \\
2 & | & 0 & 5j & 8+2j & | & 0 & 4j & 8+7j \\
2 & | & 5j & 0 & 8+2j & | & 4j & 0 & 8+7j
\end{bmatrix}.
$$

The partitions as marked in the above matrices correspond to the AICE-CRT reduction and reconstruction with regard to the modulo polynomial $P_i(u)$, $i = 1, 2, 3$, respectively.

Notes
The CRT has been studied extensively in the context of deriving fast algorithms for computing convolution of sequences defined over a field. Our objective, throughout this book has been to extend and exploit this classical result for processing data sequences defined over finite integer rings.

One can derive complex-number-theoretic-transforms (CNTTs) as a special case of Legrange Interpolation (LI) over $CZ(q)$ which in turn is a special case of AICE-CRT over $CZ(q)$ as described in this chapter, $q = p^\alpha$. The maximum number of interpolating points for the LI over $CZ(q)$ are given by $n = 2$ for $p = 2$, $n = p$ for $p = 4L + 1$, and $n = p^2$ for $p = 4L + 3$. Similarly, CNTTs over $CZ(q)$ exist for values of n given by $n = 1$ for $p = 2$, $n|(p-1)$ for $p = 4L + 1$, and $n|(p^2 - 1)$ for $p = 4L + 3$. These results can be derived in a straightforward manner.

We expect this work to find numerous applications in digital signal processing and other related areas.

This chapter is based on the research work reported in Krishna et al. [1994].

5.6 Bibliography

[5.1] C.M. Rader, "Discrete Convolutions via Mersenne Transforms," *IEEE Transactions on Computers*, vol. C-21, pp. 1269-1273, December 1972.

[5.2] R.C. Agarwal and C.S. Burrus, "Fast Convolution Using Fermat Number Transforms With Applications to Digital Filtering," *IEEE Transactions on Acoustics, Speech, and Signal Processing*, vol. ASSP-22, no. 2, pp. 87-97, April 1974.

[5.3] R.C. Agarwal and C.S. Burrus, "Number Theoretic Transforms to Implement Fast Digital Convolution," *Proceedings IEEE*, vol. 63, pp. 550-560, April 1975.

[5.4] J.H. McClellan, "Hardware Realization of a Fermat Number Transform," *IEEE Transactions on Acoustics, Speech, and Signal Processing*, vol. ASSP-24, no. 3, pp. 216-225, June 1976.

[5.5] H.J. Nussbaumer, *Fast Fourier Transform and Convolution Algorithms*, IInd Edition, Springer-Verlag, 1982.

[5.6] D.G. Myers, *Digital Signal Processing-Efficient Convolution and Fourier Transform Techniques*, Prentice Hall, Inc., 1990.

[5.7] J.H. McClellan and C.M. Rader, *Number Theory in Digital Signal Processing*, Prentice Hall, Inc., 1979.

[5.8] K.-Y. Lin, B. Krishna and H. Krishna, "Rings, Fields, the Chinese Remainder Theorem and an Extension, Part I: Theory," *IEEE Transactions on Circuits and Systems*, to appear in 1994.

[5.9] H. Krishna, K.-Y. Lin and B. Krishna, "Rings, Fields, the Chinese Remainder Theorem and an Extension, Part II: Applications to Digital Signal Processing," *IEEE Transactions on Circuits and Systems*, to appear in 1994.

[5.10] L.K. Hua, *Introduction to Number Theory*, Springer-Verlag, 1982.

[5.11] H. Krishna, K.-Y. Lin and B. Krishna, "The AICE-CRT and Digital Signal Processing Algorithms: The Complex Case," *Circuits, Systems and Signal Processing*, to appear in 1994.

Chapter 6

Fault Tolerance for Integer Sequences

In many signal processing and computing systems, one is faced with the task of processing data which is inherently defined on a ring of integers modulo an integer M, or can equivalently be converted to one. Such an arithmetic is based on the number theoretic properties of data sequences. It has fascinated and attracted the attention of mathematicians and physicists dating back to ancient times as well as modern day engineers who analyze the various problems borne of the recent technological advances. Number theory has found numerous applications in the design of computationally efficient algorithms for digital signal processing.

Along other lines, coding theory has flourished as an art and science to protect numeric data from errors, thereby improving the reliability and overall performance of data processing systems. However, by and large, with certain exceptions, coding theory is developed and studied as it applies to digital communication systems. This is evident from a large number of texts on coding theory as it relates to communication theory. Exceptions to this are the applications of coding theory concepts to fault tolerance in computing systems.

With the advent of very large scale integration (VLSI) design methodology and concepts of parallel processing, there has been tremendous thrust towards high performance computing systems that possess fault tolerance capability. Such systems provide correct results in presence of one or more faulty processors. The original idea of algorithm based fault tolerance was introduced in Huang and Abraham [1984]. It is based on some of the classical results available in coding theory. Their proposed checksum scheme can

131

detect and correct any failure within a single processor in a fault-tolerant multi-processor system under the assumption that at most one module is faulty within a given period of time. Subsequently, more applications of the algorithm based fault tolerance in signal processing and related tasks were proposed in Jou and Abraham [1986] and Chen and Abraham [1986]. A linear algebraic model of weighed checksum scheme has been introduced in Jou and Abraham [1986] and the techniques for correcting single faults were extended to correcting two faults in Anfinson and Luk [1988]. All the techniques described in these papers are valid only for processing real-valued data.

6.1   Introduction and Mathematical Preliminaries

This chapter is an effort to bring together number theory and coding theory with the objective of developing a mathematical framework for building algorithm based fault tolerance capability in vectors defined over $Z(M)$. The impetus for this work is provided by the numerous applications of integer arithmetic in certain computational environments. Lack of an underlying Galois field presents a unique challenge to this framework. We note here that the most well known application of coding theory is in digital communications where the data is defined in a Galois field of q elements, $GF(q)$, $q = p^m$. We develop the theory and algorithms for error control in data sequences defined over a ring of integers. We have used several excellent sources to strengthen our mathematical background in number theory and coding theory. The reader is referred to them for a more complete though independent treatment of these theories. They are listed in the references at the end of this chapter. Also, we refer the readers to the material in sections 2.3 and 2.4.

The mathematical preliminaries fundamental to the content of this chapter are presented in this section. A ring of integers is defined by a composite integer M which can be uniquely factored as (standard factorization of M),

$$M = \prod_{i=1}^{t} p_i^{\alpha_i} = \prod_{i=1}^{t} q_i, \tag{6.1.1}$$

where p_i is the ith prime, α_i is a positive integer exponent and $q_i = p_i^{\alpha_i}$. Integers in the range $[0, M)$ form the ring, $Z(M)$, where the arithmetic operations of '.' and '+' are defined mod M. We say that the ring $Z(M)$ has degeneracy $\{t, \alpha\}$, where $\alpha = max(\alpha_1, \alpha_2, ..., \alpha_t)$. An integer X, $0 \leq X <$

M can equivalently be represented by a residue vector having t components $X_1, X_2, \ldots, X_t$, that is,

$$X \longleftrightarrow \underline{X} = [X_1\, X_2 \ldots X_t], \qquad (6.1.2)$$

where

$$X_i \equiv X \bmod q_i, \; i = 1, 2, \ldots, t. \qquad (6.1.3)$$

Once again, integers in the range $[0, q_i)$ form a ring $Z(q_i)$, where the arithmetic operations of '.' and '+' are defined mod q_i. It is clear that $Z(q_i)$ is not the same as $GF(q_i)$ unless $\alpha_i = 1$. This is due to the reason that in $GF(q_i)$, $\alpha_i > 1$, the computations are not performed mod q_i. Of course, for $\alpha_i = 1$, $Z(q_i)$ and $GF(q_i)$ coincide and q_i is same as the prime p_i. Also, we have $(q_i, q_j) = 1$, $i \neq j$.

Given a residue vector $\underline{X}$, the corresponding integer X in $Z(M)$ can be uniquely determined by solving t congruences in (6.1.3). The uniqueness of the integer X and a procedure to compute it is established by the CRT-I.

6.2   A Framework for Fault Tolerance

The algorithm based fault control technique is an (n, k) coding scheme $\mathcal{C}$ that consists in mapping a length k data vector $\underline{A}$ over $Z(M)$ to a length n codevector $\underline{V}$ over $Z(M)$. All the length k data vectors form a vector space of dimension k. Since the coding scheme $\mathcal{C}$ being introduced is linear in nature, the corresponding length n codevectors also form a vector space of dimension k. Such a coding scheme is completely characterized by k ℓi codevectors $\underline{V}_i$, $i = 1, 2, \ldots, k$ associated with the k unit data vectors $\underline{E}_i$, $i = 1, 2, \ldots, k$. The unit data vector $\underline{E}_i$ is a vector having 1 in its ith position and 0's elsewhere. Without any loss of generality, we assume that the codevectors $\underline{V}_i$ have the form

$$\underline{V}_i = [\underline{E}_i |\ \overset{\leftarrow\, n-k\, \rightarrow}{* * \cdots * *}\ ]. \qquad (6.2.1)$$

The '$*$' in the above expression denote the $(n - k)$ parity check digits to be assigned. As a result, the codevector for an arbitrary data vector $\underline{A} = [A_1 A_2 \ldots A_k]$ can be written as $\underline{V} = \underline{A}G$, where, G, a $(k \times n)$ matrix over $Z(M)$, is called the generator matrix of the code (in systematic form) and is given by

$$G = \begin{bmatrix} \underline{V}_1 \\ \underline{V}_2 \\ \vdots \\ \underline{V}_k \end{bmatrix} = \begin{bmatrix} I_k & | & P \end{bmatrix}. \qquad (6.2.2)$$

Here $\mathbf{I}_k$ is the identity matrix of dimension k and $\mathbf{P}$ is a $k \times (n-k)$ matrix having elements in $Z(M)$. This is based on the linear nature of the coding scheme and the modulo operation, and the decomposition of $\underline{A}$ as $\underline{A} = \sum_{i=1}^{k} A_i \underline{E}_i$. We also note that the codevector $\underline{V}$ has the form $\underline{V} = [\underline{A} \mid \underline{A}\mathbf{P}]$. The first k digits of $\underline{V}$ are the same as those of $\underline{A}$ and the remaining $(n-k)$ parity check digits are appended to provide for fault tolerance. It is clear from the form of $\mathbf{G}$ in (6.2.2) that rank$(\mathbf{G}) = k$. Now consider the null space $\mathcal{U}$ of the vector space $\mathcal{V}$ created by the rows of the generator matrix $\mathbf{G}$ in (6.2.2). Let $\underline{H}_1, \underline{H}_2, \ldots, \underline{H}_{n-k}$ be the $(n-k)$ ℓi vectors that span $\mathcal{U}$. It is clear that

$$\underline{V}\mathbf{H}^t \equiv \underline{0}, \tag{6.2.3}$$

where, $\mathbf{H}$, to be termed as the parity check matrix of the code $\mathcal{C}$, is a $(n-k) \times n$ given by

$$\mathbf{H} = \begin{bmatrix} \underline{H}_1 \\ \vdots \\ \underline{H}_{n-k} \end{bmatrix}.$$

For the systematic form of $\mathbf{G}$, $\mathbf{H}$ can be written by inspection as

$$\mathbf{H} = \begin{bmatrix} -\mathbf{P}^t \mid \mathbf{I}_{n-k} \end{bmatrix}. \tag{6.2.4}$$

The *minimum distance*, d, is a fundamental parameter associated with any random error correcting code and is defined as,

$$d = min\{d(\underline{V}_i, \underline{V}_j) : \underline{V}_i, \underline{V}_j \in \mathcal{C}, \underline{V}_i \neq \underline{V}_j\}.$$

Since the coding scheme being studied is linear in nature, it is easily established that $d = min\{wt(\underline{V}), \underline{V} \in \mathcal{C}, \underline{V} \neq \underline{0}\}$. Here, $d(\underline{V}_i, \underline{V}_j)$ and $wt(\underline{V})$ denote the Hamming distance between the vectors $\underline{V}_i$ and $\underline{V}_j$ and the Hamming weight of the vector $\underline{V}$ respectively. The minimum distance, d, is intimately related to the fault detection and correction capability of the code $\mathcal{C}$.

6.2.1   Fault Detection and Correction

Given a codevector $\underline{V}$, let $\underline{V}$ be changed to a vector $\underline{R}$ due to presence of faults in computation. In this case, we write $\underline{R} = \underline{V} + \underline{E}$ or

$$R_i \equiv V_i + E_i \text{ mod } M, \; i = 1, \ldots, n, \tag{6.2.5}$$

where $\underline{E}$ is the fault vector, $\underline{E} = [E_1 \; E_2 \; \ldots \; E_n]$. If $E_i \neq 0$ then a fault is said to have taken place in the ith position. The Hamming weight of the fault vector $\underline{E}$, $wt(\underline{E})$, also denoted by γ, is called the number of faults in $\underline{R}$. Also, we have $d(\underline{R}, \underline{V}) = wt(\underline{E}) = \gamma$.

Fault detection consists in checking if $\underline{R}$ is a codevector in $\mathcal{C}$ or not. This can be accomplished by computing $\underline{R}\mathbf{H}^t$. Fault correction or decoding consists in computing the position of the faults and the fault magnitudes. The relationships between d and the fault detection and correction capabilities are straightforward to derive and are summarized in the following. The *fault detection capability* of a code $\mathcal{C}$, defined as the largest value γ for which $\underline{R}$ cannot be a codevector in $\mathcal{C}$, is $d - 1$. The *fault correcting capability*, t, of a code $\mathcal{C}$, defined as the largest value of γ for which correct decoding takes place ($\widehat{\underline{V}} = \underline{V}$), is $\lambda = \lfloor \frac{d-1}{2} \rfloor$, where $\lfloor a \rfloor$ denotes the largest integer less than or equal to a. A code $\mathcal{C}$ can correct λ faults and simultaneously detect β faults ($\beta > \lambda$), if $d \geq \lambda + \beta + 1$.

Having established the general framework for algorithm based fault control techniques over $Z(M)$, we now analyze the mathematical structures of these codes in more detail.

6.2.2 Mathematical Structure of $\mathcal{C}$ Over $Z(M)$

Consider a k-dimensional data vector $\underline{A} = [A_1 \ A_2 \ \ldots \ A_k]$, $A_j \in Z(M)$. Based on (6.1.1) and the CRT-I, it can equivalently be represented by t vectors, each of dimension k, corresponding to t distinct primes that constitute M, that is,

$$\underline{A} \leftrightarrow [\underline{A}^{(1)}\underline{A}^{(2)}\ldots\underline{A}^{(t)}], \tag{6.2.6}$$

where

$$\underline{A}^{(i)} = [A_1^{(i)} \ldots A_k^{(i)}], A_j^{(i)} \in Z(q_i), j = 1, \ldots, k. \tag{6.2.7}$$

This representation is obtained by writing $A_j \in Z(M)$ as

$$A_j \leftrightarrow \underline{A}_j = [A_j^{(1)} \ldots A_j^{(t)}] \tag{6.2.8}$$

$$A_j^{(i)} \equiv A_j \bmod q_i, \ i = 1, 2, \ldots, t \ j = 1, 2, \ldots, k. \tag{6.2.9}$$

Also, $A_j = 0$ implies that $A_j^{(i)} = 0$, $i = 1, 2, \ldots, t$, and vice versa, while $A_j \neq 0$ implies that at least one of $A_j^{(i)}$ is not equal to 0 and vice versa. The codevector $\underline{V}$ associated with $\underline{A}$ is obtained as

$$\underline{V} = [V_1 V_2 \ldots V_n] = \underline{A}\mathbf{G}. \tag{6.2.10}$$

Using the property of linearity of the coding scheme $\mathcal{C}$ and the modulo operation, we may equivalently decompose the above equation mod q_i, $i = 1, 2, \ldots t$, to get

$$\underline{V} \leftrightarrow [\underline{V}^{(1)}\underline{V}^{(2)}\ldots\underline{V}^{(t)}], \tag{6.2.11}$$

where using (6.2.10), we have

$$\underline{V}^{(i)} = [V_1^{(i)} V_2^{(i)} \ldots V_n^{(i)}] \tag{6.2.12}$$

$$\underline{V}^{(i)} = \underline{A}^{(i)} \mathbf{G}^{(i)} \bmod q_i, i = 1, 2, \ldots, t. \tag{6.2.13}$$

$\mathbf{G}^{(i)}$ being the $(k \times n)$ sub-generator matrix obtained by reducing each element of $\mathbf{G}$ modulo q_i. Thus, the process of encoding $\underline{A}$ to get the codevector $\underline{V} = \underline{A}\mathbf{G}$, can be decomposed into t parallel processes as listed below:

1. Given a length k data vector $\underline{A}$, reduce each element of $\underline{A}$ mod q_i, to get t data vectors $\underline{A}^{(1)}, \underline{A}^{(2)}, \ldots, \underline{A}^{(t)}$, each of length k, where $\underline{A}^{(i)}$ $= [A_1^{(i)} \ldots A_k^{(i)}]$ and $A_j^{(i)} \equiv A_j \bmod q_i$.

2. Encode each of these vectors independently as $\underline{V}^{(i)} \equiv \underline{A}^{(i)} \mathbf{G}^{(i)} \bmod$ q_i, $i = 1, 2, \ldots, t$.

3. Use the CRT-I to combine $V_j^{(i)}$, $i = 1, 2, \ldots, t$ to get the jth component, V_j, of the codevector $\underline{V}$, $j = 1, 2, \ldots, n$.

The expression in (6.2.13) is an important expression as it enables us to interpret the overall coding scheme $\mathcal{C}$ defined on $Z(M)$ to be constituted by t subcoding schemes $\mathcal{C}_i$ for each of the t primes that constitute M. Here, $\underline{A}^{(i)}$ is the data vector, $\underline{V}^{(i)}$ is the corresponding codevector and $\mathbf{G}^{(i)}$ is the generator matrix associated with $\mathcal{C}_i$. The encoding method outlined above has the attractive feature that if the data is available in its reduced form modulo q_i, then the encoding can be performed directly to produce the corresponding codevector in reduced form. In such a case, steps (1) and (3) are not required. The encoding method as outlined above is shown in Figure 5.1. It is clear from (6.2.2) that $\mathbf{G}^{(i)}$ is also in systematic form and is given by,

$$\mathbf{G}^{(i)} = [\mathbf{I}_k \mid \mathbf{P}^{(i)}], \ i = 1, 2, \ldots, t. \tag{6.2.14}$$

The parity check matrix, $\mathbf{H}^{(i)}$, for the subcode $\mathcal{C}_i$ can be expressed as,

$$\mathbf{H}^{(i)} = \left[-\mathbf{P}^{(i)t} \mid \mathbf{I}_{n-k} \right]. \tag{6.2.15}$$

Based on this decomposition and the CRT-I, we make several observations with respect to the coding scheme $\mathcal{C}$. These are stated in the following.

Observation 1. A length n arbitrarily chosen vector $\underline{V}^{(i)}$ having components over $Z(q_i)$ is a codevector in the subcode $\mathcal{C}_i$ *iff* it can be written in the form $\underline{V}^{(i)} = \underline{A}^{(i)} \mathbf{G}^{(i)}$, where $\underline{A}^{(i)}$ is a length k vector defined over $Z(q_i)$, or equivalently $\underline{V}^{(i)} \mathbf{H}^{(i)t} \equiv 0 \bmod q_i$, $i = 1, 2, \ldots, t$.

Observation 2. A length n arbitrarily chosen $\underline{V}$ having components over

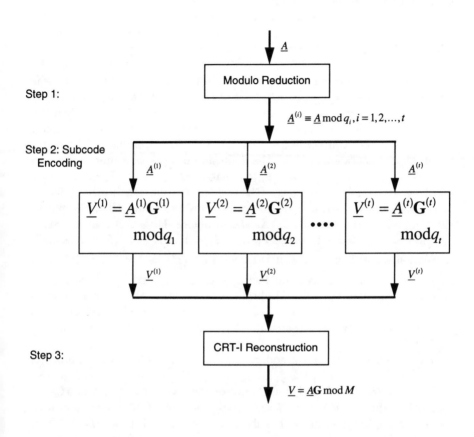

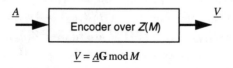

Figure 6.1: An Encoder Diagram for Encoding Over $Z(M)$.

$Z(M)$ is a codevector in the code C *iff* $\underline{V}^{(i)}$ (defined as $\underline{V}^{(i)} \equiv \underline{V}$ mod q_i) is a codevector in the subcode C_i, $i = 1, 2, \ldots, t$.

The code C defined over $M = q_1 q_2 \cdots q_t$ is equivalent to t subcodes C_i, defined over q_i, $i = 1, 2, \ldots, t$. An important question arises at this stage: How are the minimum distance d of the code C and the minimum distance d_i of the subcode C_i, $i = 1, 2, \ldots, t$ related to each other? We explore the structure and the decomposition of the C further to find an answer to this question. It is clear from the definition of ℓi of vectors that if a codevector $\underline{V}$ has non-zero component in positions $i_1, i_2, \ldots, i_a (wt(\underline{V}) = a)$, then the i_1th, i_2th,$\ldots,i_a$th columns of $\mathbf{H}$ are ℓd over $Z(M)$. Thus, if the minimum distance is d, no arbitrarily selected $d - 1$ or less columns of $\mathbf{H}$ are ℓd. Combining this statement with the definition of rank of a matrix, we can state that the rank of the $(n - k) \times j$ matrix $\mathbf{H}_j$ ($j \leq d - 1$) obtained by selecting arbitrarily j columns of $\mathbf{H}$ is j, or H_j the jth invariant factor of the matrix $\mathbf{H}_j$ satisfies $(H_j, M) = 1$. Now, examine $\mathbf{H}^{(i)}$, the parity check matrix of the subcode C_i. As stated before, $\mathbf{H}^{(i)} \equiv \mathbf{H}$ mod q_i. Since $(H_j, M) = 1$, $j \leq d - 1$, we also have $(H_j, q_i) = 1$, $i = 1, 2, \ldots, t$ as well. In Section 2.3.3, Lemma 2.2, it is shown that $(H_j^{(i)}, q_i) = 1$, $i = 1, 2, \ldots, t$, where $H_j^{(i)}$ is the jth invariant factor of the matrix $\mathbf{H}_j^{(i)}$, $\mathbf{H}_j^{(i)}$ being the $(n - k) \times j$ matrix ($j \leq d - 1$) obtained by selecting the same j columns from the matrix $\mathbf{H}^{(i)}$ as those selected from the matrix $\mathbf{H}$ to get $\mathbf{H}_j$. It is clear that no arbitrarily selected $d - 1$ or less columns of $\mathbf{H}^{(i)}$ are ℓd over $Z(q_i)$ either.

On the other hand, for a codevector $\underline{V}$ having d non-zero components in positions $j_1, j_2, \ldots, j_d$, the j_1th, j_2th,$\ldots,j_d$th columns of $\mathbf{H}$ are ℓd over $Z(M)$, thereby implying that the rank of the $(n - k) \times d$ matrix $\mathbf{H}_d$ obtained by selecting the j_1th, j_2th,$\ldots,j_d$th columns of $\mathbf{H}$ is less than d or every $d \times d$ submatrix in $\mathbf{H}_d$ has determinant Δ_d such that $(H_d, M) > 1$. Let $(H_d, M) = p_a^{i_a}$, where p_a is one of $p_1, p_2, \ldots, p_t$ and $i_a \leq \alpha_a$. Now, examine the parity check matrix, $\mathbf{H}^{(a)} \equiv \mathbf{H}$ mod q_a, of the subcode C_a. Since (H_d, M) has the factor $p_a^{i_a}$, this implies that $(H_d, q_a) = p_a^{i_a}$. In Section 2.3.3, Lemma 2.2, we show that $(H_d^{(a)}, q_a) > 1$. As a result, we conclude that the d rows $j_1, j_2, \ldots, j_d$ of $\mathbf{H}_d^{(a)}$ are ℓd over $Z(q_a)$. This implies that the codevector $\underline{V}^{(a)}$ in the code C_a corresponding to the codevector $\underline{V}$ in C has Hamming weight d or 0. This leads to the following theorem.

THEOREM 6.1 *If the minimum distance of a (n, k) code C defined over $Z(M)$ is d, $M = q_1 q_2 \cdots q_t$, then the minimum distance of the (n, k) subcode C_i defined over $Z(q_i)$ is $d_i, d_i \geq d$. Also, there is at least one subcode say C_a such that $d_a = d$.*

Also, consider the task of combining independently designed (n, k) subcodes C_i designed over $Z(q_i)$, $i = 1, 2, \ldots, t$ to obtain a (n, k) code C over $Z(M)$, $M = q_1 q_2 \cdots q_t$. If $\underline{V}^{(i)} = [V_1^{(i)} \ldots V_n^{(i)}]$ is an arbitrary codevector in C_i, $i = 1, 2, \ldots, t$ and $\underline{V} = [V_1 \ldots V_n]$ is the corresponding codevector in C, then from the CRT-I, we have $V_j \neq 0$ if one or more of $V_j^{(i)} \neq 0$. Also, if $V_j^{(i)} = 0$, $i = 1, 2, \ldots, t$, then $V_j = 0$. Noting that each of the subcodes C_i has the all-zero vector as a codevector, we get the following theorem.

THEOREM 6.2 *If the minimum distance of the subcode C_i designed over $Z(q_i)$ is d_i, $i = 1, 2, \ldots, t$, then the minimum distance of the code C defined over $Z(M)$ and obtained by combining C_i using the CRT-I, $M = q_1 q_2 \cdots q_t$, is d, where $d = min(d_1, d_2, \ldots, d_t)$.*

We now turn to the final aspect of our general framework for fault control which is the design of decoding algorithms for codes defined over $Z(M)$.

6.2.3 Decoding Algorithms Over $Z(M)$

The analysis in Section 6.2 (particularly, theorems 6.1 and 6.2) also provides for a straightforward procedure for decoding the faulty vector in (6.2.5). Since the faults in (6.2.5) corrupt the codevector in an additive manner, we outline the following decoding procedure for such a fault model:

1. Given a length n vector $\underline{R}$ defined over $Z(M)$, reduce each element of $\underline{R}$ modulo q_i to get t vectors $\underline{R}^{(i)}$, defined over $Z(q_i)$, $i = 1, 2, \ldots, t$, each of length n.

2. Decode each of $\underline{R}^{(i)}$ independently to get $\underline{E}^{(i)}$ using the decoding algorithm for C_i.

3. Combine $\underline{E}^{(i)}, i = 1, 2, \ldots, t$ using CRT-I to get the fault vector $\underline{E}$ which the decoder for C would equivalently compute.

4. Finally, $\underline{R}$ is decoded to $\widehat{\underline{V}} = \underline{R} - \underline{E}$.

It is important to note here that the validity of the above decoding procedure is based on two facts, namely (i) $wt(\underline{E}^{(i)}) \leq wt(\underline{E})$, and (ii) minimum distance d_i, of the subcode $C_i \geq d$, the minimum distance of the code C (Theorem 6.1). Thus, the complete decoding procedure over $Z(M)$ can be implemented using t independent decoders defined over $Z(q_i)$ operating in parallel. This is a very interesting and useful feature of the coding scheme that we have introduced here. It exposes the inherent parallelism in the encoding and the decoding algorithms which may be exploited to obtain suitable VLSI architectures. A block diagram for the decoder is shown in Figure 6.2. Also, we note that each of the subcodes C_i are defined over a much smaller size alphabet as compared to the original code.

Given a minimum distance d (n,k) code C defined over $Z(M), M = q_1 q_2 \cdots q_t$, consider the problem of designing a minimum distance d code C' defined over $Z(N)$, $N = M \cdot q_b$, $q_b = p_b^{\alpha_b}$, $(q_i, q_b) = 1$, $i = 1, 2, \ldots, t$. Based on theorems 6.1 and 6.2, it is clear that we need to design a code C_b over $Z(q_b)$ having minimum distance $d_b \geq d$. The generator matrix of the code C' is obtained by combining the generator matrices of the coding scheme C and C_b using the CRT-I. The encoder for C' is obtained by including one more branch in parallel to the encoder to C. Interestingly, the decoder for C' is obtained by combining the decoders for C and C_b in parallel as well.

Since the design of a coding scheme over $Z(M)$ is equivalent to the design of t coding schemes C_i over $Z(q_i)$, $i = 1, 2, \ldots, t$, we now turn to the design of a coding scheme C over $Z(q)$, $q = p^\alpha$. We reassert that $Z(q)$ is not the same as (or an automorphism of) $GF(q)$ due to modulo computation.

6.3   Coding Techniques Over $Z(q)$

For $\alpha = 1$, $Z(q) = Z(p) = GF(p)$. Therefore, for $\alpha = 1$, we can apply all the well-known coding theory techniques to the design of codes of interest here. As a result, we are motivated to establish a close correspondence between codes defined over $GF(p)$ and codes defined over a ring of integers, $Z(q)$, $q = p^\alpha$. Such a correspondence is in two parts, part I is the characterization of the codes in terms of the generator matrices and minimum distance and part II is the decoding algorithms.

Consider a (n,k) code C over $Z(q)$ having generator matrix $\mathbf{G}$, parity check matrix $\mathbf{H}$, and minimum distance d, and an equivalent code C_e over $GF(p)$ having generator matrix $\mathbf{G}_e$, parity check matrix $\mathbf{H}_e$, and minimum distance d_e. The equivalence between the codes C and C_e is defined in terms of their generator matrices $\mathbf{G}$ and $\mathbf{G}_e$ as

$$\mathbf{G}_e \equiv \mathbf{G} \bmod p. \tag{6.3.1}$$

We now proceed to establish a relationship between d and d_e. Once again, minimum distance of C equal to d implies that all i ($i \leq d-1$) arbitrarily selected columns of $\mathbf{H}$ are ℓi, that is, the $(n-k) \times i$ matrix formed by these i columns has the ith invariant factor H_i such that $(H_i, p^\alpha) = 1$. In other words, it has rank i. If the same i columns were selected from $\mathbf{H}_e$ ($\mathbf{H}_e \equiv \mathbf{H} \bmod p$ from (6.3.1)) to obtain the $(n-k) \times i$ matrix formed by these columns, it will have ith invariant factor $H_i^{(e)}$ such that $(H_i^{(e)}, p) = 1$, or equivalently the i columns in $\mathbf{H}_e$ are ℓi (refer to Section 2.3.3, Lemma

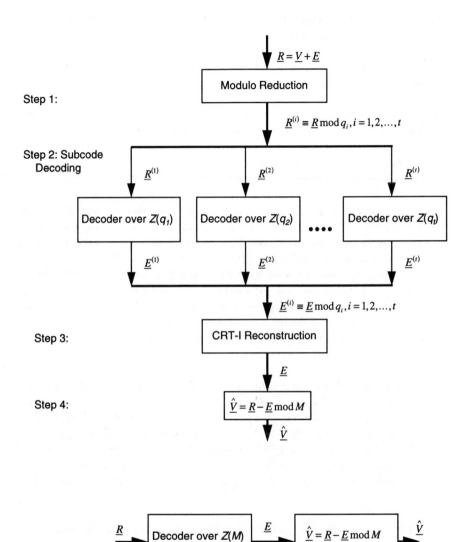

Figure 6.2: The Decoder Diagram for Decoding Over $Z(M)$.

2.3 for a proof). Also, for $i = d$, there is at least one set of d columns in
H (there is at least one codeword in C having Hamming weight d) such
that the $(n - k) \times d$ matrix formed by these columns has rank less than
d. Clearly, these d columns are ℓd when reduced modulo p. This implies
existence of a codeword in C_e having Hamming weight d or 0. This leads
to the following theorem.

THEOREM 6.3 *For every (n, k) coding scheme C defined over $Z(q)$,
$q = p^\alpha$, having minimum distance d, there is an equivalent coding scheme
C_e in $GF(p)$ having minimum distance d. The generator matrix $\mathbf{G}_e$ of C_e
is obtained by reducing the generator matrix $\mathbf{G}$ of C modulo p.*

Using similar arguments as above, we can show that if we are given a
coding scheme C over $GF(p)$, having minimum distance d, then the generator matrix for C can also be used to obtain a coding scheme over $Z(q)$,
$q = p^\alpha$ having minimum distance d (refer to Lemma 2.4 in Section 2.3.3).
A consequence of the analysis is that one only needs to design a code over
$GF(p)$ which can also be used over $Z(q)$, $q = p^\alpha$. There is an abundance
of good linear codes over $GF(p)$ available in coding theory texts and any
of them can be used in the present context. Therefore, we do not need
to pursue the topic of design of linear codes defined over $Z(q)$, $q = p^\alpha$.
Combining the above analysis and the analysis in Section 6.2.2, we get the
following fundamental theorem:

THEOREM 6.4 *The generator matrix of a coding scheme defined over
$Z(M), M = q_1 q_2 \cdots q_t, q_i = p_i^{\alpha_i}$ having minimum distance d is equivalent to the generator matrix of a coding scheme defined over $Z(M_e), M_e =
p_1 p_2 \cdots p_t$, having minimum distance d. The generator matrix of a coding
scheme defined over $Z(M_e)$, M_e having minimum distance d is obtained
by combining the generator matrices of coding schemes C_i over $GF(p_i)$
having minimum distances $d_i > d$ using the CRT-I. In addition, $d =
min(d_1, d_2, \ldots, d_t)$.*

In addition to finding the generator matrix of a coding scheme over $Z(q)$,
$q = p^\alpha$, we have the issue of the decoding algorithm for the code over
$Z(q)$ which is originally designed over $GF(p)$. Once again, the decoding
algorithms for linear codes over $GF(p)$ are well established. Consequently,
there is a strong impetus to derive the decoding algorithm for the code
over $Z(q)$ in terms of the decoding algorithm for the code over $GF(p)$. In
the following, it is shown that the decoding algorithm over $GF(p)$ can be
used recursively to decode a received vector over $Z(q)$. This is based on
the observation that a linear code defined over $Z(q)$, $q = p^\alpha$ is also a linear

code over $Z(p^j)$, $j = 1, 2, \ldots, \alpha$. It is worthwhile to mention here that the code over $Z(p^j)$ is obtained by reducing the code over $Z(p^i)$, $j < i$, modulo p^j.

6.3.1 Decoding Algorithms Over $Z(q)$

The decoding over $GF(p)$ is performed in three steps:

1. Compute the $(n - k)$ length vector called the syndrome as $\underline{s} = \underline{r}\mathbf{H}^t$;

2. Define a mapping called the standard array from the vector $\underline{s}$ over $GF(p)$ to a length n fault vector $\underline{e}$ over $GF(p)$; and

3. Obtain the estimate $\widehat{\underline{v}}$ of the codevector as $\widehat{\underline{v}} = \underline{r} - \underline{e}$.

Observe that the decoding procedure as outlined here does not require division, thereby alleviating the need for a finite field. Therefore, it may also be used to decode over $Z(q)$, $q = p^\alpha$, where the operation of division by a non-zero scalar is not always available. However, except certain simple cases, any practical realization of step (2) must employ division in order to decode in a reasonable amount of time by computing $\underline{e}$ for a given $\underline{s}$ rather than storing $\underline{e}$ for every possible $\underline{s}$ in a memory. Note that the size of the memory for a memory-based realization of step (2) is $p^{(n-k)}$. It is assumed that the previous statements are valid in the present context also, that is, given $\underline{S} = \underline{R}\mathbf{H}^t$ over $Z(q)$, we would like to compute $\underline{E}$ over $Z(q)$ in order to perform decoding over $Z(q)$. The size of the memory for a memory-based realization of step (2) for a decoder over $Z(q)$ will be $p^{\alpha(n-k)}$. We will denote the decoder over $GF(p)$ by $\mathcal{D}$.

Given $\underline{R} = \underline{V} + \underline{E}$ over $Z(q)$, we may write $R_i = V_i + E_i$, $i = 1, 2, \ldots, n$; $R_i, V_i, E_i \in Z(q)$. Since R_i, V_i, and E_i are defined mod p^α, we can express them in the p-adic number representation as,

$$R_i = \sum_{m=0}^{\alpha-1} r_{i,m} p^m, \ 0 \leq r_{i,m} < p$$

$$V_i = \sum_{m=0}^{\alpha-1} v_{i,m} p^m, \ 0 \leq v_{i,m} < p$$

$$E_i = \sum_{m=0}^{\alpha-1} e_{i,m} p^m, \ 0 \leq e_{i,m} < p. \tag{6.3.2}$$

Note that

$$r_{i,0} \equiv v_{i,0} + e_{i,0} \bmod p, \tag{6.3.3}$$

while, in general, $r_{i,m} \neq (v_{i,m} + e_{i,m}) \bmod p, m > 0$. This is due to the non-linear nature of the arithmetic operation of '+'. Given $\underline{R}$, the recursive algorithm to compute $\underline{E}$ consists in computing $e_{i,m}$, $i = 1, \ldots, n$, $m = 0, 1, \ldots, \alpha - 1$ recursively by employing $\mathcal{D}$. Based on (6.3.2) and the expression, $\underline{R} = \underline{V} + \underline{E}$, we can write

$$
\begin{aligned}
\underline{R} &= \sum_{m=0}^{\alpha-1} \underline{r}_m p^m \\
&= \sum_{m=0}^{\alpha-1} \underline{v}_m p^m + \sum_{m=0}^{\alpha-1} \underline{e}_m p^m,
\end{aligned} \tag{6.3.4}
$$

where the vectors $\underline{r}_m$, $\underline{v}_m$ and $\underline{e}_m$ are given by

$$
\begin{aligned}
\underline{r}_m &= [r_{1,m}\, r_{2,m} \ldots r_{n,m}], \\
\underline{v}_m &= [v_{1,m}\, v_{2,m} \ldots v_{n,m}],
\end{aligned}
$$

and

$$
\underline{e}_m = [e_{1,m}\, e_{2,m} \ldots e_{n,m}]. \tag{6.3.5}
$$

As $\underline{V}$ is a code vector in a code over $Z(q)$,

$$
\underline{V}_m \equiv \underline{V} \bmod p^{m+1}, \; m = 0, 1, \ldots, \alpha - 1, \tag{6.3.6}
$$

is a codevector in a code over $Z(p^{m+1})$. Corresponding to (6.3.6), we may write

$$
\begin{aligned}
\underline{R}_m &\equiv \underline{R} \bmod p^{m+1} \\
&\equiv \underline{V}_m + \underline{E}_m \bmod p^{m+1} \\
\underline{E}_m &\equiv \underline{E} \bmod p^{m+1}.
\end{aligned} \tag{6.3.7}
$$

Recall that the first step in decoding $\underline{R}$ is the computation of the syndrome as

$$
\underline{S} \equiv \underline{R}\mathbf{H}^t = \underline{E}\mathbf{H}^t, \tag{6.3.8}
$$

thereby, implying that the syndrome $\underline{S}_m$ corresponding to the received vector $\underline{R}_m$ can be obtained as

$$
\begin{aligned}
\underline{S}_m &\equiv \underline{R}\mathbf{H}^t \bmod p^{m+1} \\
&\equiv \underline{S} \bmod p^{m+1}, \; m = 0, 1, 2, \ldots, \alpha - 1.
\end{aligned} \tag{6.3.9}
$$

The decoder $\mathcal{D}$ over $GF(p)$ consists in a computational procedure to obtain $\underline{e}$, given the syndrome $\underline{s}$ and the relation $\underline{s} = \underline{e}\mathbf{H}^t$ over $GF(p)$. We can convert the above equation to α relations of the type $\underline{s} = \underline{e}\mathbf{H}^t$ over $GF(p)$ and then employ the decoder over $GF(p)$ as follows. Let

$$
\begin{aligned}
\underline{s}_0 &= \underline{S}_0 \equiv \underline{R}_0\mathbf{H}^t \bmod p \\
&= \underline{e}_0\mathbf{H}^t \bmod p.
\end{aligned}
\tag{6.3.10}
$$

Given $\underline{s}_0$, $\mathcal{D}$ is used to find $\underline{e}_0$. We note that if $wt(\underline{E})$ is less than $\lfloor \frac{d-1}{2} \rfloor$, the fault correcting capability of the code over $Z(q)$, then $wt(\underline{E}_m)$ and $wt(\underline{e}_m)$ are less than the fault correcting capability of the code over $Z(p^{m+1})$. This is due to the fact that $wt(\underline{E}) \geq wt(\underline{E}_m) \geq wt(\underline{e}_m)$ while the fault correcting capability of the codes over $Z(q)$ and $Z(p^{m+1})$ are the same.

Once $\underline{e}_0$ is known, consider the expression for $\underline{S}_1$ as

$$
\begin{aligned}
\underline{S}_1 &\equiv \underline{R}_1\mathbf{H}^t \bmod p^2 \\
&\equiv \underline{E}_1\mathbf{H}^t \bmod p^2 \\
&= (\underline{e}_0 + \underline{e}_1 p)\,\mathbf{H}^t \bmod p^2.
\end{aligned}
\tag{6.3.11}
$$

Rearranging various terms, we get

$$
\begin{aligned}
\underline{s}_1 &= \{(\underline{S}_1 - \underline{e}_0\mathbf{H}^t)p^{-1}\} \bmod p \\
&\equiv \underline{e}_1\mathbf{H}^t \bmod p.
\end{aligned}
$$

Clearly, given $\underline{S}$ and $\underline{e}_0$, $\underline{s}_1$ can be computed and $\mathcal{D}$ can be used to obtain $\underline{e}_1$. The recursive nature of the decoding algorithm is clear now. Let us say that $\underline{e}_0, \underline{e}_1, \ldots, \underline{e}_{m-1}$ are determined. The computations to determine $\underline{e}_m$ proceeds as follows, let

$$
\begin{aligned}
\underline{s}_m &= \left\{ \left(\underline{S}_m - \sum_{j=0}^{m-1} \underline{e}_j p^j \mathbf{H}^t\right) p^{-m} \right\} \bmod p \\
&\equiv \underline{e}_m\mathbf{H}^t \bmod p.
\end{aligned}
\tag{6.3.12}
$$

Given $\underline{s}_m$, $\mathcal{D}$ is employed to compute $\underline{e}_m$, $m = 1, \ldots, \alpha - 1$. Once the fault vectors $\underline{e}_0, \underline{e}_1, \ldots, \underline{e}_{\alpha-1}$ (or equivalently, the vector $\underline{E}$) are determined, the decoder decodes $\underline{R}$ to a code vector $\hat{\underline{V}}$ where

$$
\hat{\underline{V}} \equiv \underline{R} - \underline{E} \bmod q.
\tag{6.3.13}
$$

A step-by-step description of the decoding algorithm over $Z(q)$ is given in the following

1. Given $\underline{R}$, compute the syndrome $\underline{S} \equiv \underline{R}\,\mathbf{H}^t$.

2. Let $\underline{s}_0 \equiv \underline{S} \bmod p = \underline{e}_0\mathbf{H}^t$.

3. Employ $\mathcal{D}$ to determine $\underline{e}_0$.

4. Let $m \leftarrow 1$.

5. Compute $\underline{s}_m \equiv \underline{e}_m \mathbf{H}^t \bmod p$ by using (6.3.12).

6. Employ $\mathcal{D}$ to determine $\underline{e}_m$.

7. $m \leftarrow m + 1$. If $m < \alpha$, go to step (5).

8. Decode $\underline{R}$ to the codevector $\widehat{\underline{V}}$ using (6.3.13).

A flowchart for this algorithm is shown in Figure 6.3. Of course, this algorithm leads to correct decoding provided that $wt(\underline{E}) \leq \lfloor \frac{d-1}{2} \rfloor$. A most interesting feature of this algorithm is that it also leads to correct decoding for all those fault vectors $\underline{E}$ for which $wt(\underline{e}_m) \leq \lfloor \frac{d-1}{2} \rfloor$, $m = 0, 1, \ldots, \alpha - 1$, though $wt(\underline{E})$ exceeds $\lfloor \frac{d-1}{2} \rfloor$. Clearly, if $wt(\underline{E}) \leq \lfloor \frac{d-1}{2} \rfloor$, then $wt(\underline{e}_m) \leq \lfloor \frac{d-1}{2} \rfloor$, but if $wt(\underline{e}_m) \leq \lfloor \frac{d-1}{2} \rfloor$, then it *does not* imply that $wt(\underline{E}) \leq \lfloor \frac{d-1}{2} \rfloor$, $m = 0, 1, 2, \ldots, \alpha - 1$. We conclude this section by noting that one cannot decode directly the vector $\underline{r}_m$, the mth coefficient in the p-adic representation of $\underline{R}, m = 0, 1, \ldots, \alpha - 1$ as the corresponding vector $\underline{v}_m$, though defined over $GF(p)$, is *not* a codevector in any code. Also, a divide-and-conquer technique based fast algorithm can be described for performing the decoding over $Z(q)$, $q = p^\alpha$. This algorithm is described in the following section.

6.3.2   A Fast Algorithm For Decoding Over $Z(q)$

In this subsection, we outline a divide-and-conquer technique based fast algorithm for decoding over $Z(q)$, $q = p^\alpha$. It assumes that α is an integer of the type $\alpha = 2^\beta$ and recursively decodes the received vector $\underline{R}$ over $Z(p^m)$, $m = 1, 2, 4, \ldots, 2^\beta$. Since this algorithm is similar in approach to the algorithm described in Section 6.3.1, we simply summarize it in the following

Decoder for $m = 1$

$\underline{s}_{m-1} = \underline{S} \bmod p^m$.

Use decoder over $Z(p^m)$ to get $\underline{e}_0, \underline{e}_1, \ldots, \underline{e}_{m-1}$.

Decoder for $m = 2, 4, \ldots, 2^\beta$

$$\underline{s}_{m/2} \equiv \left\{ (\underline{S}_m - \sum_{j=0}^{m/2-1} \underline{e}_j p^j \mathbf{H}^t) p^{-m/2} \right\} \bmod p^{m/2}$$

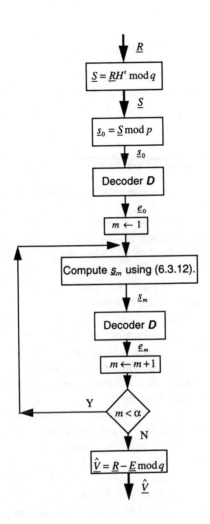

Figure 6.3: A Flowchart for the Decoder Over $Z(q)$, $q = p^{\alpha}$.

$$\equiv \left\{ \sum_{j=m/2}^{m-1} \underline{e}_j p^{j-m/2} \mathbf{H}^t \right\} \bmod p^{m/2}.$$

Use decoder over $Z(p^{m/2})$ to get $\underline{e}_{m/2}, \ldots, \underline{e}_{m-1}$.

We have kept the details of this algorithm to a minimum as in most applications α would be a small integer, and for a small value of α, the order-recursive algorithm described in Section 6.3.1 may perform better than this divide-and-conquer technique based algorithm due to its simple mathematical structure.

6.3.3   Some General Remarks

Sections 6.2 and 6.3 establish a complete mathematical framework for fault control techniques in data sequences defined over $Z(M)$. It is established that the generator matrix $\mathbf{G}$ of a linear code defined over $Z(M), M = q_1 q_2 \cdots q_t, q_i = p_i^{\alpha_i}$ is equivalent to the generator matrix of a linear code defined over $Z(M_e)$, $M_e = p_1 p_2 \cdots p_t$. Also, the generator matrix of a linear code defined over $Z(M)$ is obtained by combining the generator matrices of linear codes C_i defined over $GF(p_i)$ using the CRT-I. The decoder over $Z(M)$ can be implemented using t decoders operating in parallel; here the ith decoder performs decoding over $Z(q_i)$. The decoder over $Z(q_i)$ can be implemented by using recursively the decoder over $GF(p_i)$ α_i times. Consequently, the degeneracy of M, $\{t, \alpha\}$, plays a vital role in establishing the overall realization and the decoding time, t being the number of parallel branches such a realization will have and α being the maximum number of recursions along each of the branches.

In some cases, it may not be possible to directly derive (n, k) codes over $Z(p_i^{\alpha_i})$ having minimum distance d. An easier task is to design (n_i, k) codes C_i over $Z(p_i^{\alpha_i})$ having minimum distance $d_i \geq d$. Then, an (n, k) code over $Z(M)$ having minimum distance d can be obtained such that $n = max(n_1, n_2, \ldots, n_t)$. This is accomplished by appending $n - n_i$ zeros to every codeword in C_i and then combining the resulting codewords using the CRT-I. The configuration of the decoder remains essentially unaltered.

We now present an example and specific details of minimum distance 4 single fault correcting-double fault detecting (SFC-DFD) codes.

6.4   Examples and SFC-DFD Codes

Let $M = 110,852,311 = 31^3 \cdot 61^2 = q_1 \cdot q_2$, $q_1 = 31^3$, $q_2 = 61^2$. Clearly, we have $p_1 = 31$, $\alpha_1 = 3$, $p_2 = 61$, and $\alpha^2 = 2$. Here, M has degeneracy

{2,3}. Suppose we are interested in designing a (n, k) minimum distance 3 SFC code over $Z(M)$. This is accomplished by designing (n, k) SFC codes over $Z(29{,}791)$ or $GF(31)$ and $Z(3{,}721)$ or $GF(61)$, respectively. A SFC code over $GF(p)$ is a Hamming code having the largest value of n as $(p^m - 1)/(p - 1)$, $m = n - k$. Thus, a SFC code over $Z(M)$ can have the largest value of n as

$$\begin{aligned}
n &= max(n_1, n_2, \ldots, n_t), \\
n_i &= (p_i^m - 1)/(p_i - 1), \ i = 1, 2, \ldots, t, \\
m &= n - k.
\end{aligned}$$

In our case, let $m = n - k = 2$. Then $n_1 = 32$, $n_2 = 62$ and $n = 62$. If we are interested in a $(10, 8)$ SFC code over $Z(110{,}852{,}311)$, then we need to obtain the generator matrices $\mathbf{G}^{(1)}$ and $\mathbf{G}^{(2)}$ of the $(10,8)$ shortened Hamming codes over $GF(31)$ and $GF(61)$ respectively. One such set of generator matrices is

$$\mathbf{G}^{(1)} = \begin{bmatrix} & & & & \mathbf{I_8} & & & \\ 1 & 1 & 1 & 1 & 1 & 1 & 1 & 1 \\ 1 & 2 & 3 & 4 & 5 & 6 & 7 & 8 \end{bmatrix}^t,$$

and

$$\mathbf{G}^{(2)} = \begin{bmatrix} & & & & \mathbf{I_8} & & & \\ 1 & 1 & 1 & 1 & 1 & 1 & 1 & 1 \\ 1 & 2 & 3 & 4 & 5 & 40 & 7 & 8 \end{bmatrix}^t. \tag{6.4.1}$$

Combining $\mathbf{G}^{(1)}$ and $\mathbf{G}^{(2)}$ using the CRT-I, we get the overall generator matrix as

$$\mathbf{G} = \begin{bmatrix} & & & & \mathbf{I_8} & & & \\ 1 & 1 & 1 & 1 & 1 & 1 & 1 & 1 \\ 1 & 2 & 3 & 4 & 5 & 77{,}158{,}696 & 7 & 8 \end{bmatrix}^t.$$

Let the data vector be $\underline{A} = [50{,}000\ 29\ 60\ 2\ 5\ 6\ 110\ 25]$. The corresponding code vector in $\mathcal{C}_1$ having $\mathbf{G}^{(1)}$ as the generator matrix is computed as $\underline{V}^{(1)} = \underline{A}^{(1)} \mathbf{G}^{(1)} \bmod 29{,}791$, where $\underline{A}^{(1)} \equiv \underline{A} \bmod 29{,}791 = [20{,}209\ 29\ 60\ 2\ 5\ 6\ 110\ 25]$. The codevector $\underline{V}^{(1)} = [20{,}209\ 29\ 60\ 2\ 5\ 6\ 110\ 25\ 20{,}446\ 21{,}486]$. Similarly, $\underline{A}^{(2)} = \underline{A} \bmod 3{,}721 = [1{,}627\ 29\ 60\ 2\ 5\ 6\ 110\ 25]$ and $\underline{V}^{(2)} = [1{,}627\ 29\ 60\ 2\ 5\ 6\ 110\ 25\ 1{,}864\ 3{,}108]$. The codevector $\underline{V}$ may be obtained by combining $\underline{V}^{(1)}$ and $\underline{V}^{(2)}$ using the CRT-I or directly by using the expression $\underline{V} = \underline{A}\mathbf{G}$. In this example, $\underline{V} = [50{,}000\ 29\ 60\ 2\ 5\ 6\ 110\ 25\ 50{,}237\ 19{,}594{,}173]$. Let the vector $\underline{R} = [50{,}000\ 29\ 60\ 1{,}222\ 5\ 6\ 110\ 25\ 50{,}237\ 19{,}594{,}173]$. Computing the syndrome, we get $\underline{S} = \underline{R}\mathbf{H}^t = [110{,}851{,}091$

110,847,431]. Since $\underline{S} \neq \underline{0}$, we assume that one fault has taken place and proceed to find it. It can be shown that for a SFC code, $\underline{S} = E_i \ \underline{H}_i$ where i is the fault location, E_i is the fault value and $\underline{H}_i$ is the ith row of the parity check matrix $\mathbf{H}^t$. This relation can be employed directly to compute i and E_i as $i = 4$ and $E_i = 1,220$. However, we would like to demonstrate the parallel and recursive nature of the decoder.

Since the degeneracy of M is $\{2,3\}$, the overall decoder can be implemented in parallel using two decoders over $Z(29,791)$ and $Z(3,721)$ respectively. The first step is to compute the syndrome over $Z(29,791)$ and $Z(3,271)$. They are given by

$$
\begin{aligned}
\underline{S}^{(1)} &\equiv \underline{R}^{(1)}\mathbf{H}^t \bmod 29{,}791 \\
&\equiv \underline{S} \bmod 29{,}791 = [28{,}571 \ 24{,}911], \\
\underline{S}^{(2)} &\equiv \underline{R}^{(2)}\mathbf{H}^t \bmod 3{,}721 \\
&\equiv \underline{S} \bmod 3{,}721 = [2{,}501 \ 2{,}562].
\end{aligned}
\tag{6.4.2}
$$

Given $\underline{S}^{(1)}$ and $\underline{S}^{(2)}$, the decoder will now proceed to determine $\underline{E}^{(1)}$ and $\underline{E}^{(2)}$ using the decoders over $Z(29,751)$ and $Z(3,721)$ respectively, and the recursive approach as described in Section 6.3. At the heart of these decoders are two SFC decoders, one that operates in $GF(31)$ using $\mathbf{G}^{(1)}$ as the generator matrix and the second that operates in $GF(61)$ using $\mathbf{G}^{(2)}$ as its generator matrix.

In Table 6.1, we show the results of these recursive computations. Note that the first decoder performs three recursions while the second decoder performs two recursions. To simplify the notations (and hoping that it will not confuse the reader), we delete the parenthesized superscript of (1) or (2) from the table. Based on Table 6.1, we conclude that the fault is in the fourth location. The fault magnitude is computed as follows: $E_4^{(1)} \equiv E_4$ mod $29,791$. The coefficients in the 31-adic representation of $E_4^{(1)}$ are $11, 8$ and 1 respectively, implying that

$$
E_4^{(1)} = 11 + 8 \times 31 + 1 \times 31^2 = 1{,}220.
\tag{6.4.3}
$$

Similarly, $E_4^{(2)} \equiv E_4$ mod $3,721$ and the coefficients in the 61-adic representation of $E_4^{(2)}$ are 0 and 20 respectively, implying that

$$
E_4^{(2)} = 0 + 20 \times 61 = 1{,}220.
\tag{6.4.4}
$$

Table 6.1: Results of Computations for the Example in Section 6.4.

Computations recursions	$\underline{S}_m$	$\underline{s}_m$	Decoder over Fault Location	GF(31) Fault Value
$m = 0$	(20 18)	(20 18)	4	11
$m = 1$	(702 886)	(23 30)	4	8
$m = 2$	(28,571 24,911)	(30 24)	4	1

A. Steps in decoding over $Z(29, 751)$ $\underline{S} = [28, 571\ 24, 911]$, $\alpha = 3$.

Computations recursions	$\underline{S}_m$	$\underline{s}_m$	Decoder over Fault Location	GF(61) Fault Value
$m = 0$	(0 0)	(0 0)	no errors	0
$m = 1$	(2,501 2,562)	(41 42)	4	20

B. Steps in decoding over $Z(3, 721)$, $\underline{S} = [2, 501\ 2, 562]$, $\alpha = 2$.

Combining (6.4.3) and (6.4.4) using the CRT-I, we get $E_4 = 1,220$. Finally the decoder decodes $\underline{R}$ to $\underline{\widehat{V}}$ where

$$\widehat{V}_i = R_i, \ i = 1, 2, \ldots, 10, \ i \neq 4$$

and

$$\widehat{V}_4 = R_4 - E_4 = 2.$$

A SFC-DFD code is obtained by adding an overall parity digit to every codevector in the SFC code. For example, the (11, 8) SFC-DFD code over $Z(110,852,311)$ is obtained by combining the generator matrices of the SFC-DFD codes over $GF(31)$ and $GF(61)$ respectively. For the SFC generator matrices in (6.4.1), the corresponding SFC-DFD generator matrices (in systematic form) are as follows,

$$\mathbf{G}^{(1)} = \begin{bmatrix} & & & \mathbf{I_8} & & & \\ 1 & 1 & 1 & 1 & 1 & 1 & 1 & 1 \\ 1 & 2 & 3 & 4 & 5 & 6 & 7 & 8 \\ 3 & 4 & 5 & 6 & 7 & 8 & 9 & 10 \end{bmatrix}^t,$$

$$\mathbf{G}^{(2)} = \begin{bmatrix} & & & \mathbf{I_8} & & & \\ 1 & 1 & 1 & 1 & 1 & 1 & 1 & 1 \\ 1 & 2 & 3 & 4 & 5 & 40 & 7 & 8 \\ 3 & 4 & 5 & 6 & 7 & 42 & 9 & 10 \end{bmatrix}^t.$$

The decoding for SFC-DFD code is performed in a manner similar to the decoding for the SFC code. However, if any one of the t decoders operating in parallel detects the presence of two faults or the fault vector $\underline{E}$ has two non-zero components, then the decoder declares the presence of two faults.

In the next section, we describe the algebraic construction of cyclic codes based on NTTs.

6.5   NTT Based Cyclic Codes

It was demonstrated in Section 3.8 that number theoretic transforms (NTTs) are a special case of LI over $Z(M)$. NTTs lead to an efficient technique for computing the cyclic convolution over $Z(M)$ and are based on the factorization of $u^n - 1$ as

$$u^n - 1 = \prod_{i=1}^{n} (u - G^i) \tag{6.5.1}$$

where G is the generator over $Z(M)$. Also, $G^n \equiv 1 \bmod M$, where as $(G^i - 1, M) = 1$, $i = 1, 2, \ldots, n - 1$. An NTT of length n exists provided that

$$n|(p_j - 1), \ j = 1, 2, \ldots, t. \tag{6.5.2}$$

These properties also form the basis of NTT based cyclic codes described in this section.

DEFINITION 6.1 An (n, k) code $\mathcal{C}$ is called *cyclic* provided that every cyclic shift of a codeword is also a codeword. In other words, if $\underline{V} = [V_0 \ V_1 \ \ldots \ V_{n-1}]$ is a codeword, then so is $[V_{n-1} \ V_0 \ \ldots \ V_{n-2}]$.

In the case of cyclic codes, it is mathematically convenient to express vectors as polynomials. The vector $\underline{V} = [V_0 \ V_1 \ \ldots \ V_{n-1}]$ can equivalently be expressed as the polynomial $V(u) = \sum_{i=0}^{n-1} V_i \ u^i$. Let $G(u)$ be the *monic* code polynomial of the lowest degree in the cyclic code $\mathcal{C}$. Such a polynomial always exists and is called the generator polynomial of $\mathcal{C}$. A number of properties associated with $G(u)$ are listed in the following:

1. $deg(G(u)) = n - k$,

2. $G(u)|(u^n - 1)$,

3. Every monic polynomial $G(u)$, such that $G(u)|(u^n - 1)$ generates a cyclic code and vice versa.

4. A polynomial $V(u)$ of degree $n - 1$ or less is a codepolynomial *iff* $G(u)|V(u)$, that is, every codepolynomial $V(u)$ can be expressed as $V(u) = U(u)G(u)$ and vice versa, $deg(U(u)) \leq k - 1$.

Based on the definition of a cyclic code and the factorization of $u^n - 1$ in (6.5.1), we see that for every value of n that satisfies (6.5.2), a cyclic code can be obtained. This is done by selecting $n - k$ distinct powers of G, say $G^{i_1}, G^{i_2}, \ldots, G^{i_{n-k}}$ and forming the generator polynomial $G(u)$ as,

$$G(u) = \prod_{j=1}^{n-k} (u - G^{i_j}). \tag{6.5.3}$$

A particular choice of the powers of G leads to Bose-Chaudhuri-Hocquengham (BCH) or Reed-Solomon (RS) type codes over $Z(M)$. These codes are maximum-distance-separable (MDS) over $Z(M)$.

THEOREM 6.5 *If the $n-k$ distinct powers of G are consecutive in nature, then the resulting generating polynomial $G(u)$ gives rise to a cyclic code $\mathcal{C}$ having minimum distance d,*

$$d = n - k + 1. \tag{6.5.4}$$

PROOF. Without any loss of generality, we assume that the $n-k$ consecutive powers of G are $1, 2, \ldots, n-k$. The resulting generator polynomial $G(u)$ becomes,

$$G(u) = \prod_{j=1}^{n-k} (u - G^j). \qquad (6.5.5)$$

Thus, $V(u) = \sum_{i=0}^{n-k} V_i\, u^i$ is a code polynomial *iff*

$$V(G^j) = 0, \; j = 1, 2, \ldots, n-k, \qquad (6.5.6)$$

or

$$[V_0 V_1 \ldots V_{n-1}] \begin{bmatrix} 1 & 1 & \cdots & 1 \\ G & G^2 & & G^{n-k} \\ G^2 & G^4 & & \vdots \\ \vdots & \vdots & & \\ G^{n-1} & G^{2(n-1)} & & G^{(n-k)(n-1)} \end{bmatrix} = \underline{0}. \qquad (6.5.7)$$

The parity check matrix for this code is

$$\mathbf{H}^t = \begin{bmatrix} 1 & 1 & \cdots & 1 \\ G & G^2 & & G^{n-k} \\ G^2 & G^4 & & G^{2(n-k)} \\ \vdots & \vdots & & \vdots \\ G^{n-1} & G^{2(n-1)} & & G^{(n-k)(n-1)} \end{bmatrix}. \qquad (6.5.8)$$

The minimum distance, d, of this code is $n-k+1$, provided that $d-1 = n-k$ arbitrarily selected rows of $\mathbf{H}^t$ are ℓi over $Z(M)$. Let us select rows $i_1, i_2, \ldots, i_a$, $a \le n-k$ of $\mathbf{H}^t$ and form the reduced $a \times (n-k)$ matrix $\mathbf{H}'$ as

$$\mathbf{H}' = \begin{bmatrix} G^{i_1} & G^{2i_1} & \cdots & G^{(n-k)i_1} \\ G^{i_2} & G^{2i_2} & \cdots & G^{(n-k)i_2} \\ G^{i_a} & G^{2i_a} & & G^{(n-k)i_a} \end{bmatrix}. \qquad (6.5.9)$$

In order to show that $\text{rank}(\mathbf{H}') = a$, $a \le n-k$, it is sufficient to show that there is a square $(a \times a)$ submatrix in $\mathbf{H}'$ having determinant Δ_a such that $(\Delta_a, M) = 1$ (Lemma 2.1). Consider the $(a \times a)$ submatrix of $\mathbf{H}'$ consisting of the first a rows and a columns. Such a matrix is given by

$$\mathbf{H}'' = \begin{bmatrix} G^{i_1} & G^{2i_1} & \cdots & G^{ai_1} \\ G^{i_2} & G^{2i_2} & \cdots & G^{ai_2} \\ G^{i_a} & G^{2i_a} & & G^{ai_a} \end{bmatrix} \qquad (6.5.10)$$

$$= \begin{bmatrix} G_1 & G_1^2 & \cdots & G_1^a \\ G_2 & G_2^2 & \cdots & G_2^a \\ \vdots & \vdots & & \vdots \\ G_a & G_a^2 & \cdots & G_a^a \end{bmatrix} \quad (6.5.11)$$

$$= G_1 G_2 \ldots G_a \begin{bmatrix} 1 & G_1 & \cdots & G_1^{a-1} \\ 1 & G_2 & \cdots & G_2^{a-1} \\ \vdots & & & \\ 1 & G_a & \cdots & G_a^{a-1} \end{bmatrix} \quad (6.5.12)$$

$$= G_1 G_2 \ldots G_a \mathbf{V}, \quad (6.5.13)$$

where

$$G_j = G^{ij}. \quad (6.5.14)$$

In the above expression, $\mathbf{V}$ is a Vandermonde matrix. The determinant of $\mathbf{H}''$ is given by (refer to (3.8.3)),

$$\begin{aligned} \Delta &= det(\mathbf{H}'') = G_1 G_2 \ldots G_a det(\mathbf{V}) \\ &= G_1 G_2 \ldots G_a \prod_{j=1}^{a-1} \prod_{k=j+1}^{a} (G_j - G_k) \quad (6.5.15) \end{aligned}$$

$$= G_1 G_2 \ldots G_a \prod_{j=1}^{a-1} \prod_{k=j+1}^{a} G^{i_k}(G^{i_j - i_k} - 1). \quad (6.5.16)$$

It is clear from the properties of the generator G of the NTT that $(\Delta, M) = 1$. This proves that rank$(\mathbf{H}') = a$, $a \leq n - k$.

The general mathematical framework for the decoding algorithm for codes defined over $Z(M)$ was derived in Section 6.2. It is applicable in the present situation as well. It requires decoders for decoding BCH codes of minimum distance $n - k + 1$ over $GF(p_i)$, $i = 1, 2, \ldots, t$. Given the BCH decoder over $GF(p)$, the decoding algorithm over $Z(p^\alpha)$ is the same as the one described in Section 6.3. Once again, the overall decoder consists of t decoders operating in parallel.

NOTES

In this work, our goal has been to bring together the richness of results available in the number theory and the theory of error control coding. The importance of this work lies in the suitability of modulo arithmetic in certain computational environments. A complete mathematical framework is described which may be used for designing schemes for algorithm based fault tolerant computing over a ring of integers.

Linearity of the modulo operation and the coding scheme leads to very interesting and useful properties for both the encoding and the decoding algorithms. The encoding operation can be performed on the data vector directly or equivalently on the modulo reduced vectors associated with it. The decoding is performed by combining several decoders operating in parallel. All the computations required for fault correction are converted to equivalent computations on a Galois field. This enables us to distinguish between the computations that may be performed in parallel and computations that are intrinsically serial in nature. Also, the emphasis in this chapter is on the mathematical structure of the various algorithms. The issues related to computational efficiency are not being treated directly. We expect that this work will motivate further research on the application of coding theory techniques to algorithm based fault tolerant computing systems.

There are a number of texts on coding theory and its applications to communication systems. We have used the texts Peterson and Weldon [1972], Blahut [1983], McElice [1977], MacWilliams and Sloane [1977], and Lin and Costello [1983] quite extensively. Application of coding theory concepts to fault tolerance in computing systems have been described in Wakerly [1978], Rao and Fujiwara [1989], Huang and Abraham [1984], Jou and Abraham [1986], Chen and Abraham [1986], and Anfinson and Luk [1988].

It is worthwhile to describe the earlier work on codes constructed over integer rings in Blake [1972], Blake [1975], Spiegel [1977] and Spiegel [1978]. In Blake [1972], cyclic codes of block length n over $Z(M)$ were constructed by combining cyclic (n, k_i) cyclic codes over $GF(p_i)$, where M is the product of distinct primes and the integers n and M have no factors in common. This approach was extended in Blake [1975] and Spiegel [1977] to derive cyclic codes over $Z(p^\alpha)$ with focus on Hamming and Reed-Solomon codes. These ideas were further extended in Spiegel [1978] to derive BCH codes over $Z(p^\alpha)$.

Our work departs from the above papers in a significant manner. We study the construction of linear codes, *not necessarily cyclic*, over $Z(M)$ having length n and dimension k. The notion of dimension is not present in earlier papers. No a priori assumption is made here about the relationship between n and M. More importantly, given the standard factorization of M as in (6.1.1), we establish a complete equivalence between a linear code over $Z(M)$ and the resulting codes over $Z(p_i^{\alpha_i})$, $i = 1, 2, \ldots, t$, (refer to Theorem 6.1) and vice versa. The earlier work has focused only on the equivalence between the (n, k_i) linear cyclic codes over $Z(p_i^{\alpha_i})$ and the resulting linear cyclic code over $Z(M)$ (refer to Theorem 6.2). The most important difference is the lack of a decoding strategy in the earlier papers for the codes considered therein. We describe a complete step-by-step decoding method for decoding linear codes over $Z(M)$ and study its inherently serial and parallel nature. Finally, the various group and ring theoretic results required in Blake [1972], Blake [1975], Spiegel [1977] and Spiegel [1978] are far more abstract than the elementary number theory employed here in order to describe the mathematical construction of the codes.

This chapter is based on the research work reported in Krishna [1994].

6.6   Bibliography

[6.1] W.W. Peterson and E.J. Weldon, Jr., *Error Correcting Codes*, IInd Edition, MIT Press, 1972.

[6.2] R.E. Blahut, *Theory and Practice of Error Control Codes*, Addison-Wesley Publishing Co., 1983.

[6.3] R.J. McElice, *The Theory of Information and Coding*, Addison-Wesley Publishing Co., 1977.

[6.4] F.J. MacWilliams and N.J.A. Sloane, *The Theory of Error Correcting Codes*, North Holland Publishing Co., 1997.

[6.5] S. Lin and D.J. Costello, Jr., *Error Control Coding: Fundamentals and Applications*, Prentice Hall, 1983.

[6.6] J. Wakerly, *Error Detecting Codes, Self-checking Circuits and Applications*, North Holland, 1978.

[6.7] T.R.N. Rao and E. Fujiwara, *Error Control Coding for Computer Systems*, Prentice Hall, 1989.

[6.8] M.A. Soderstrand, W.K. Jenkins, G.A. Jullien and F.J. Taylor, *Residue Number System Arithmetic: Modern Applications in Digital Signal Processing*, IEEE Press, 1986.

[6.9] K.H. Huang and J.A. Abraham, "Algorithm-Based Fault Tolerance for Matrix Operations," *IEEE Transactions on Computers*, vol. C-33, no. 6, pp. 518-528, June 1984.

[6.10] J.Y. Jou and J.A. Abraham, "Fault-Tolerant Matrix Arithmetic and Signal Processing on Highly Concurrent Computing Structures," *Proceedings IEEE*, no. 5, pp. 721-741, May 1986.

[6.11] C.Y. Chen and J.A. Abraham, "Fault Tolerance Systems for the Computation of Eigen Values and Singular Values," *SPIE, Advanced Algorithms and Architectures for Signal Processing*, vol. 696, pp. 228-237, 1986.

[6.12] C.J. Anfinson and F.T. Luk, "A Linear Algebraic Model of Algorithm Based Fault Tolerance," *IEEE Transactions on Computers*, vol. 37, no. 12, pp. 1599-1604, Dec. 1988.

[6.13] I.F. Blake, "Codes Over Certain Rings," *Information and Control*, vol. 20, pp. 396-404, 1972.

[6.14] I.F. Blake, "Codes Over Integer Residue Rings," *Information and Control*, vol. 29, pp. 295-300, 1975.

[6.15] E. Spiegel, "Codes Over Z_m," *Information and Control*, vol. 35, pp. 48-51, 1977.

[6.16] E. Spiegel, "Codes Over Z_m, Revised," *Information and Control*, vol. 37, pp. 100-104, 1978.

[6.17] H. Krishna, "A Mathematical Framework for Algorithm Based Fault Tolerant Computing Over a Ring of Integers," *Circuits, Systems and Signal Processing*, to appear in 1994.

PART III

ERROR CONTROL TECHNIQUES IN RESIDUE NUMBER
SYSTEMS

Theory, Algorithms and Implementation

Chapter 7

Fault Control in Residue Number Systems

In the last decade, fault-tolerant techniques in RNS have been studied extensively for the protection of arithmetic operations and data transmissions in digital filters as well in general purpose computers. If a digital filter is implemented using an RNS with appropriately selected redundancy, the resulting RNS structure can detect and correct errors in the processed data. Since carry-free operations are inherent in RNS, fault-tolerant RNS may provide capability to perform fast arithmetic in a reliable manner.

In this chapter, we develop a coding theory approach to fault control in RNS. The basic idea is to introduce redundancy into the RNS arithmetic in order to make it fault tolerant. There are two methods for adding redundancy to an RNS, namely, redundant residue number systems (RRNS) and residue number system product codes (RNS-PC). These two methods are studied from a coding theory point of view. The concepts of Hamming weight, minimum distance, and error detection and correction capabilities are introduced. The necessary and sufficient conditions for the desired error control capability are derived from the minimum distance of the code. A special case will generate the maximum-distance-separable (MDS) RRNS. The following chapters will deal with the application of this approach to deriving computationally efficient algorithms for the purpose of fault control (detection and correction of errors and erasures).

7.1   Background and Terminology

Consider an RNS II defined by a set of n pairwise relatively prime moduli $m_1, m_2, \ldots, m_k, m_{k+1}, \ldots, m_n$. Let the product $\prod_{i=1}^{n} m_i$ be denoted by M_T. An integer X in the range $[0, M_T)$ can be uniquely represented by the n-tuple, $X \leftrightarrow \underline{x}$, $\underline{x} = [x_1 \ x_2 \ \ldots \ x_n]$, of its residue modulo m_i,

$$x_i \equiv X \bmod m_i, \ i = 1, 2, \ldots, n. \tag{7.1.1}$$

According to the CRT-I, for any given n-tuple $[x_1 \ x_2 \ \ldots \ x_n]$, where $0 \leq x_i < m_i$, there exists one and only one integer X such that $0 \leq X < M_T$ and $x_i \equiv X \pmod{m_i}$. The numerical value of X may be computed using

$$X \equiv \sum_{i=1}^{n} x_i T_i' M_i' (\bmod M_T), \tag{7.1.2}$$

where

$$M_i' = \frac{M_T}{m_i}, \tag{7.1.3}$$

and the integers T_i' are computed a priori by solving the congruences,

$$T_i' M_i' \equiv 1 (\bmod m_i), \ i = 1, 2, \ldots, n. \tag{7.1.4}$$

The MRC is a useful operation in RNS arithmetic. It consists in representing an integer X, $0 \leq X < M_T$ in the form

$$X = \sum_{\ell=1}^{n} a_\ell \prod_{i=1}^{\ell-1} m_i, \tag{7.1.5}$$

where $0 \leq a_\ell < m_\ell$, $\ell = 1, 2, \ldots, n$, and $\prod_{i=1}^{0} m_i = 1$. MRC has been extensively studied in the literature in the context of single error correcting RNS.

Consider an arbitrary vector $\underline{x} = [x_1 \ x_2 \ \ldots \ x_n]$. It is clear from (7.1.1) that x_i is zero *iff* X is a multiple of m_i. Therefore, if the vector $\underline{x}$ has non-zero components in positions $i_1, i_2, \ldots, i_\alpha$ and zero in all other positions, then X is an integer of the form

$$X = \frac{M_T}{m_{i_1} m_{i_2} \cdots m_{i_\alpha}} X', \tag{7.1.6}$$

where $0 \leq X' < m_{i_1} m_{i_2} \cdots m_{i_\alpha}$. Given $\underline{x}$, the integer X' can be computed by solving the following congruences

$$\frac{M_T}{m_{i_1} m_{i_2} \cdots m_{i_\alpha}} X' \equiv x_{i_j} \pmod{m_{i_j}}, \ j = 1, 2, \ldots, \alpha. \tag{7.1.7}$$

7.2 A Coding Theory Framework for RRNS

An RRNS is defined in the RNS II as follows. Among the n moduli of the RNS II, the first k moduli form the set of nonredundant moduli and their product represents the legitimate range, M, of the RRNS,

$$M = \prod_{i=1}^{k} m_i. \qquad (7.2.1)$$

The remaining $n - k = r$ moduli form the set of redundant moduli that allows for error detection and correction in the RRNS. Let M_R be the product of the redundant moduli; that is,

$$M_R = \prod_{i=k+1}^{n} m_i. \qquad (7.2.2)$$

It is clear from (7.2.1) and (7.2.2) that $M_T = M \cdot M_R$. In RRNS, every integer X in the range $[a, M + a)$ is represented by a residue vector or sequence $\underline{x}$ having n components $x_1, x_2, \ldots, x_n$, that is,

$$X \longleftrightarrow [x_i]_{i=1}^{n} = [x_1 \ x_2 \ \ldots \ x_n] \qquad (7.2.3)$$

where

$$x_i \equiv X \pmod{m_i}. \qquad (7.2.4)$$

The integer x_i is the ith residue digit corresponding to the ith modulus m_i and satisfies the inequality $0 \le x_i < m_i$. Without any loss of generality, we assume a to be zero. The residue vector $\underline{x}$ can be divided into two parts; the first k residue digits corresponding to the k nonredundant moduli are called the information digits and the remaining r residue digits corresponding to the r redundant moduli are called the parity digits. The parity digits represent the redundancy introduced in the RNS to provide it with the prespecified fault control capability. We call the RRNS to be a code in the systematic form as the integer X, $0 \le X < M$, can be uniquely determined from the first k components of the residue vector $\underline{x}$.

In the following, we explore the mathematical structure of an RRNS using certain well-known algebraic results in coding theory. To begin with, we define the RRNS as an (n, k) semi-linear code Ω. All the integers in the range $[0, M)$ are valid and are referred to as code-integers subsequently. The corresponding residue vectors are said to constitute the k-dimensional code space. Note that all the n-tuple residue representations form an n-dimensional vector space. Every residue representation in the code space is a *codevector* which consists of an information part (the first k residue digits) and a parity part (the remaining r residue digits). Strictly speaking, the RRNS is not a linear code as a Galois field defining the code does not exist. We now present a coding theory interpretation of an RRNS.

7.2.1   Minimum Distance of RRNS

The minimum distance is a fundamental parameter associated with any random error-correcting code.

DEFINITION 7.1 The *Hamming weight* of a vector $\underline{x}$, $wt(\underline{x})$, is defined as the number of non-zero components of $\underline{x}$.

DEFINITION 7.2 The *Hamming distance* between two vectors $\underline{x}_i$ and $\underline{x}_j$, $d(\underline{x}_i, \underline{x}_j)$, is the number of places in which $\underline{x}_i$ and $\underline{x}_j$ differ.

DEFINITION 7.3 The *minimum distance* d of a code Ω is defined as,

$$d = min\{d(\underline{x}_i, \underline{x}_j) : \underline{x}_i, \underline{x}_j \in \Omega, \ \underline{x}_i \neq \underline{x}_j\}. \qquad (7.2.5)$$

In subsequent analysis, the intimate relationship between d and the error detection and correction capability of the RRNS will be established. Also, if the cardinality of Ω is high, it may not be feasible to compute d using (7.2.5).

Consider two code-integers X_1 and X_2 lying in the range $[0, M)$. The corresponding codevectors are $\underline{x}_1$ and $\underline{x}_2$ in the (n, k) RRNS code. If $M \leq X_1 + X_2 < M_T$, then the RRNS representation of the integer $X_1 + X_2$, that is, the vector $\underline{x}_1 + \underline{x}_2$, is not a codevector. Clearly, this violates the definition of a linear code. Similarly, for a given α, if $M \leq \alpha X < M_T$, where $X \in [0, M)$, then the RRNS representation of αX, that is, the vector $\alpha \underline{x}$, is not a codevector. Since most results in coding theory are available for linear codes, we will circumvent this problem of non-linearity in RRNS by defining the term *semi-linearity* or *conditional-linearity* (that is, linearity under additional conditions).

DEFINITION 7.4 A code is called *semi-linear* if its codevectors satisfy the property of linearity under certain appropriately predefined conditions.

For an RRNS code, given two code integers X_1 and X_2 and the corresponding codevectors $\underline{x}_1$ and $\underline{x}_2$ respectively, it is clear that for $X = \alpha_1 X_1 + \alpha_2 X_2$, if $0 \leq X < M$, then $\alpha_1 \underline{x}_1 + \alpha_2 \underline{x}_2$ is a codevector. Thus, an RRNS code is a linear code under the condition that all the quantities be in the range $[0, M)$; or it is a semi-linear code. An important result for such a code is established in the following lemma.

LEMMA 7.1 *For the semi-linear RRNS code, the minimum distance d, is the same as the Hamming weight of the codevector in Ω having the smallest positive Hamming weight,*

$$d = min\{wt(\underline{x}) : \underline{x} \in \Omega, \underline{x} \neq \underline{0}\}. \tag{7.2.6}$$

PROOF. It is clear from the definition of $d(\underline{x}_i, \underline{x}_j)$ that

$$d(\underline{x}_i, \underline{x}_j) = wt(\underline{x}_i - \underline{x}_j)$$

or

$$d(\underline{x}_i, \underline{x}_j) = wt(\underline{x}_j - \underline{x}_i). \tag{7.2.7}$$

Since $\underline{x}_i \neq \underline{x}_j$ in (7.2.5), we must have either $X_i > X_j$ or $X_j > X_i$ where X_i and X_j are the code-integers associated with the codevectors $\underline{x}_i$ and $\underline{x}_j$ respectively. Without any loss of generality, let us assume that $X_i > X_j$. Therefore, $X = X_i - X_j$ is a non-zero code-integer in the range $[0, M)$ and the corresponding vector $\underline{x}$ is a codevector in Ω. Also, since $X = X_i - X_j$, we have

$$\underline{x} = \underline{x}_i - \underline{x}_j. \tag{7.2.8}$$

Combining (7.2.7) and (7.2.8), we get

$$d(\underline{x}_i, \underline{x}_j) = wt(\underline{x}), \ \underline{x} \neq \underline{0}.$$

This proves the lemma.

In the following theorem, we derive necessary and sufficient conditions which the redundant moduli must satisfy for an RRNS code to have minimum distance equal to d. The relationship between the minimum distance and the error correction and detection capabilities will be described in the next section.

THEOREM 7.1 *The minimum distance of an RRNS code is d iff the product of the redundant moduli satisfies the following relation,*

$$max\{\prod_{i=1}^{d} m_{j_i}\} > M_R \geq max\{\prod_{i=1}^{d-1} m_{j_i}\}, \ 1 \leq j_i \leq n. \tag{7.2.9}$$

PROOF. For an RRNS code to have minimum distance d, the following conditions must be satisfied: (i) there is no codevector of Hamming weight $d-1$ or less (except for the all-zero codevector), and (ii) there is at least one codevector of Hamming weight d. Given an arbitrary vector $\underline{x} = [x_1 \ x_2 \ \ldots \ x_n]$, the first condition implies that if $wt(\underline{x}) = \alpha \leq d-1$, then the corresponding integer X satisfies $X \geq M$. Recall that if $X < M$, then the corresponding vector $\underline{x}$ is a codevector. Let $wt(\underline{x}) = \alpha$, and $\underline{x}$ be non-zero in positions $j_1, j_2, \ldots, j_\alpha$, and zero elsewhere. As a result, X is a multiple of m_j, $j = 1, 2, \ldots, n; j \neq j_1, j_2, \ldots, j_\alpha$. Thus, we have

$$X = X' \prod_{\substack{j=1 \\ j \neq j_1, j_2, \ldots, j_\alpha}}^{n} m_j, \qquad (7.2.10)$$

where X' is an arbitrary integer satisfying,

$$0 < X' < m_{j_1} m_{j_2} \cdots m_{j_\alpha}. \qquad (7.2.11)$$

The first condition is satisfied *iff* for $\alpha \leq d - 1, X \geq M$. From (7.2.10), the smallest value of X is obtained by setting $X' = 1$, and by including the smallest of $n - d + 1$ moduli among the n moduli in the product in (7.2.10). Consequently, the moduli must satisfy,

$$min\{ \prod_{\substack{j=1 \\ j \neq j_1, j_2, \ldots, j_{d-1}}}^{n} m_j \} \geq M$$

or

$$\frac{M M_R}{max\{\prod_{i=1}^{d-1} m_{j_i}\}} \geq M$$

or

$$M_R \geq max\{\prod_{i=1}^{d-1} m_{j_i}\}. \qquad (7.2.12)$$

Using a similar argument, we can show that there is at least one codevector of Hamming weight d *iff*

$$M_R < max\{\prod_{i=1}^{d} m_{j_i}\}. \qquad (7.2.13)$$

Combining (7.2.12) and (7.2.13), we get the statement of the theorem.

COROLLARY 7.1 *The minimum distance of the RRNS is 3 iff*

$$max\{m_{j_1} m_{j_2} m_{j_3}\} > M_R \geq max\{m_{j_1} m_{j_2}\}, \ 1 \leq j_1, j_2, j_3 \leq n. \qquad (7.2.14)$$

EXAMPLE 7.1 Consider the RRNS based on the moduli

$$[m_1 \ m_2 \ m_3 \ m_4 \ m_5 \ m_6] = [11 \ 16 \ 17 \ 3 \ 7 \ 13].$$

For $d = 3$, we have the condition, $3536 > M_R \geq 272$. Note that $max\{m_{j_1} m_{j_2} m_{j_3}\} = 3536$, while $max\{m_{j_1} m_{j_2}\} = 272$. Therefore, an RRNS having minimum distance 3 can be obtained using any one of the following sets of moduli as the redundant moduli: $\{m_2, m_3; M_R = 272\}$, $\{m_4, m_5, m_6; M_R = 273\}$, $\{m_2, m_4, m_5; M_R = 336\}$ and so on. Also, it is easy to verify that if m_1, m_2, m_4 and m_5 are chosen to be the redundant moduli ($M_R = 3696$), then the resulting RRNS has minimum distance equal to 4.

It is seen from the form of (7.2.9) and the above example that the choice of redundant moduli is non-unique for a given set of moduli (that constitutes the RRNS) and minimum distance d. An important question arises at this stage – "What is the optimal set of redundant moduli for an RRNS ?" The corresponding RRNS is termed the optimal RRNS. In the present context, we say that a given RRNS is optimal if it minimizes M_R for a given minimum distance. Note that minimizing M_R equivalently implies that M is maximized, as $M \cdot M_R = M_T$, where M_T is a constant. Maximizing M, in turn, implies the range of values that a code-integer X can take is maximized.

The smallest value of M_R for minimum distance d is obtained by setting

$$M_R = max\{\prod_{i=1}^{d-1} m_{j_i}\}, \ 1 \leq j_i \leq n \tag{7.2.15}$$

in (7.2.9). The *lhs* inequality in (7.2.9) is satisfied trivially in this case. This implies that an optimal RRNS having minimum distance d has the largest $d - 1$ moduli as the redundant moduli. In this case,

$$d - 1 = n - k$$

or

$$n = k + d - 1. \tag{7.2.16}$$

Using the coding theory terminology, we will refer to an RRNS that satisfies (7.2.15) as the maximum-distance-separable RRNS (MDS-RRNS). The justification for this name is provided by the fact that the minimum distance of an (n, k) code satisfies the inequality

$$d \leq n - k + 1.$$

Clearly, the MDS-RRNS has the largest possible minimum distance for a given redundancy. Since most literature on RRNS deals with the MDS-RRNS, we will also focus on the error correction and detection for the MDS-RRNS throughout this work. In Example 7.1, if the moduli m_2 and m_3 are chosen as the redundant moduli, then we obtain the MDS-RRNS having minimum distance equal to 3. The useful range of the resulting MDS-RRNS is $[0, 3003)$.

7.2.2   Error Detection and Correction in RRNS

Consider the RNS Π defined by the moduli $m_i, i = 1, 2, \ldots, n$, and the RRNS Ω defined by the nonredundant moduli $m_i, i = 1, 2, \ldots, k$ and the redundant moduli $m_i, i = k + 1, \ldots, n$. In a multiprocessor environment, where each component of the codevector is processed on one processor, if the ith processor is faulty, it will introduce an error in the ith component, x_i, of the codevector $\underline{x}$. The output of the ith processor, therefore, will be y_i, where,

$$y_i \equiv x_i + e_i \quad (\mathrm{mod}\ m_i),\ 0 \le e_i < m_i, \qquad (7.2.17)$$

e_i being the error value. If $e_i = 0$, then the ith processor is fault free. In general, the combined output can be expressed in the vector form as,

$$\underline{y} = \underline{x} + \underline{e}, \qquad (7.2.18)$$

where $\underline{e}$ is the error vector and $\underline{y}$ is the output vector. Clearly, $\underline{x} \in \Omega$, while, in general, $\underline{y}, \underline{e} \in \Pi$. The Hamming weight of the error vector $\underline{e}$, $wt(\underline{e})$, also denoted by α, is called the number of errors in $\underline{y}$. Also, we have $d(\underline{y}, \underline{x}) = wt(\underline{e}) = \alpha$. For a given $\underline{y}$, one of the two fundamental problems arises:

(1) perform error detection, that is, check if $\underline{y} \in \Omega$, and

(2) perform error correction, that is, use a decoding procedure to decode $\underline{y}$ to a unique codevector $\hat{\underline{x}} \in \Omega$ (an estimate of $\underline{x}$).

In the following, these problems are examined in more detail, error detection and correction capabilities are defined, and their relationship to the minimum distance d of the RRNS is established. The following lemma (triangular inequality) is crucial in establishing the relationship between the minimum distance and the error detection and error correction capabilities of an RRNS.

LEMMA 7.2 *Given any three vectors $\underline{a}$, $\underline{b}$, and $\underline{c}$ in an RNS Π, the Hamming distance among $\underline{a}$, $\underline{b}$ and $\underline{c}$, satisfies the triangular inequality:*

$$d(\underline{a}, \underline{b}) + d(\underline{b}, \underline{c}) \ge d(\underline{a}, \underline{c}).$$

PROOF. Let A, B and C be the integers associated with the vectors $\underline{a}$, $\underline{b}$, and $\underline{c}$ respectively. Without any loss of generality, let us assume that $C > A$. Thus,

$$d(\underline{a}, \underline{c}) = wt(\underline{c} - \underline{a}).$$

Depending on the value of B, three distinct cases arise: (1) $C > B > A$, (2) $C > A > B$, and (3) $B > C > A$. For case (1), we may write

$$
\begin{aligned}
d(\underline{a}, \underline{c}) &= wt(\underline{c} - \underline{a}) \\
&= wt[(\underline{c} - \underline{b}) + (\underline{b} - \underline{a})] \\
&= wt(\underline{c} - \underline{b}) + wt(\underline{b} - \underline{a}) - \Theta_1 - 2\Theta_2,
\end{aligned}
$$

where Θ_1 is the number of places in which the non-zero residue digits of $(\underline{c} - \underline{b})$ and $(\underline{b} - \underline{a})$ add to a non–zero number, and Θ_2 is the number of places in which the non-zero residue digits of $(\underline{c} - \underline{b})$ and $(\underline{b} - \underline{a})$ add to zero. Clearly,

$$
wt(\underline{c} - \underline{b}) + wt(\underline{b} - \underline{a}) \geq wt(\underline{c} - \underline{a}).
$$

For case (2), we may write

$$
\begin{aligned}
d(\underline{a}, \underline{c}) &= wt(\underline{c} - \underline{a}) \\
&= wt(\underline{c} - \underline{a} + \underline{b} + \underline{m} - \underline{b}) \\
&= wt[(\underline{c} - \underline{b}) + \underline{m} - (\underline{a} - \underline{b})] \\
&= wt(\underline{c} - \underline{b}) + wt[\underline{m} - (\underline{a} - \underline{b})] - \Theta_3 - 2\Theta_4 \\
&\leq wt(\underline{c} - \underline{b}) + wt[\underline{m} - (\underline{a} - \underline{b})] \\
&= wt(\underline{c} - \underline{b}) + wt(\underline{a} - \underline{b}),
\end{aligned}
$$

where $\underline{m} = [m_1 \ m_2 \ \cdots \ m_n]$, Θ_3 is the number of places in which the non-zero residue digits of $(\underline{c} - \underline{b})$ and $[\underline{m} - (\underline{a} - \underline{b})]$ add to a non–zero number, and Θ_4 is the number of places in which the non-zero residue digits of $(\underline{c} - \underline{b})$ and $[\underline{m} - (\underline{a} - \underline{b})]$ add to zero. Case (3) is similar to case (2). This proves the lemma.

Error detection in RRNS

A conceptually simple but computationally impractical method to perform error detection is to compare $\underline{y}$ with every possible codevector in Ω. If $\underline{y} \neq \underline{x}$, but $\underline{y} \in \Omega$, then undetectable errors are said to have occurred. If $\underline{y}$ is not in Ω, then errors are detected. To assess the performance of this scheme, we define a parameter called the error detecting capability.

DEFINITION 7.5 The *error detecting capability*, l, of a code Ω is the largest number of errors that may occur for which $\underline{y}$ is not in Ω.

In other words, it is the largest value of α for which $\underline{y}$ is not in Ω. As defined earlier, α is the number of errors present in the received vector. The following lemma establishes the relation between the parameter l and

d of the RRNS.

LEMMA 7.3 *The value of l is $d-1$ for an RRNS having minimum distance equal to d.*

PROOF. For an RRNS having minimum distance d, no error vector $\underline{e}$ having $wt(\underline{e}) = \alpha$, $0 < \alpha < d$, can change one codevector into another. Note that if $d(\underline{y}, \underline{x}) = \alpha$, $0 < \alpha < d$, then from Lemma 7.2, $d(\underline{y}, \underline{x}_i) \geq d - \alpha > 0$ for all $\underline{x}_i \in \Omega$, and $\underline{x}_i \neq \underline{y}$. Therefore, $\underline{y}$ can not be a codevector. Also, there exist at least two codevectors $\underline{x}_1$ and $\underline{x}_2$ such that $d = d(\underline{x}_1, \underline{x}_2)$. If $\underline{x} = \underline{x}_1$ and $\underline{e} = \underline{x}_2 - \underline{x}_1$ (Assume that $X_2 > X_1$), then $\underline{y} = \underline{x}_2$. In such a case $\underline{y} \in \Omega$. Therefore, not all error vectors having $wt(\underline{e}) = d$ are detectable, thereby, proving the statement of the lemma.

Error correction in RRNS

The decoding procedure for performing error correction depends on the criterion of performance. If errors occur independently, all errors are equiprobable, and the criterion of performance is to minimize the probability of decoding error (maximum likelihood decoding or MLD), then the decoding procedure can be summarized as — decode $\underline{y}$ to a codevector $\hat{\underline{x}}$ which differs from $\underline{y}$ in the least number of places. We will use this decoding procedure in our subsequent analysis. A justification for this MLD-based decoding procedure is that the probability of occurrence of i errors $<$ probability of occurrence of j errors, $i > j$, under the above stated conditions.

A conceptually simple but computationally impractical method to perform error correction is to compare $\underline{y}$ with every codevector in Ω. The vector $\underline{y}$ is decoded to that codevector $\hat{\underline{x}}$ which satisfies the condition

$$d(\underline{y}, \hat{\underline{x}}) \leq d(\underline{y}, \underline{z}), \ \forall \underline{z} \in \Omega, \ \underline{z} \neq \hat{\underline{x}}. \qquad (7.2.19)$$

If there are two codevectors, say $\underline{x}_1$ and $\underline{x}_2$, in Ω such that $d(\underline{y}, \underline{x}_1) = d(\underline{y}, \underline{x}_2)$, then the decoder may declare a decoding failure or it may decode $\underline{y}$ to any one of $\underline{x}_1$ or $\underline{x}_2$.

DEFINITION 7.6 The *error correcting capability*, λ, of a code Ω is the largest number of errors that may occur for which correct decoding ($\hat{\underline{x}} = \underline{x}$) takes place.

In other words, it is the largest value of α for which correct decoding takes place. The following lemma establishes the relationship between the parameters λ and d of an RRNS.

LEMMA 7.4 *The value of λ for an RRNS is $\lambda = \lfloor \frac{d-1}{2} \rfloor$, where $\lfloor a \rfloor$ denotes the largest integer less than or equal to a.*

PROOF. Let $\underline{x}_i$ be a codeword other than $\underline{x}$ in Ω. The Hamming distance among $\underline{x}$, $\underline{x}_i$ and $\underline{y}$ satisfies the triangular inequality

$$d(\underline{y}, \underline{x}) + d(\underline{y}, \underline{x}_i) \geq d(\underline{x}, \underline{x}_i).$$

Since $d(\underline{x}, \underline{x}_i) \geq d$ and $d(\underline{y}, \underline{x}) = \alpha$, we have

$$d(\underline{y}, \underline{x}_i) \geq (d - \alpha).$$

If $\alpha \leq \lfloor \frac{d-1}{2} \rfloor$, then $d - \alpha > \lfloor \frac{d-1}{2} \rfloor \geq \alpha$. Thus,

$$d(\underline{y}, \underline{x}_i) > d(\underline{y}, \underline{x}),$$

thereby, implying that $\underline{y}$ is closer (in Hamming distance) to $\underline{x}$ than any other codevector in Ω. On the other hand, when $\alpha > \lambda$, we can show that there exists at least one codevector $\underline{x}_i$ such that

$$d(\underline{y}, \underline{x}) > d(\underline{y}, \underline{x}_i).$$

In case $d(\underline{y}, \underline{x}) > d(\underline{y}, \underline{x}_i)$, incorrect decoding will take place. Similarly, in case $d(\underline{y}, \underline{x}) = d(\underline{y}, \underline{x}_i)$, $\underline{x}_i \neq \underline{x}$, incorrect decoding may take place. This proves the lemma.

In some RRNS, one may be interested in performing simultaneous error detection and correction. The following lemma deals with such a situation.

LEMMA 7.5 *An RRNS is capable of correcting λ or fewer errors and simultaneously detecting β ($\beta > \lambda$) or fewer errors if $d \geq \lambda + \beta + 1$.*

PROOF. It follows from Lemma 7.4 that for $\beta > \lambda$, an RRNS with $d \geq \lambda + \beta + 1 > 2\lambda + 1$ can correct λ or fewer errors. Now consider the situation where more than λ but fewer than $\beta + 1$ errors occur. If the decoder makes a decoding error, that is, $\underline{y}$ is decoded to $\hat{\underline{x}}$ which is a codevector but $\hat{\underline{x}} \neq \underline{x}$, then

$$d(\underline{y}, \hat{\underline{x}}) \leq \lambda, \quad d(\underline{y}, \underline{x}) \leq \beta,$$

and

$$d(\underline{y}, \hat{\underline{x}}) + d(\underline{y}, \underline{x}) \leq \lambda + \beta < d.$$

However, from Lemma 7.2 the Hamming distance among $\underline{x}$, $\hat{\underline{x}}$, and $\underline{y}$ satisfies the triangular inequality $d(\underline{y}, \hat{\underline{x}}) + d(\underline{y}, \underline{x}) \geq d(\hat{\underline{x}}, \underline{x})$. Therefore, $d(\hat{\underline{x}}, \underline{x}) < d$, that is, $\hat{\underline{x}}$ can not be a codevector. This contradicts the original assumption. This proves the lemma.

7.2.3   Weight Distribution of the MDS-RRNS Code

Let W_j be the number of codevectors having Hamming weight j, $j = 0, 1, \ldots, n$. These numbers constitute the weight distribution of the code. For an MDS code defined over the finite field, $GF(q)$, W_j can be computed using the expression,

$$W_j = \binom{n}{j} \sum_{h=0}^{j-1-(n-k)} (-1)^h \binom{j}{h} (q^{j-h-(n-k)} - 1), \qquad (7.2.20)$$

for $n - k + 1 \leq j \leq n$. Of course, $W_j = 0$, $1 \leq j \leq n - k$ and $W_0 = 1$. There is no finite field underlying the algebra of RRNS. However, since the RRNS are being studied as a coding scheme with emphasis on MDS-RRNS, we hope to find, at least in approximation, their weight distribution in a manner similar to MDS codes defined over a finite field. It is derived in the following theorem.

THEOREM 7.2 *For an (n, k) MDS-RRNS code having k nonredundant moduli and $r = (n - k)$ redundant moduli, if all the moduli are close to a mean value $\overline{m}$, that is, $m_1 \simeq m_2 \simeq \cdots \simeq m_n \simeq \overline{m}$, then the weight distribution of the MDS-RRNS can be approximated by,*

$$W_j \simeq \binom{n}{j} \sum_{h=0}^{j-1-(n-k)} (-1)^h \binom{j}{h} (\overline{m}^{j-h-(n-k)} - 1), \qquad (7.2.21)$$

for $n - k + 1 \leq j \leq n$, $W_j = 0$ for $1 \leq j \leq n - k$ and $W_0 = 1$.

PROOF. Let $N(i_1, i_2, \ldots, i_h)$ be the set of codevectors belonging to Ω whose i_1th, i_2th, ..., i_hth digits are zero. For a codevector $\underline{x}$, if its ith component is 0, that is, $x_i = 0$, it implies that the integer X is a multiple of m_i. Similarly, if a codevector $\underline{x}$ has zero in i_1th, i_2th, ..., i_hth positions, it implies that the corresponding integer X is a multiple of $\prod_{j=1}^{h} m_{i_j} \simeq \overline{m}^h$. Thus, the number of codevectors in $N(i_1, i_2, \ldots, i_h)$ is $\overline{m}^{k-h}$ for $h \leq k - 1$.

If W_j is the number of codevectors with weight j, then by definition, W_j is the number of codevectors which have exactly $n - j$ zero components. Let U_s be the sum of the numbers of codevectors in all the sets $N(i_1, i_2, \ldots, i_s)$ which have zeros in s or more positions. Let $|N(i_1, i_2, \ldots, i_s)|$ represents the number of vectors in $N(i_1, i_2, \ldots, i_s)$. Then

$$\begin{aligned} U_s &= \sum_{1 \leq i_1, i_2, \cdots, i_s \leq n} |N(i_1, i_2, \ldots, i_s)| \\ &\simeq \binom{n}{s} \overline{m}^{k-s}, \end{aligned}$$

for $k \geq s$, while $U_s = \binom{n}{s} 1$, for $k < s$. Note that U_s is the sum of the number of codevectors which have zeros in s or more positions. It implies that codevectors of weight less than $n - s$ are counted more than once. In fact,

$$U_s = W_{n-s} + \binom{s+1}{1} W_{n-(s+1)} + \binom{s+2}{2} W_{n-(s+2)} + \cdots + \binom{n}{n-s} W_0.$$

Using these expressions, a closed-form expression can be derived using the principle of inclusion and exclusion. This yields

$$W_j = \binom{n}{j} \sum_{h=0}^{j-1-(n-k)} (-1)^h \binom{j}{h} (\overline{m}^{j-h-(n-k)} - 1),$$

for $n - k + 1 \leq j \leq n$. For $1 \leq j \leq n - k$, $W_j = 0$ and $W_0 = 1$. This proves the theorem.

The approximated weight distribution may be helpful to analyze the performance of MDS-RRNS coding scheme. Also, a lower and upper bound on the value of W_j can be obtained by setting $\overline{m} = m_1$ and $\overline{m} = m_k$ respectively in (7.2.21). We now consider an example.

EXAMPLE 7.2 Given an $(8, 4)$ MDS-RRNS Ω with nonredundant moduli 23, 25, 27, 29 and redundant moduli 31, 32, 37, 41. The geometrically averaged modulus $\overline{m} = 26$. The value $\overline{m}$ can be chosen to be the integer nearest to the kth-root of the product of nonredundant moduli. Recall that $\prod_{i=1}^{k} m_i = M$. Thus, $\overline{m} \simeq M^{1/k}$. The weight distributions W_i, $0 \leq i \leq 8$ for both exact and approximated cases are shown below:

Weight	Exact	Approximated
W_0	1	1
W_1	0	0
W_2	0	0
W_3	0	0
W_4	0	0
W_5	928	1,400
W_6	11,485	14,700
W_7	95,767	107,000
W_8	342,044	333,875
Total	450,225	456,976

The exact weight distribution was computed by exhaustively generating all the 450,225 codevectors in Ω.

7.3   Consistency Checking for RRNS

The first step in any error correction procedure (and the only step in the error detection procedure) is to check if the given residue vector y is a code-vector in the RRNS. A simple technique to accomplish this is to compute the corresponding integer Y and check if $0 \leq Y < M$. If $0 \leq Y < M$, then y is a codevector. However, the computation of Y from y may require processing large valued integers. A suitable method to avoid this is the BEX method which is also the first step of the error correcting algorithms described here.

The residues $y_1, y_2, \cdots, y_n$ of an integer Y consist of two parts : $y_i, i = 1, 2, \ldots, k$ correspond to the information part and $y_{k+i}, i = 1, 2, \ldots, n-k$, correspond to the parity part. First, we compute the integer $\overline{Y}$ using $y_i, i = 1, 2, \ldots, k$, that is,

$$\overline{Y} \equiv \sum_{i=1}^{k} y_i M_i T_i \pmod{M} \tag{7.3.1}$$

$$\equiv \sum_{i=1}^{k} y_i' M_i \pmod{M}, \tag{7.3.2}$$

where $M_i = M/m_i$, $y_i' \equiv y_i T_i \pmod{m_i}$, and the predetermined integers T_i satisfy the congruences $T_i M_i \equiv 1 \pmod{m_i}$ $i = 1, 2, \ldots, k$. Note that the CRT-I is used here only to simplify the presentation. Then, based on $\overline{Y}$, the parity digits are recomputed to get

$$\overline{y}_{k+i} \equiv \overline{Y} \pmod{m_{k+i}}, \ i = 1, 2, \ldots, n-k. \tag{7.3.3}$$

The MRC approach is a useful method for performing BEX. Given the nonredundant residues $[y_1 \ y_2 \ \ldots \ y_k]$ for an integer $\overline{Y}$, $\overline{Y} \in [0, M)$, the redundant residues $[\overline{y}_{k+1} \ \overline{y}_{k+2} \ \ldots \ \overline{y}_n]$ can be computed using BEX based on MRC algorithm as given in Table 7.1. Next, the test quantities (to be called syndrome digits henceforth) Δ_i are formed as

$$\Delta_i \equiv (\overline{y}_{k+i} - y_{k+i}) \pmod{m_{k+i}}, \ \ i = 1, 2, \ldots, n-k. \tag{7.3.4}$$

A fundamental property of the syndrome digits is stated in the following theorem.

THEOREM 7.3 *A residue vector y is a codevector iff all the syndrome digits are zero.*

Table 7.1: Base Extension Method Based on MRC Algorithm

Based on $[y_1 \ y_2 \ \ldots \ y_k]$, compute the mixed radix digits

$$
\begin{aligned}
a_1 &\equiv y_1 \pmod{m_1} \\
a_2 &\equiv m_1^{-1}(y_2 - a_1) \pmod{m_2} \\
a_3 &\equiv m_2^{-1}(m_1^{-1}(y_3 - a_1) - a_2) \pmod{m_3} \\
&\vdots
\end{aligned}
$$

Compute each term of the following equation as
soon as the mixed radix digits are available.
$$\overline{y}_{k+r} \equiv a_k m_{k-1} \cdots m_2 m_1 + a_{k-1} m_{k-2} \cdots m_2 m_1$$
$$+ \cdots + a_2 m_1 + a_1 \pmod{m_{k+r}}, \ r = 1, 2, \ldots, n - k$$
Note that $m_1^{-1} \pmod{m_2}$ denotes the multiplicative inverse of
m_1 modulo m_2. The advantage of using the above equations is
that the calculation of $\overline{y}_{k+r}$ can be overlapped with the
computation of the mixed radix digits, by making use of the
mixed radix digits as soon as they are available.

PROOF. It is clear from (7.3.4) that $\Delta_i = 0$ implies $\overline{y}_{k+i} = y_{k+i}$ and vice
versa. If all the syndrome digits are zero, then the integer $\overline{Y}$ computed in
(7.3.1) has the residues $y_i, i = 1, \ldots, n$, and satisfies the condition $0 \leq \overline{Y} <$
M. Therefore, $\underline{y}$ is a codevector.

For the case when one or more syndrome digits are non-zero, let the
syndrome digits $\Delta_{i_j}, j = 1, \ldots, \alpha$ be non-zero. In this case, we may write

$$y_{k+i_j} \equiv (\overline{y}_{k+i_j} - \Delta_{i_j}) \bmod m_{k+i_j}, \ j = 1, \ldots, \alpha, \qquad (7.3.5)$$

$$y_{k+i} = \overline{y}_{k+i}, \ i = 1, \ldots, n - k; \ i \neq i_j, j = 1, \ldots, \alpha. \qquad (7.3.6)$$

Eq.(7.3.5) can equivalently be written as

$$y_{k+i_j} \equiv (\overline{y}_{k+i_j} + (m_{k+i_j} - \Delta_{i_j})) \bmod m_{k+i_j}, \qquad (7.3.7)$$

or

$$y_{k+i_j} \equiv (\overline{y}_{k+i_j} + \Delta'_{i_j}) \bmod m_{k+i_j}, \qquad (7.3.8)$$

where

$$\Delta'_{i_j} = (m_{k+i_j} - \Delta_{i_j}).$$

The vector $[y_1 \ y_2 \ \cdots \ y_k \ \bar{y}_{k+1} \cdots \bar{y}_n]$ corresponds to the integer $\bar{Y}$, $0 \le \bar{Y} < M$, and therefore, is a codevector. The integer Y can be written as,

$$Y = (\bar{Y} + Y') \bmod M_T. \tag{7.3.9}$$

The integer Y' corresponds to the vector having zeros in all positions except $k + i_j$, the components in the position $k + i_j$ being Δ'_{i_j}, $j = 1, 2, \ldots, \alpha$. Recalling (7.1.6), Y' is an integer of this type

$$Y' = (\prod_{j=1}^{\alpha} m_{k+i_j})^{-1} M_T Y'', \tag{7.3.10}$$

where

$$0 < Y'' < \prod_{j=1}^{\alpha} m_{k+i_j}. \tag{7.3.11}$$

Since Y' is a multiple of M, $Y' > M$. The minimum value of Y' is

$$Y'_{min} = (\prod_{j=1}^{\alpha} m_{k+i_j})^{-1} M_T,$$

and the maximum value of Y' is

$$\begin{aligned} Y'_{max} &= M_T - (\prod_{j=1}^{\alpha} m_{k+i_j})^{-1} M_T \\ &= M_T - Y'_{min}. \end{aligned}$$

It is clear from the above expressions that

$$Y'_{min} > M,$$

and

$$Y'_{max} + M < M_T.$$

Substituting the above expressions in (7.3.9) leads to

$$M < \bar{Y} + Y' < M_T$$

or

$$M < Y < M_T.$$

Consequently, the corresponding vector y cannot be a codevector. This proves the theorem.

An unfortunate feature of this consistency checking is that the errors in the information digits and the errors in the parity digits effect the syndromes differently, and therefore, must be treated differently. Based on this theorem, two important observations can be made.

<u>Observation 1</u>. For a λ error-correcting RRNS, if the number of non-zero syndromes is λ or less, say the syndromes $\Delta_{i_j}, j = 1, \ldots, \alpha$ are non-zero, $\alpha \leq \lambda$, then the error-correcting algorithm declares the digits $y_{k+i_j}, j = 1, \ldots, \alpha$ in error and the vector y is decoded to the codevector $\widehat{x} = [\widehat{x}_1 \, \widehat{x}_2 \cdots \widehat{x}_n]$, where

$$\widehat{x}_i = y_i, \quad i = 1, 2, \ldots, n, \; i \neq k + i_j, \; j = 1, \ldots, \alpha,$$

$$\widehat{x}_{k+i_j} = \overline{y}_{k+i_j}, \quad j = 1, 2, \ldots, \alpha.$$

In this case, all the α errors are assumed to be in the positions for which the syndrome digits are non-zero. Also, all the errors are in the parity digits.

<u>Observation 2</u>. For a λ error-correcting RRNS, if up to λ errors occur in a way that at least one of the information digits is in error, then the number of non-zero syndromes is greater than λ.

Based on the above analysis for consistency checking, Chapter 8 will deal with the fast and superfast algorithms for single error correction in RRNS. Multiple error correction for RRNS will be the focus of attention in Chapter 9. We now establish a coding theory framework for the RNS-PC in the following section.

7.4 A Coding Theory Framework for RNS-PC

The RNS-PC Ω is defined in the RNS II as follows. All the code-integers (legitimate numbers), X, in RNS-PC must be of the form AX', $0 \leq AX' < M_T$, where A is a given positive integer, called the *code generator*. In other words, an integer X, $0 \leq X < M_T$, is a code-integer, *iff* $X \equiv 0 \pmod{A}$. Associated with each code-integer X, is the vector of residues $\underline{x} = [x_1 \, x_2 \, \ldots \, x_n]$, which is termed as the corresponding codevector. A number $\overline{Y}$ in the range $[0, M_T)$, such that $\overline{Y} \not\equiv 0 \pmod{A}$ is said to be an illegitimate number. It is clear that the integer X' is in the range $[0, R)$, where $R = [M_T/A] + 1$, where $[M_T/A]$ denotes the integer part of M_T/A. Given a codevector $\underline{x}$, we must be able to obtain the integer $X' \leftrightarrow \underline{x}' = [x_1' \, x_2' \, \ldots \, x_n']$ by solving the congruences $Ax_i' \equiv x_i \bmod m_i$,

$i = 1, 2,\ldots,n$. This is possible *iff* $gcd(A, m_i) = 1, i = 1, 2,\ldots,n$. Therefore, throughout this work A is always selected such that $gcd(A, m_i) = 1$, $i = 1, 2,\ldots, n$.

In the following, we explore the mathematical structure of an RNS-PC by defining the RNS-PC as a semi-linear code Ω. All the integers having the form $X = AX'$, $0 \leq X < M_T$, are valid, and the corresponding residue vectors are said to constitute the code space. Every residue representation in the code space is a codevector. Strictly speaking, a Galois field defining the code does not exist. Also, the RNS-PC is not a linear code. For example, consider two code-integers $X_1 = AX'_1$ and $X_2 = AX'_2$ in RNS-PC, the corresponding codevectors being $\underline{x}_1$ and $\underline{x}_2$, respectively. Let $\overline{X} = X_1 + X_2$ and $X \equiv \overline{X} \pmod{M_T}$. If $M_T \leq \overline{X} < 2M_T$, then $X = X_1 + X_2 - M_T$ and $X \equiv -M_T \not\equiv 0 \pmod{A}$, and therefore, the vector $\underline{x}_1 + \underline{x}_2$ is not a codevector. Clearly, this violates the definition of a linear code. Similarly, for a given scalar α_1, if $M_T \leq \alpha_1 X_1 < AM_T$, then the vector $\alpha_1\underline{x}_1$ is not a code vector.

All the fundamental results established for the RRNS in Section 7.2 are also valid for the RNS-PC. The minimum distance of the RNS-PC, d, is defined as in Definition 7.3. Once again, we term the RNS-PC as a semi-linear code. For an RNS-PC defined by the code generator A, given two code-integers $X_1 = AX'_1$ and $X_2 = AX'_2$, where $0 \leq X_1, X_2 < M_T$, the corresponding codevectors being $\underline{x}_1$ and $\underline{x}_2$, respectively, it is clear that for $X = \alpha_1 X_1 + \alpha_2 X_2$, if $0 \leq X < M_T$, then $\alpha_1\underline{x}_1 + \alpha_2\underline{x}_2$ is a codevector. Thus, an RNS-PC is a linear code under the condition that all the quantities, X, be in the range $[0, M_T)$ and $X \equiv 0 \pmod{A}$; or it is a semi-linear code.

LEMMA 7.6 *For the semi-linear RNS-PC, d, is the same as the Hamming weight of the codevector in Ω having the smallest positive Hamming weight, that is,*

$$d = min\{wt(\underline{x}) : \underline{x} \in \Omega, \underline{x} \neq \underline{0}\}. \tag{7.4.1}$$

PROOF. By Definition 7.2,

$$d(\underline{x}_i, \underline{x}_j) = wt(\underline{x}_i - \underline{x}_j); \text{ if } X_i > X_j \tag{7.4.2}$$

or

$$d(\underline{x}_i, \underline{x}_j) = wt(\underline{x}_j - \underline{x}_i); \text{ if } X_j > X_i.$$

Without any loss of generality, let us assume that $X_i > X_j$. Therefore, $0 < X = X_i - X_j = AX'_i - AX'_j = A(X'_i - X'_j) < M_T$ and $\underline{x}$ is a codevector in Ω. Also,

$$\underline{x} = \underline{x}_i - \underline{x}_j. \tag{7.4.3}$$

Combining (7.4.2) and (7.4.3), we get

$$d(\underline{x}_i, \underline{x}_j) = wt(\underline{x}), \ \underline{x} \neq \underline{0}.$$

This proves the lemma.

In the following we derive the necessary and sufficient conditions on the moduli and code generator for an RNS-PC having minimum distance equal to d.

THEOREM 7.4 *The minimum distance of an RNS-PC is d iff the code generator A satisfies the following relation,*

$$max\{ \prod_{\substack{i=1 \\ 1 \leq j_i \leq n}}^{d} m_{j_i} \} > A \geq max\{ \prod_{\substack{i=1 \\ 1 \leq j_i \leq n}}^{d-1} m_{j_i} \}, \ d \geq 2. \qquad (7.4.4)$$

PROOF. For an RNS-PC to have minimum distance d, the following conditions must be satisfied: (i) there is no codevector of Hamming weight $d-1$ or less (except for the all-zero codevector), and (ii) there is at least one codevector of Hamming weight d. Given an arbitrary vector $\underline{x} = [x_1 \ x_2 \ldots \ x_n]$, the first condition implies that if $wt(\underline{x}) = c \leq d-1$, then $X \not\equiv 0 \ (\text{mod } A)$. Recall that if $X < M_T$ and $X \equiv 0 \ (\text{mod } A)$, then the corresponding vector $\underline{x}$ is a codevector. Let $wt(\underline{x}) = c$, and $\underline{x}$ be non-zero in positions $j_1, j_2, \ldots, j_c$, and zero elsewhere. As a result, X is a multiple of m_j, $j = 1, 2, \ldots, n; \ j \neq j_1, j_2, \ldots, j_c$. Thus, we have

$$X = X'A\frac{M_T}{\prod_{i=1}^{c} m_{j_i}}, \qquad (7.4.5)$$

where X' is an arbitrary integer satisfying,

$$0 < X'A < \prod_{i=1}^{c} m_{j_i}. \qquad (7.4.6)$$

The first condition is satisfied *iff* the smallest non-zero values of $X \geq M_T$ for $c \leq d-1$. From (7.4.5), the smallest value of X is obtained by setting $X' = 1$, and by letting the product of c moduli be maximum in (7.4.5). Consequently, the moduli must satisfy

$$\frac{AM_T}{max\{\prod_{\substack{i=1 \\ 1 \leq j_i \leq n}}^{d-1} m_{j_i}\}} \geq M_T$$

or

$$A \geq max\{ \prod_{\substack{i=1 \\ 1 \leq j_i \leq n}}^{d-1} m_{j_i} \}. \qquad (7.4.7)$$

Using a similar argument, we can show that there is at least one codevector of Hamming weight d *iff*

$$A < max\{ \prod_{\substack{i=1 \\ 1 \le j_i \le n}}^{d} m_{j_i} \}. \qquad (7.4.8)$$

Combining (7.4.7) and (7.4.8), we get the statement of the theorem.

EXAMPLE 6.3 Consider an RNS-PC based on the moduli $m_1 = 2$, $m_2 = 3$, $m_3 = 5$, $m_4 = 7$, and the code generator $A = 31$. Since $m_4 < A < m_3 m_4$, by Theorem 7.4, this RNS-PC has minimum distance 2. Now consider the same moduli set and change $A = 31$ to $A = 37$. Since $m_3 m_4 < A < m_2 m_3 m_4$, by Theorem 7.4, the new RNS-PC has minimum distance 3. The code spaces for $A = 31$ and $A = 37$ are given as follows.

RNS-PC $A = 31$ $(d = 2)$						RNS-PC $A = 37$ $(d = 3)$				
X	x_1	x_2	x_3	x_4		X	x_1	x_2	x_3	x_4
0	0	0	0	0		0	0	0	0	0
31	1	1	1	1		37	1	1	2	2
62	0	2	2	6		74	0	2	4	4
93	1	0	3	2		111	1	0	1	6
124	0	1	4	5		148	0	1	3	1
155	1	2	0	1		185	1	2	0	3
186	0	0	1	4						

7.4.1   Error Detection and Correction in RNS-PC

Once again, consider the RNS Π defined by the moduli m_i, $i = 1, 2, \ldots, n$, and the RNS-PC Ω defined by the moduli m_i, $i = 1, 2, \ldots, n$, and code generator A. If the ith residue digit x_i is in error, then the altered residue digit will be y_i, where

$$y_i \equiv x_i + e_i \pmod{m_i}, \ 0 < e_i < m_i, \qquad (7.4.9)$$

e_i being the error value. In general, the altered residue vector $\underline{y}$ can be expressed as the sum of the codevector $\underline{x}$ and error vector $\underline{e}$,

$$\underline{y} = \underline{x} + \underline{e}. \qquad (7.4.10)$$

Clearly, $\underline{x} \in \Omega$, while, in general, $\underline{y}, \underline{e} \in \Pi$. The Hamming weight of the error vector $\underline{e}$, $wt(\underline{e})$, denoted by c, is called the number of errors in $\underline{y}$. Also, we have $d(\underline{y}, \underline{x}) = wt(\underline{e}) = c$. For a given $\underline{y}$, one of the two fundamental problems arises:

(1) perform error detection, that is, check if $\underline{y} \in \Omega$, and

(2) perform error correction, that is, use a decoding procedure to decode $\underline{y}$ to a unique codevector $\underline{\hat{x}} \in \Omega$ (an estimate of $\underline{x}$).

The error detecting capability, l, and the error correcting capability, λ, of a code Ω were defined in Definition 7.5 and 7.6 respectively. The error detecting capability, l, of an RNS-PC is $d-1$. In other words, any l or less residue errors in an RNS-PC of the code generator A are detectable *iff* the following condition is satisfied,

$$A > max\{ \prod_{\substack{i=1 \\ 1 \le j_i \le n}}^{l} m_{j_i} \}. \qquad (7.4.11)$$

Given a vector $\underline{y}$, it is decoded to that codevector $\underline{\hat{x}}$ which satisfies the condition in (7.2.19). If this condition is satisfied with equality, it implies that there exist more than one codevectors equidistant from $\underline{y}$. In such a case, $\underline{y}$ can be decoded to any one of them. The error correcting capability, λ, of an RNS-PC is $\lambda = \lfloor \frac{d-1}{2} \rfloor$. In other words, any λ or less residue errors in an RNS-PC of the code generator A are correctable *iff* the following condition is satisfied,

$$A > max\{ \prod_{\substack{i=1 \\ 1 \le j_i \le n}}^{2\lambda} m_{j_i} \}. \qquad (7.4.12)$$

Finally, a RNS-PC of the code generator A is capable of correcting λ or fewer residue errors and simultaneously detecting β ($\beta > \lambda$) or fewer residue errors *iff* $d \ge \lambda + \beta + 1$. In such a case, the following condition is satisfied

$$A > max\{ \prod_{\substack{i=1 \\ 1 \le j_i \le n}}^{\lambda+\beta} m_{j_i} \}. \qquad (7.4.13)$$

The expressions in (7.4.11), (7.4.12) and (7.4.13) for RNS-PC can be established using the proofs for the lemmas 7.3, 7.4 and 7.5, and Theorem 7.4.

7.5 Consistency Checking for RNS-PC

Under the assumption that no more than β residue errors occur, given a number $\overline{Y}$ to be tested, a simple procedure for error detection is as follows:

Step 1: $\Delta \equiv \overline{Y} \pmod{A}$, called syndrome, is computed.

Step 2: Check if $\Delta \equiv 0 \pmod{A}$, if yes, no error occurs; otherwise, one or more errors are detected.

Now, given $\overline{Y} \equiv \overline{X} \pmod{M_T}$, let $\overline{X} = AX_1' + AX_2'$, the sum of two code-integers $0 \leq AX_1', AX_2' < M_T$. If no *additive overflow* occurs, (that is, $0 \leq \overline{X} < M_T$), $\overline{Y}$ is also a code-integer. If an additive overflow occurs, then $\overline{Y} = \overline{X} - M_T$ is illegitimate since $\overline{Y} \equiv -M_T \pmod{A} \not\equiv 0$. From the above analysis, we see that the above procedure is not suitable for simultaneously detecting residue errors and additive overflow, since in both cases, illegitimate numbers are generated. A procedure for simultaneously detecting additive overflow and multiple residue errors is analyzed and derived in the following.

Given an integer $X \in [0, M_T)$, where $X \equiv \overline{X} \pmod{M_T}$, $\overline{X} = AX_1' + AX_2'$ and $0 \leq AX_1', AX_2' < M_T$, if c $(1 \leq c \leq \beta)$ errors occur, then the altered residue vector can be represented as

$$\overline{Y} \equiv X + E \pmod{M_T} \tag{7.5.1}$$

$$
\begin{aligned}
\longleftrightarrow \underline{y} &= [y_1 \; y_2 \; \ldots \; y_n] \\
&\equiv \underline{x} + \underline{e} = [x_1 \; x_2 \; \ldots \; x_n] \\
&+ \quad [0 \ldots 0 \, e_{i_1} \, 0 \ldots 0 \, e_{i_2} \, 0 \ldots 0 \, e_{i_c} \, 0 \ldots 0], \quad (7.5.2)
\end{aligned}
$$

where

$$
\begin{aligned}
X &\longleftrightarrow \underline{x} = [x_1 \; x_2 \; \ldots \; x_n] \\
E &\longleftrightarrow \underline{e} = [0 \ldots 0 \, e_{i_1} \, 0 \ldots 0 \, e_{i_2} \, 0 \ldots 0 \, e_{i_c} \, 0 \ldots 0] \\
& \qquad y_i = x_i, \; i \neq i_\alpha, \; \alpha = 1, 2, \ldots, c, \\
& \qquad y_{i_\alpha} \equiv x_{i_\alpha} + e_{i_\alpha} \pmod{m_{i_\alpha}}, \; 1 \leq i_\alpha \leq n, \; \alpha = 1, 2, \ldots, c.
\end{aligned}
$$

Since $E \equiv 0 \pmod{m_i}$ for all $i \neq i_\alpha$, $\alpha = 1, 2, \ldots, c$, and $E \not\equiv 0 \pmod{m_i}$ for $i = i_\alpha$, $\alpha = 1, 2, \ldots, c$, the number E is a multiple of all the moduli except $m_{i_1}, m_{i_2}, \ldots, m_{i_c}$, that is,

$$
\begin{aligned}
E &= e' \frac{M_T}{\prod_{\alpha=1}^{c} m_{i_\alpha}} \\
&\equiv e_{i_\alpha} \pmod{m_{i_\alpha}}, \; \alpha = 1, 2, \ldots, c, \tag{7.5.3}
\end{aligned}
$$

where $0 < e' < \prod_{\alpha=1}^{c} m_{i_\alpha}$. Based on $\underline{y}$ and BEX which avoids processing large valued integers, the syndrome digit Δ is computed

$$\Delta \equiv \overline{Y} \pmod{A}. \tag{7.5.4}$$

Since both X and E in (7.5.1) are less than M_T, we consider the following four cases

(i) $X + E < M_T$ and $X = \overline{X}$. In this case,

$$\overline{Y} = X + E = X + e' \frac{M_T}{\prod_{\alpha=1}^{c} m_{i_\alpha}}, \tag{7.5.5}$$

and

$$\Delta \equiv e' \frac{M_T}{\prod_{\alpha=1}^{c} m_{i_\alpha}} \pmod{A}. \tag{7.5.6}$$

(ii) $M_T \leq X + E < 2M_T$ and $X = \overline{X}$. In this case,

$$\overline{Y} = X + E - M_T = X + \left(e' - \prod_{\alpha=1}^{c} m_{i_\alpha}\right) \frac{M_T}{\prod_{\alpha=1}^{c} m_{i_\alpha}}, \tag{7.5.7}$$

and

$$\Delta \equiv \left(e' - \prod_{\alpha=1}^{c} m_{i_\alpha}\right) \frac{M_T}{\prod_{\alpha=1}^{c} m_{i_\alpha}} \pmod{A}. \tag{7.5.8}$$

(iii) $X + E < M_T$ and $X = \overline{X} - M_T$. In this case,

$$\begin{aligned} \overline{Y} &= X + E = \overline{X} - M_T + E \\ &= \overline{X} + \left(e' - \prod_{\alpha=1}^{c} m_{i_\alpha}\right) \frac{M_T}{\prod_{\alpha=1}^{c} m_{i_\alpha}}, \end{aligned} \tag{7.5.9}$$

and

$$\Delta \equiv \left(e' - \prod_{\alpha=1}^{c} m_{i_\alpha}\right) \frac{M_T}{\prod_{\alpha=1}^{c} m_{i_\alpha}} \pmod{A}. \tag{7.5.10}$$

(iv) $M_T \leq X + E < 2M_T$ and $X = \overline{X} - M_T$. In this case,

$$\begin{aligned} \overline{Y} &= X + E - M_T = \overline{X} + E - 2M_T \\ &= \overline{X} + \left(e' - 2\prod_{\alpha=1}^{c} m_{i_\alpha}\right) \frac{M_T}{\prod_{\alpha=1}^{c} m_{i_\alpha}}, \end{aligned} \tag{7.5.11}$$

and

$$\Delta \equiv \left(e' - 2\prod_{\alpha=1}^{c} m_{i_\alpha}\right) \frac{M_T}{\prod_{\alpha=1}^{c} m_{i_\alpha}} \pmod{A}. \tag{7.5.12}$$

Under an additional constraint on the form of the moduli and code generator A, additive overflow and multiple errors can be distinguished and concurrently detected. The result is established in the following theorem.

THEOREM 7.5 *Assume that no more than β residue errors occur. If the moduli and code generator A of an RNS-PC satisfy the condition*

$$A > max\{2 \prod_{\substack{i=1 \\ 1 \le j_i \le n}}^{\beta} m_{j_i}\}, \qquad (7.5.13)$$

then β or fewer errors, affecting either an arbitrary legitimate number or the sum of two arbitrary legitimate numbers, are detectable and generate illegitimate numbers such that $\Delta \not\equiv 0 \pmod{A}$ and $\Delta \not\equiv -M_T \pmod{A}$.

PROOF. Consider (i) for $X + E < M_T$ and $X = \overline{X}$. If

$$\Delta \equiv e' \frac{M_T}{\prod_{\alpha=1}^{c} m_{i_\alpha}} \pmod{A} \equiv 0,$$

then e' is a multiple of A. This is not possible as $e' < \prod_{\alpha=1}^{c} m_{i_\alpha} < A$, $c \le \beta$. Now, if

$$\Delta \equiv e' \frac{M_T}{\prod_{\alpha=1}^{c} m_{i_\alpha}} \pmod{A} \equiv -M_T,$$

then

$$(e' + \prod_{\alpha=1}^{c} m_{i_\alpha}) \frac{M_T}{\prod_{\alpha=1}^{c} m_{i_\alpha}} \pmod{A} \equiv 0.$$

This is not possible as $e' + \prod_{\alpha=1}^{c} m_{i_\alpha} < 2\prod_{\alpha=1}^{c} m_{i_\alpha} < A$, $c \le \beta$. The cases (ii), (iii), and (iv) can also be analyzed in an analogous manner. This proves the theorem.

Based on Theorem 7.5, we can add one more step to the procedure for error detection (given at the beginning of this section) to obtain the procedure for simultaneous detection of errors and additive overflow. The additional step is that if $\Delta \equiv -M_T \pmod{A}$, then declare overflow detected; and if $\Delta \not\equiv 0 \pmod{A}$ and $\Delta \not\equiv -M_T \pmod{A}$, then declare one or more errors detected. When errors are detected, the procedure to determine error location and error value for single errors is described in the following chapter. Algorithms for double and multiple error detection and correction will be described in chapters 9-11.

NOTES

Among the earliest researchers, Szabo and Tanaka [1967] have briefly sketched a method for single-error detection or single-error correction for use with an RNS. However, the error correction procedure given in Szabo and Tanaka

[1967] is computationally inefficient. Also, it appears to be quite complicated for implementation. Watson and Hastings [1966] have constructed an RRNS to detect or correct single errors. They proposed a consistency-checking procedure in which the digits contained in the information part are used to obtain new values for the parity checking digits through BEX. The new redundant residues can then be compared with the previously existing redundant residues to provide for error detection and correction. However, their method for error correction needs a correction table which may require large memory space, thereby making it impractical for the correction of more than single residue errors. Therefore, multiple-residue-error correction has not been investigated by them.

Mandelbaum [1972] showed how single error correction can be accomplished in an RRNS code and established that two redundant moduli with redundancy less than the redundancy in Watson and Hastings [1966] are necessary for single residue digit error correction. Later, necessary and sufficient conditions for minimal redundancy allowing the correction of the whole class of single residue errors were derived from the concept of modulus projections by Barsi and Maestrini [1973].

The concept of RNS-PC was developed in Barsi and Maestrini [1974], a concept that was earlier suggested by Mandelbaum [1972]. Mandelbaum [1976a, 1976b, 1978, 1984] gave the sufficient condition on redundancy for correcting multiple errors and derived a decoding procedure for multiple errors by using continued fraction expansion and Euclid's algorithm with a slight increase in redundancy. Later, Barsi and Maestrini [1978a, 1978b] showed that a multiple-error-correcting procedure based on convergents of continued fraction can be extended to the RNS-PC. Since the above continued fractions expansion based decoding methods have to process large valued integers and use an iterative process, they appear to be more suitable for a general purpose computer.

Yau and Liu [1973] designed two error-correcting algorithms, one for single residue-error correction and the other for burst residue-error correction. Basically, the method of Yau and Liu is Watson's method with the error-correction table replaced by an iterative computation. Consequently, their implementation needs memory space which is much smaller than that required in Watson and Hastings [1966]. However, their method becomes computationally inefficient due to the iterative computations. Ramachandran [1983] has proposed a method to correct single errors which establishes a tradeoff between computational complexity and the number of extra redundant moduli.

A number of papers by Jenkins and his associates [1980, 1983, 1984, 1988] applied the MRC algorithm and the concept of modulus projections to digital filters and residue error checkers. The error checker is emphasized for the property of self-checking. However, after determining the location of the residue digit error, error correction requires some additional circuitry to obtain the correct residue digit through BEX. Su and Lo [1990] have used the redundant mixed radix digits as entries to construct a lookup table for correcting single residue error. Using this algorithm, an error correction circuit with scaling is designed by them. However, their algorithm becomes impractical for correcting multiple residue errors due to the large memory space required by lookup tables, even if it remains valid for multiple error correction.

From the above description, it is seen that all the literature published thus far deals almost exclusively with single error correction.

This chapter is based on the research work reported in Krishna et al. [1992] and Krishna et al. [1993].

7.6   Bibliography

[7.1] N.S. Szabo and R.I. Tanaka, *Residue Arithmetic and Its Applications to Computer Technology*, McGraw Hill, New York, 1967.

[7.2] R.W. Watson and C.W. Hastings, "Self-Checked Computation Using Residue Arithmetic," *Proceeding IEEE*, no. 72, pp. 1920-1931, 1966.

[7.3] D. Mandelbaum, "Error Correction in Residue Arithmetic," *IEEE Transactions on Computers*, vol. C-21, no. 6, pp. 538-545, 1972.

[7.4] F. Barsi and P. Maestrini, "Error Correcting Properties of Redundant Residue Number systems," *IEEE Transactions on Computers*, vol. C-22, no. 3, pp. 307-375, 1973.

[7.5] F. Barsi and P. Maestrini, "Error Detection and Correction by Product Codes in Residue Number Systems," *IEEE Transactions on Computers*, vol. C-23, pp. 915-924, 1974.

[7.6]D.M. Mandelbaum, "On a Class of Arithmetic Codes and a Decoding Algorithm," *IEEE Transactions on Information Theory*, vol. IT-22, pp. 85-88, 1976a.

[7.7] D.M. Mandelbaum, "On a Class of Nonlinear Arithmetic Codes That Are Easy to Decode," *Information and Control*, vol. 30, pp. 151-168, 1976b.

[7.8] D.M. Mandelbaum, "Further Results on Decoding Arithmetic Residue Codes," *IEEE Transactions on Information Theory*, vol. IT-24, pp. 643-644, 1978.

[7.9] D.M. Mandelbaum, "An Approach to an Arithmetic Analog of Berlekamp's Algorithm," *IEEE Transactions on Information Theory*, vol. IT-30, pp. 758-762, 1984.

[7.10] F. Barsi and P. Maestrini, "Improved Decoding Algorithm for Arithmetic Residue Codes," *IEEE Transactions on Information theory*, vol. IT-24, pp. 640-643, 1978a.

[7.11] F. Barsi and P. Maestrini, "A Class of Multiple-Error-Correcting Arithmetic Residue Codes," *Information and Control*, vol. 36, pp. 28-41, 1978b.

[7.12] S.S.-S. You and Y.-C. Lin, "Error Correction in Redundant Residue Number Systems," *IEEE Transactions on Computers*, vol. C-22, pp. 5-11, 1973.

[7.13] V. Ramachandran, "Single Residue Error Correction in Residue Number Systems," *IEEE Transactions on Computers*, vol. C-32, pp. 504-507, 1983.

[7.14] M.H. Etzel and W.K. Jenkins, "Redundant Residue Number Systems for Error Detection and Correction in Digital Filters," *IEEE Transactions on Acoustics, Speech, and Signal Processing*, no. 5, pp. 538-544, 1980.

[7.15] W.K. Jenkins, "The Design of Error Checkers for Self-Checking Residue Number Arithmetic," *IEEE Transactions on Computers*, vol. C-32, pp. 388-396, 1983.

[7.16] M.J. Bell, Jr. and W.K. Jenkins, "A Residue to Mixed Radix Converted and Error Checker for a Five-Moduli Residue Number System," *Proceedings IEEE International Conference on Acoustics, Speech, and Signal Processing*, vol. 3, 1984.

[7.17] W.K. Jenkins and E.J. Altman, "Self-Checking Properties of Residue Number Error Checkers Based on Mixed Radix conversion," *IEEE Transactions on Circuits and Systems*, vol. 35, pp. 159-167, 1988.

[7.18] C.-C. Su and H.-Y. Lo, "An Algorithm for Scaling and Single Residue Error Correction in Residue Number Systems," *IEEE Transactions on Computers*, vol. C-39, pp. 1053-1064, 1990.

[7.19] H. Krishna, K.-Y. Lin, and J.-D. Sun, "A Coding Theory Approach to Error Control in Redundant Residue Number Systems, Part I: Theory and Single Error Correction," *IEEE Transactions on Circuits and Systems*, vol. 39, pp. 8-17, 1992.

[7.20] H. Krishna and J.-D. Sun, "On Theory and Fast Algorithms for Error Correction in Residue Number System Product Codes," *IEEE Transactions on Computers*, vol. 42, pp. 840-852, 1993.

Chapter 8

Single Error Correction in RNS

In the next two chapters, we describe several algorithms for single and multiple error correction and detection in RRNS and the RNS-PC. This chapter is devoted to single error correction in RNS. Based on the concept of consistency checking and the coding theory framework, fast and superfast algorithms are derived for single error correction. The algorithms derived here are superior to the previous algorithms by at least an order of magnitude. Based on these algorithms, two architectures are designed for their hardware implementation.

8.1 Single Error Correction in RRNS

In an RRNS, for $d = 3$, we can correct all single errors in residue digits. In this case, we have two syndromes Δ_1 and Δ_2 computed from the given residue vector $\underline{y}$ using the MRC based BEX method (refer to Table 7.1). If only one of them is non-zero, then Observation 1 in Section 7.3 applies and the corresponding parity digit is declared to be in error. If only the ith information digit is in error, that is, $y_i \equiv x_i + e_i \pmod{m_i}$ and $y_j = x_j$, $j = 1, 2, \ldots, k$, $j \neq i$, then $\overline{Y}$ in (7.3.1) becomes

$$\overline{Y} \equiv (X + E) \pmod{M}, \tag{8.1.1}$$

where

$$E = e_i' M_i, \ M_i = M/m_i, \ e_i' \equiv e_i T_i \pmod{m_i}, \tag{8.1.2}$$

and $0 < e_i, e_i' < m_i$. Note that e_i can be uniquely determined if e_i' is known and vice versa. Since both X and E in (8.1.1) are less than M, we consider

189

the following two situations:
(i) $X + E < M$. In this case

$$\overline{Y} = (X + E) = X + e_i' \frac{M}{m_i}, \tag{8.1.3}$$

and

$$\Delta_1 \equiv e_i' \frac{M}{m_i} \pmod{m_{k+1}}, \tag{8.1.4}$$

$$\Delta_2 \equiv e_i' \frac{M}{m_i} \pmod{m_{k+2}}. \tag{8.1.5}$$

(ii) $M \le X + E < 2M$. In this case

$$\overline{Y} = X + e_i' \frac{M}{m_i} - M = X + (e_i' - m_i)\frac{M}{m_i}, \tag{8.1.6}$$

and

$$\Delta_1 \equiv (e_i' - m_i)\frac{M}{m_i} \pmod{m_{k+1}}, \tag{8.1.7}$$

$$\Delta_2 \equiv (e_i' - m_i)\frac{M}{m_i} \pmod{m_{k+2}}. \tag{8.1.8}$$

Given Δ_1 and Δ_2, based on (8.1.1–8.1.8), a procedure to determine i (error location) and e_i (error value) can be outlined as follows: For $j = 1, 2, \ldots, k$, solve the congruences

$$\begin{cases} \Delta_1 \equiv e_j^{(1,1)} \frac{M}{m_j} \pmod{m_{k+1}} \\ \Delta_2 \equiv e_j^{(1,2)} \frac{M}{m_j} \pmod{m_{k+2}}, \end{cases} \tag{8.1.9}$$

and

$$\begin{cases} \Delta_1 \equiv (e_j^{(2,1)} - m_j)\frac{M}{m_j} \pmod{m_{k+1}} \\ \Delta_2 \equiv (e_j^{(2,2)} - m_j)\frac{M}{m_j} \pmod{m_{k+2}}, \end{cases} \tag{8.1.10}$$

to obtain the values $e_j^{(1,1)}, e_j^{(1,2)}, e_j^{(2,1)}$ and $e_j^{(2,2)}$ respectively. Compare to check if (a) $e_j^{(1,1)} = e_j^{(1,2)}$ and $e_j^{(1,1)} < m_j$, or (b) $e_j^{(2,1)} = e_j^{(2,2)}$ and $e_j^{(2,1)} < m_j$. Let us say that either of the two conditions (a) or (b) is satisfied for $j = l$. Then, the lth information digit is declared to be in error, the value of the error being $e_l \equiv e_j^{(1,1)} T_l^{-1} \pmod{m_l}$ if $e_j^{(1,1)} = e_j^{(1,2)}$ and $e_l \equiv e_j^{(2,1)} T_l^{-1} \pmod{m_l}$ if $e_j^{(2,1)} = e_j^{(2,2)}$. Here T_l^{-1} is the multiplicative inverse of $T_l \bmod m_l$, that is, $T_l^{-1} \equiv \frac{M}{m_l} \pmod{m_l}$. The correct value of the lth residue digit is $(y_l - e_l) \pmod{m_l}$, and $\underline{y}$ is decoded to the codevector $\hat{\underline{x}}$, where

$$\hat{x}_j = y_j, \ j = 1, 2, \ldots, n, \ j \neq l, \tag{8.1.11}$$

$$\hat{x}_l \equiv y_l - e_l \pmod{m_l}. \tag{8.1.12}$$

It is clear from (8.1.3–8.1.5) and (8.1.9) that if one error takes place in the ith information digit such that $X + E < M$, then for $j = i$, $e_i' = e_j^{(1,1)} = e_j^{(1,2)}$ and $l = i$. Similarly, if one error takes place in the ith information digit such that $M \leq X + E < 2M$, then for $j = i$, $e_i' = e_j^{(2,1)} = e_j^{(2,2)}$ and $l = i$ from (8.1.6–8.1.8) and (8.1.10). However, it remains to be shown that if one error takes place in the ith information digit, then for $j \neq i$ and $0 < e_j^{(1,1)}, e_j^{(1,2)}, e_j^{(2,1)}, e_j^{(2,2)} < m_j$, $e_j^{(1,1)} \neq e_j^{(1,2)}$ and $e_j^{(2,1)} \neq e_j^{(2,2)}$. Under an additional constraint on the form of the moduli, the result is established in the following theorem.

THEOREM 8.1 *If the moduli* $m_j, j = 1, 2, \ldots, k$ *are such that there do not exist integers* $n_i, n_j, 0 < n_i < m_i, 0 < n_j < m_j$, *that satisfy*

$$n_i m_j + n_j m_i = m_{k+1} m_{k+2}, \ j = 1, 2, \ldots, k, \tag{8.1.13}$$

then for $j \neq i$ *and* $0 < e_j^{(1,1)}, e_j^{(1,2)}, e_j^{(2,1)}, e_j^{(2,2)} < m_j$, $e_j^{(1,1)} \neq e_j^{(1,2)}$ *and* $e_j^{(2,1)} \neq e_j^{(2,2)}$, *where the* ith *information digit is received in error.*

PROOF. For $X + E < M$ and $j \neq i$, if $e_j^{(1,1)} = e_j^{(1,2)} = e_j' < m_j$ (say), then comparing (8.1.4, 8.1.5) to (8.1.9) we get

$$\frac{M}{m_i m_j}(e_j' m_i - e_i' m_j) \equiv 0 \pmod{m_{k+1} m_{k+2}},$$

that is, $e_j' m_i - e_i' m_j$ is a multiple of $m_{k+1} m_{k+2}$. This is not possible as $-m_{k+1} m_{k+2} < e_j' m_i - e_i' m_j < m_{k+1} m_{k+2}$. Similarly, for $X + E < M$ and $j \neq i$, if $e_j^{(2,1)} = e_j^{(2,2)} = e_j' < m_j$ (say), then comparing (8.1.4, 8.1.5) to (8.1.10) we get

$$\frac{M}{m_i m_j}(e_i' m_j + (m_j - e_j') m_i) \equiv 0 \pmod{m_{k+1} m_{k+2}}.$$

This cannot hold due to (8.1.13). The case for $M \leq X + E < 2M$, can also be analyzed in an analogous manner. This proves the theorem.

We observe that if $m_{k+1} m_{k+2} > 2 m_i m_j - m_i - m_j$, then the condition stated in Theorem 8.1 is satisfied in a trivial manner. It is interesting to note here that there can be at most one solution to (8.1.13). Based on the concepts developed above and the assumption that the moduli $m_i, i = 1, 2, \ldots, k$ satisfy the condition in Theorem 8.1, the algorithm to correct a single error in RRNS is given in Table 8.1. A flowchart for this algorithm is given in Figure 8.1.

8.2 Computational Analysis and Examples

Once Δ_1 and Δ_2 are known, the remaining steps in the decoding algorithm require $2k + 1$ modular multiplications (MMULT) and $2k + 1$ modular additions (MADD). Also, all of these operations in the algorithm (Step 4) are independent of each other, and therefore, can be performed in parallel. If BEX is based on the MRC approach, the computation of Δ_1 and Δ_2 requires $k^2/2 + 3k/2 - 2$ MMULT and $k^2/2 + 3k/2$ MADD. The total number of MMULT and MADD for the algorithm derived here are $k^2/2 + 7k/2 - 1$ and $k^2/2 + 7k/2 + 1$, respectively. A detailed computational complexity analysis is given in Appendix A. A comparison of this algorithm with those in Yau and Liu [1973] and Jenkins and Altman [1988] in terms of the requirement of MMULT and MADD is shown in Table 8.2. From Table 8.2, we note that it is superior to the algorithms in Yau and Liu [1973] and Jenkins and Altman [1988] by an order of magnitude. This, we feel, is a significant improvement in the computational complexity for correcting single errors in RRNS. Finally, the algorithms in Yau and Liu [1973] and Jenkins and Altman [1988] remain valid even for the case when the moduli do not satisfy the conditions stated in Theorem 8.1. In the following we present three examples to illustrate the use of the decoding algorithm.

EXAMPLE 8.1 Consider the (8,6) RRNS code based on the moduli $m_1 = 23$, $m_2 = 25$, $m_3 = 27$, $m_4 = 29$, $m_5 = 31$, $m_6 = 32$, $m_7 = 41$ and $m_8 = 47$, where m_7 and m_8 are the redundant moduli. Clearly, $m_7 m_8 = 1927 > 2m_i m_j - m_i - m_j, i, j = 1, 2, \ldots, 6$, and therefore, the condition in Theorem 8.1 is satisfied trivially. The legitimate range is

$$M = \prod_{i=1}^{6} m_i = 446, 623, 200.$$

(1). Let $X = 21, 600$, then $\underline{x}$ is represented by [3 0 0 24 24 0 34 27]. Since $0 \leq X < M$, $\underline{x}$ is a codevector. Assume that one error takes place in the ith position during data processing (say $i = 2$), and the received vector is

$$\underline{y} = [3\ 1\ 0\ 24\ 24\ 0\ 34\ 27].$$

Table 8.1: An Algorithm for Single Error Correction in RRNS

Step 1:	According to the received vector, compute the syndromes using BEX as given in Table 7.1 and $\Delta_r \equiv \bar{y}_{k+r} - y_{k+r} \pmod{m_{k+r}}$, $r = 1, 2, \dots, n - k$.
Step 2:	Check how many syndromes are zero: (1). If $\Delta_1 = \Delta_2 = 0$, then no error occurs. Stop. (Theorem 7.3) (2). If only one syndrome is non-zero, then the corresponding parity digit is declared to be in error. The estimate of erroneous parity digit is $\bar{y}_{k+r}$. Stop. (Theorem 7.3, Observation 1) (3). If all the syndromes are non-zero, then go to Step 3.
Step 3:	Let $j = 1$.
Step 4:	Perform the single-error consistency-checking for the nonredundant modulus m_j: (1). Compute $e_j^{(1,1)}, e_j^{(1,2)}, e_j^{(2,1)}, e_j^{(2,2)}$ using (8.1.9) and (8.1.10). (2). Check the consistency of the solutions, $e_j^{(1,1)} = e_j^{(1,2)} < m_j$ or $e_j^{(2,1)} = e_j^{(2,2)} < m_j$. If it is consistent, then go to Step 5. If it is not consistent, then $j = j + 1$. Go to Step 4 for $j \leq k$. For $j = k + 1$, go to Step 6.
Step 5:	Declare only one error in the jth position. Correct it using (8.1.12) and stop.
Step 6:	Declare more than one error detected and stop.

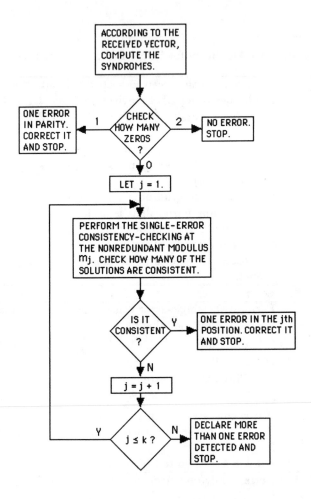

Figure 8.1: A Decoder Flowchart for Single Error Correction in RRNS

Table 8.2: MMULT and MADD Required for Single Error Correction in RRNS

Yau and Liu's Algorithm [1973]	Jenkins and Altman's Algorithm [1988]	New Algorithm	Modular Operations
$\frac{1}{4}k^3 + \frac{5}{4}k^2 + \frac{1}{2}k - 2$	$\frac{1}{2}k^3 + \frac{3}{2}k^2 + 2k$	$\frac{1}{2}k^2 + \frac{7}{2}k - 1$	MMULT
$\frac{1}{4}k^3 + \frac{5}{4}k^2 + \frac{3}{2}k$	$\frac{1}{2}k^3 + \frac{3}{2}k^2 + 2k$	$\frac{1}{2}k^2 + \frac{7}{2}k + 1$	MADD

Based on the received information part $[3\,1\,0\,24\,24\,0]$ and the BEX, we compute the syndrome digits $\Delta_1 \equiv 7 \pmod{41}$ and $\Delta_2 \equiv 22 \pmod{47}$. Following the decoding algorithm, we check the consistency for $j = 1, 2, \ldots, k$, based on $e_j^{(1,1)}$, $e_j^{(1,2)}$, $e_j^{(2,1)}$ and $e_j^{(2,2)}$ as computed from (8.1.9) and (8.1.10). It shows that when $j = 2$

$$e_2^{(1,1)} \equiv \Delta_1 m_2 M^{-1} \pmod{41} \equiv 17,$$

$$e_2^{(1,2)} \equiv \Delta_2 m_2 M^{-1} \pmod{47} \equiv 17,$$

$$e_2^{(2,1)} \equiv (\Delta_1 m_2 M^{-1} + m_2) \pmod{41} \equiv 1,$$

$$e_2^{(2,2)} \equiv (\Delta_2 m_2 M^{-1} + m_2) \pmod{47} \equiv 42.$$

The identity $e_2^{(1,1)} = e_2^{(1,2)} = 17$ declares that one error occurs in the second position and the error digit is 1 obtained from $e_2 \equiv e_2^{(1,1)} \frac{M}{m_2} \pmod{m_2}$. (2). Consider the case when $X + E > M$. Let $X = 400,000,000$, then codevector $\underline{x}$ is represented by $[8\,0\,22\,13\,25\,0\,23\,14]$. Assume that one error takes place in the ith position during data processing (say $i{=}4$), and the received vector is

$$\underline{y} = [8\,0\,22\,18\,25\,0\,23\,14].$$

Note that $X + E = 661,813,600 > M$. Based on the received information part $[8\,0\,22\,18\,25\,0]$ and the BEX, we compute the syndrome digits $\Delta_1 \equiv 32$ $\pmod{41}$ and $\Delta_2 \equiv 40$ $\pmod{47}$. Following the decoding algorithm, we check the consistency for $j = 1, 2, \ldots, 6$, based on $e_j^{(1,1)}$, $e_j^{(1,2)}$, $e_j^{(2,1)}$ and

$e_j^{(2,2)}$ as computed from (8.1.9) and (8.1.10). It shows that when $j = 4$

$$e_4^{(1,1)} \equiv \Delta_1 m_4 M^{-1} \pmod{41} \equiv 29,$$

$$e_4^{(1,2)} \equiv \Delta_2 m_4 M^{-1} \pmod{47} \equiv 35,$$

$$e_4^{(2,1)} \equiv (\Delta_1 m_4 M^{-1} + m_4) \pmod{41} \equiv 17,$$

$$e_4^{(2,2)} \equiv (\Delta_2 m_4 M^{-1} + m_4) \pmod{47} \equiv 17.$$

The identity $e_4^{(2,1)} = e_4^{(2,2)} = 17$ declares that one error occurs in the fourth position and the error digit is 5 obtained from $e_4 \equiv e_4^{(2,1)} \frac{M}{m_4} \pmod{m_4}$.

EXAMPLE 8.2 In this example, the condition $m_{k+1} m_{k+2} > 2 m_i m_j - m_i - m_j$, $i, j = 1, 2, \ldots, k$, is not satisfied. Consider the (8,6) RRNS code based on the moduli $m_1 = 23$, $m_2 = 25$, $m_3 = 27$, $m_4 = 29$, $m_5 = 31$, $m_6 = 32$, $m_7 = 37$, $m_8 = 47$, where m_7 and m_8 are the redundant moduli. Clearly, $m_7 m_8 = 1739 < 2 m_5 m_6 - m_5 - m_6 = 1921$, but the condition in Theorem 8.1 is satisfied. The legitimate range is

$$M = \prod_{i=1}^{6} m_i = 446,623,200.$$

Let $X = 21,600$, then $\underline{x}$ is represented by [3 0 0 24 24 0 29 27]. Since $0 \leq X < M$, $\underline{x}$ is a codevector. Assume that one error takes place in the ith position during data processing (say $i = 2$), and the received vector is

$$\underline{y} = [3\ 1\ 0\ 24\ 24\ 0\ 29\ 27].$$

Based on the received information part [3 1 0 24 24 0] and the BEX, we compute the syndrome digits $\Delta_1 \equiv 6 \pmod{37}$ and $\Delta_2 \equiv 22 \pmod{47}$. Following the decoding algorithm, we check the consistency for $j = 1, 2, \ldots, 6$, based on computing $e_j^{(1,1)}$, $e_j^{(1,2)}$, $e_j^{(2,1)}$ and $e_j^{(2,2)}$ from (8.1.9) and (8.1.10). It shows that when $j = 2$

$$e_2^{(1,1)} \equiv \Delta_1 m_2 M^{-1} \pmod{37} \equiv 17,$$

$$e_2^{(1,2)} \equiv \Delta_2 m_2 M^{-1} \pmod{47} \equiv 17,$$

$$e_2^{(2,1)} \equiv (\Delta_1 m_2 M^{-1} + m_2) \pmod{37} \equiv 5,$$

$$e_2^{(2,2)} \equiv (\Delta_2 m_2 M^{-1} + m_2) \pmod{47} \equiv 42.$$

The identity $e_2^{(1,1)} = e_2^{(1,2)} = 17$ declares that one error occurs in the second position and the error digit is 1 obtained from $e_2 \equiv e_2^{(1,1)} \frac{M}{m_2}$ (mod m_2).

EXAMPLE 8.3 Consider the (6,4) RRNS code based on the moduli $m_1 = 7, m_2 = 9$, $m_3 = 11$, $m_4 = 13$, $m_5 = 16$ and $m_6 = 17$, where m_5 and m_6 are the redundant moduli. Clearly, $m_5 m_6 = 272 > 2m_3 m_4 - m_3 - m_4 = 262$. The legitimate range is $[0, 9,009)$. Let $X = 5,000$, then $\underline{x} = [2\ 5\ 6\ 8\ 8\ 2]$. Assume that one error takes place in the 4th position during data processing, and the received vector is $\underline{y} = [2\ 5\ 6\ 6\ 8\ 2]$. We compute the syndrome digits $\Delta_1 \equiv 13$ (mod 16) and $\Delta_2 \equiv 11$ (mod 17). Following the decoding algorithm, it is seen that when $j = 3, 4$

$$e_3^{(1,1)} = e_3^{(1,2)} = 15 > m_3,$$

$$e_4^{(2,1)} = e_4^{(2,2)} = 6 < m_4.$$

The values $e_4^{(2,1)} = e_4^{(2,2)} = 6$ imply that one error occurs in the 4th position. Note that $e_3^{(1,1)} = e_3^{(1,2)} = 15 > m_3$, which implies that even though $e_3^{(1,1)} = e_3^{(1,2)}$, it is not a consistent solution.

In Chapter 9, we derive computationally efficient algorithms for simultaneous correction and detection of multiple errors, burst errors, and erasures in a given RRNS coding scheme.

8.3 Single-Error Correction in RNS-PC

Once again, for single error correction in RNS-PC, $d = 3$. Using (7.4.12) and Theorem 7.5, one can show that an RNS-PC with $A > max\{2m_{j_1}, m_{j_2}\}$, $1 \leq j_1, j_2 \leq n$, can simultaneously correct single errors and detect additive overflow. Assume that only the pth residue digit is in error, then (7.5.5–7.5.12) can be analyzed as follows.
(i) $X + E < M_T$ and $X = \overline{X}$. In this case,

$$\overline{Y} = X + E = X + e' \frac{M_T}{m_p}, \quad 0 < e' < m_p, \qquad (8.3.1)$$

and

$$\Delta \equiv e' \frac{M_T}{m_p} \pmod{A}. \qquad (8.3.2)$$

(ii) $M_T \leq X + E < 2M_T$ and $X = \overline{X}$. In this case,

$$\overline{Y} = X + E - M_T = X + (e' - m_p) \frac{M_T}{m_p}, \quad 0 < e' < m_p, \qquad (8.3.3)$$

and

$$\Delta \equiv (e' - m_p)\frac{M_T}{m_p} \quad (\mathrm{mod}\ A). \qquad (8.3.4)$$

(iii) $X + E < M_T$ and $X = \overline{X} - M_T$. In this case,

$$\overline{Y} = X + E = \overline{X} + E - M_T = \overline{X} + (e' - m_p)\frac{M_T}{m_p},\ 0 < e' < m_p, \quad (8.3.5)$$

and

$$\Delta \equiv (e' - m_p)\frac{M_T}{m_p} \quad (\mathrm{mod}\ A). \qquad (8.3.6)$$

(iv) $M_T \le X + E < 2M_T$ and $X = \overline{X} - M_T$. In this case,

$$\begin{aligned} \overline{Y} &= X + E - M_T = \overline{X} + E - 2M_T \\ &= \overline{X} + (e' - 2m_p)\frac{M_T}{m_p},\ 0 < e' < m_p, \qquad (8.3.7) \end{aligned}$$

and

$$\Delta \equiv (e' - 2m_p)\frac{M_T}{m_p} \quad (\mathrm{mod}\ A). \qquad (8.3.8)$$

Given Δ, based on (8.3.1–8.3.8) a procedure to determine p (error location) and e_p (error value) can be outlined as follows:

For $j = 1, 2, \ldots, n$, solve the congruences

$$\Delta \equiv e_j^{(1)}\frac{M_T}{m_j} \quad (\mathrm{mod}\ A), \qquad (8.3.9)$$

$$\Delta \equiv (e_j^{(2)} - m_j)\frac{M_T}{m_j} \quad (\mathrm{mod}\ A), \qquad (8.3.10)$$

$$\Delta \equiv (e_j^{(3)} - 2m_j)\frac{M_T}{m_j} \quad (\mathrm{mod}\ A), \qquad (8.3.11)$$

to obtain the values $e_j^{(1)}$, $e_j^{(2)}$ and $e_j^{(3)}$. Check if the solutions are consistent, that is, one of the following three conditions is satisfied:
Condition (1) $e_j^{(1)} < m_j$, $e_j^{(2)} > m_j$ and $e_j^{(3)} > m_j$.
Condition (2) $e_j^{(1)} > m_j$, $e_j^{(2)} < m_j$ and $e_j^{(3)} > m_j$.
Condition (3) $e_j^{(1)} > m_j$, $e_j^{(2)} > m_j$ and $e_j^{(3)} < m_j$.
Let us say that one of the three conditions is satisfied, for $j = l$, then the lth residue digit is declared in error, the value of the error being $e_l \equiv e\frac{M_T}{m_l}$ (mod m_l), where $e = e_l^{(1)}$, $e_l^{(2)}$, or $e_l^{(3)}$ depending on which of the

conditions (1), (2), or (3) is satisfied. The correct value of the lth residue digit is $(y_l - e_l) \bmod m_l$, and $\underline{y}$ is decoded to $\widehat{\underline{x}}$, where

$$
\begin{aligned}
\widehat{\underline{x}} &= [\widehat{x}_1 \, \widehat{x}_2 \ldots \widehat{x}_n], \\
\widehat{x}_j &= y_j, \; j = 1, 2, \ldots, n, \; j \neq l, &\text{(8.3.12)} \\
\widehat{x}_l &\equiv y_l - e_l \pmod{m_l}. &\text{(8.3.13)}
\end{aligned}
$$

It is clear from (8.3.1, 8.3.2, 8.3.9) that if one error takes place in the pth residue digit such that $X + E < M_T$ and $X = \overline{X}$, then for $j = p$, $e_j^{(1)} = e' < m_p$ and $l = p$. Similarly, if one error takes place in the pth residue digit such that $M_T \le X + E < 2M_T$ and $X = \overline{X}$, then for $j = p$, $e_j^{(2)} = e' < m_p$ and $l = p$ from (8.3.3, 8.3.4, 8.3.10); if one error takes place in the pth residue digit such that $X + E < M_T$ and $X = \overline{X} - M_T$, then for $j = p$, $e_j^{(2)} = e' < m_p$ and $l = p$ from (8.3.5, 8.3.6, 8.3.10); if one error takes place in the pth residue digit such that $M_T \le X + E < 2M_T$ and $X = \overline{X} - M_T$, then for $j = p$, $e_j^{(3)} = e' < m_p$ and $l = p$ from (8.3.7, 8.3.8, 8.3.11). Based on the above analysis, if Condition (2) is satisfied, then after correcting the error, we recompute the syndrome, denoted by Δ'. If $\Delta' \equiv -M_T \pmod{A}$ then an additive overflow is declared. However, the following two cases have to be shown. Case (1): if only one error occurs in the pth residue digit and $j = p$, then only one of the above three conditions is satisfied. Case (2): if only one error occurs in the pth residue digit and $j \neq p$, then none of the above three conditions is satisfied. Under an additional constraint on the moduli and the code generator A, the result is established in the following theorem.

THEOREM 8.2 *If the moduli and code generator A of an RNS-PC are such that there do not exist integers n_p, n_j, $0 < n_p < m_p$, $0 < n_j < m_j$, that satisfy either one of the following equations*

$$
n_p m_j + n_j m_p = A \qquad\qquad \text{(8.3.14)}
$$

$$
m_p m_j + n_p m_j + n_j m_p = A, \; 1 \le j \le k, \qquad \text{(8.3.15)}
$$

and the pth residue digit is received in error, then (i) only one of conditions (1), (2), or (3) is satisfied for $j = p$, and (ii) none of conditions (1), (2), or (3) is satisfied for $j \neq p$.

PROOF. (i) For $j = p$, if $X + E < M_T$ and $X = \overline{X}$, then $e_j^{(1)} = e'$ satisfies

$$
\Delta \left(\frac{M_T}{m_j} \right)^{-1} \bmod A \equiv e_j^{(1)} = e' < m_j,
$$

where $\left(\frac{M_T}{m_j}\right)^{-1}$ is the multiplicative inverse of $\left(\frac{M_T}{m_j}\right)$ mod A. Also (8.3.10) and (8.3.11) lead to

$$\left(\Delta\left(\frac{M_T}{m_j}\right)^{-1} + m_j\right) \text{ mod } A \equiv e_j^{(2)} > m_j,$$

$$\left(\Delta\left(\frac{M_T}{m_j}\right)^{-1} + 2m_j\right) \text{ mod } A \equiv e_j^{(3)} > m_j.$$

Similarly, for $j = p$, if $M_T \leq X + E < 2M_T$ and $X = \overline{X}$, then $e_j^{(2)} = e'$ is a solution to (8.3.10), that is,

$$\left(\Delta\left(\frac{M_T}{m_j}\right)^{-1} + m_j\right) \text{ mod } A \equiv e_j^{(2)} < m_j.$$

and from (8.3.9) and (8.3.11),

$$\left(\Delta\left(\frac{M_T}{m_j}\right)^{-1} \text{ mod } A\right) \equiv A - m_j + e_j^{(2)} = e_j^{(1)} > m_j,$$

$$\left(\Delta\left(\frac{M_T}{m_j}\right)^{-1} + 2m_j \text{ mod } A\right) \equiv m_j + e_j^{(2)} = e_j^{(3)} > m_j.$$

The case for $X + E < M_T$ and $X = \overline{X} - M_T$, and the case for $M_T \leq X + E < 2M_T$ and $X = \overline{X} - M_T$ can also be analyzed in an analogous manner.

(ii) For $j \neq p$, $X + E < M_T$ and $X = \overline{X}$, if $e_j^{(1)} < m_j$, then comparing (8.3.2) and (8.3.9), we obtain

$$\frac{M_T}{m_j m_p}(e'm_j - e_j^{(1)}m_p) \equiv 0 \pmod{A},$$

that is, $e'm_j - e_j^{(1)}m_p$ is a multiple of A. This is not possible as

$$|e'm_j - e_j^{(1)}m_p| < m_j m_p < A.$$

Similarly, for $X + E < M_T$ and $X = \overline{X}$, if $e_j^{(2)} < m_j$, then comparing (8.3.2) and (8.3.10), we obtain

$$\frac{M_T}{m_j m_p}[e'm_j + (m_j - e_j^{(2)})m_p] \equiv 0 \pmod{A}.$$

This can not hold due to (8.3.14). For $X + E < M_T$ and $X = \overline{X}$, if $e_j^{(3)} < m_j$, then comparing (8.3.2) and (8.3.11), we obtain

$$\frac{M_T}{m_j m_p}[e'm_j + (2m_j - e_j^{(3)})m_p] \equiv 0 \pmod{A}.$$

This can not hold due to (8.3.15). The cases (1) $M_T \leq X + E < 2M_T$ and $X = \overline{X}$, (2) $X + E < M_T$ and $X = \overline{X} - M_T$, and (3) $M_T \leq X + E < 2M_T$ and $X = \overline{X} - M_T$ can also be analyzed in an analogous manner. This proves the theorem.

LEMMA 8.1 *If the moduli and code generator A of an RNS-PC satisfy the condition*

$$A > max\{3m_{j_1} m_{j_2} - m_{j_1} - m_{j_2}\}, \quad 1 \leq j_1, j_2 \leq n, \qquad (8.3.16)$$

and only one error occurs in the pth residue digit, then (i) only one of conditions (1), (2), or (3) is satisfied for $j = p$, and (ii) none of conditions (1), (2), or (3) is satisfied for $j \neq p$.

PROOF. The condition is sufficient since, if the moduli and code generator A satisfy (8.3.16), then the conditions in Theorem 8.2 are satisfied trivially. This completes the proof.

Based on the concepts developed above and under the assumption that the moduli m_j, $j = 1, 2, \ldots, n$, and the code generator A satisfy the constraints in Theorem 8.2 (or Lemma 8.1), the algorithm to correct a single error and simultaneously detect additive overflow in RNS-PC is given in Table 8.3. A flowchart for this algorithm is given in Figure 8.2. We observe here that the structure of this algorithm is identical to the structure of the algorithm for single error correction, multiple error and additive overflow detection (to be established in Chapter 11). Therefore we have included "multiple error" in the title of Table 8.3 and Figure 8.2.

EXAMPLE 8.4 Consider an RNS-PC based on the moduli $m_1 = 3$, $m_2 = 5$, $m_3 = 7$, $m_4 = 11$, $m_5 = 13$, $m_6 = 17$, $m_7 = 19$, $m_8 = 23$, and the code generator $A = 1073$, which satisfy the conditions in Theorem 8.2. Here, $-M_T \bmod A = 499$. Let $X = 10,730 < M_T = \prod_{i=1}^{8} m_i = 111,546,435$. Therefore, X is a code-integer in this RNS-PC as $X = 0 \bmod A$. The residue representation of X is $\underline{x} = [1\ 0\ 6\ 5\ 5\ 3\ 14\ 12]$. Assume that X is altered by an error in the first residue digit, and $E \longleftrightarrow \underline{e} = [2\ 0\ 0\ 0\ 0\ 0\ 0\ 0]$. Then the received vector is $\underline{y} = [0\ 0\ 6\ 5\ 5\ 3\ 14\ 12]$. Based on the received vector and BEX, we compute the syndrome digit $\Delta \equiv 25 \pmod{A}$. Clearly, $\Delta \not\equiv 0 \pmod{A}$ and $\Delta \not\equiv -M_T \pmod{A}$.

Table 8.3: An Algorithm for Correcting Single Error and Simultaneously Detecting Multiple Errors and Additive Overflow in RNS-PC

Step 1:	According to the received vector, compute the syndrome Δ using BEX given in Table 7.1.
Step 2:	If $\Delta \equiv 0 \,(\mathrm{mod}\ A)$, then no error occurs. Stop. If $\Delta \equiv -M_T \,(\mathrm{mod}\ A)$, then no error occurs and an additive overflow is detected. Stop. Otherwise, go to next step.
Step 3:	Let $j = 1$.
Step 4:	Perform the single-error consistency-checking for the modulus m_j: (1). Compute $e_j^{(1)}, e_j^{(2)}, e_j^{(3)}$ using (8.3.9, 8.3.10, 8.3.11). (2). Check the consistency of the solutions, $$e_j^{(1)} < m_j \text{ or } e_j^{(2)} < m_j \text{ or } e_j^{(3)} < m_j.$$ If it is consistent, then go to Step 6. If it is inconsistent, then go to next step.
Step 5:	Let $j = j + 1$. If $j \le n$, go to Step 4. Go to Step 7 for $j > n$.
Step 6:	Declare one error in the jth position. Correct it using (8.3.13). If $e_j^{(3)} < m_j$, an additive overflow is detected. Stop. If $e_j^{(2)} < m_j$, recompute the new syndrome Δ' as in Step 1. If $\Delta' \equiv -M_T \,(\mathrm{mod}\ A)$, then an additive overflow is detected as well. Stop. Otherwise, stop.
Step 7:	Declare more than one error detected and stop.

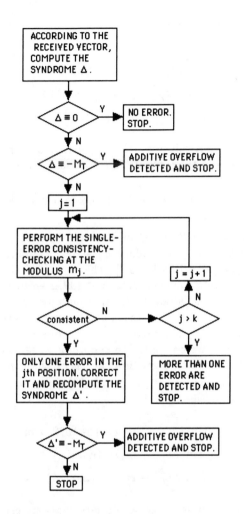

Figure 8.2: A Decoder Flowchart for Single Error Correction and Multiple Error and Additive Overflow Detection in RNS-PC

Therefore, following the decoding algorithm, we check the consistency for $j = 1, 2, \ldots, 8$. This is based on $e_j^{(1)}$, $e_j^{(2)}$ and $e_j^{(3)}$ as computed from (8.3.9–8.3.11). It is seen that when $j = 1$

$$
\begin{aligned}
e_1^{(1)} &\equiv \Delta m_1 M_T^{-1} \pmod{A} \equiv 2 < m_1, \\
e_1^{(2)} &\equiv (\Delta m_1 M_T^{-1} + m_1) \pmod{A} \equiv 5 > m_1, \\
e_1^{(3)} &\equiv (\Delta m_1 M_T^{-1} + 2m_1) \pmod{A} \equiv 8 > m_1.
\end{aligned}
$$

Since $e_1^{(1)}$ is a consistent solution, we declare that one error has occurred in the first position and the estimate of erroneous digit is

$$
\widehat{x}_1 \equiv y_1 - e_1^{(1)} \frac{M_T}{m_1} \pmod{m_1} \equiv 1.
$$

EXAMPLE 8.5 In the RNS-PC of Example 8.4, let $X \equiv \overline{X} = X_1 + X_2 \pmod{M_T}$, where $X_1 = 107,300,000 \longleftrightarrow \underline{x}_1 = [2\,0\,3\,5\,2\,1\,2\,8\,9]$ and $X_2 = 4,246,934 \longleftrightarrow \underline{x}_2 = [2\,4\,6\,10\,3\,11\,16\,7]$. Then $\underline{x} = [1\,4\,2\,4\,5\,6\,5\,16]$. Based on $\underline{x}$ and BEX, we compute the syndrome $\Delta \equiv 499 \equiv -M_T \pmod{A}$. In this case an additive overflow is detected. Now, assume that a single residue digit error affects the number X considered above, the error vector being $\underline{e} = [2\,0\,0\,0\,0\,0\,0\,0]$. Then the received vector becomes $\underline{y} = [0\,4\,2\,4\,5\,6\,5\,16]$. Based on $\underline{y}$ and BEX, we compute the syndrome $\Delta \equiv 524 \not\equiv -M_T \pmod{A}$. Following the decoding algorithm, we check the consistency for $j = 1, 2, \ldots, 8$. This is based on $e_j^{(1)}$, $e_j^{(2)}$ and $e_j^{(3)}$ as computed from (8.3.9–8.3.11). It is seen that when $j = 1$,

$$
\begin{aligned}
e_1^{(1)} &\equiv \Delta m_1 M_T^{-1} \pmod{A} \equiv 1072 > m_1, \\
e_1^{(2)} &\equiv (\Delta m_1 M_T^{-1} + m_1) \pmod{A} \equiv 2 < m_1, \\
e_1^{(3)} &\equiv (\Delta m_1 M_T^{-1} + 2m_1) \pmod{A} \equiv 5 > m_1.
\end{aligned}
$$

Since $e_1^{(2)}$ is a consistent solution, we declare that one error has occurred in the first position and the estimated value of erroneous digit is

$$
\widehat{x}_1 \equiv y_1 - e_1^{(2)} \frac{M_T}{m_1} \pmod{m_1} \equiv 1.
$$

Based on the corrected vector $[1\,4\,2\,4\,5\,6\,5\,16]$ and BEX, we recompute the syndrome $\Delta' \equiv 499 \equiv -M_T \pmod{A}$. Now, an additive overflow is detected.

EXAMPLE 8.6 In the RNS-PC of Example 8.4, let $X = 21,288,320$ and $\underline{x} = [2\ 0\ 4\ 9\ 1\ 2\ 17\ 3]$. Assume that X is altered by two residue errors, the error vector being $\underline{e} = [0\ 0\ 0\ 0\ 0\ 0\ 8\ 5]$. Then the received vector is $\underline{y} = [2\ 0\ 4\ 9\ 1\ 2\ 6\ 8]$. Based on $\underline{y}$ and BEX, we compute the syndrome $\Delta \equiv 524 \pmod{A}$. Following the decoding algorithm, we check the consistency for $j = 1, 2, \ldots, 8$; This is based on $e_j^{(1)}$, $e_j^{(2)}$ and $e_j^{(3)}$ as computed from (8.3.8–8.3.10). It is seen that when $j = 1$,

$$e_1^{(1)} \equiv \Delta m_1 M_T^{-1} \pmod{A} \equiv 1072 > m_1,$$

$$e_1^{(2)} \equiv (\Delta m_1 M_T^{-1} + m_1) \pmod{A} \equiv 2 < m_1,$$

$$e_1^{(3)} \equiv (\Delta m_1 M_T^{-1} + 2m_1) \pmod{A} \equiv 5 > m_1.$$

So, it is declared that one error has occurred in the first position. It is obvious that the decoder makes a decoding error in this case.

The computational complexity of the algorithm for single error correction and overflow detection in RNS-PC can be analyzed in a manner similar to the analysis of the algorithm for single error correction in RRNS. The details of such an analysis are omitted here. Based on the MRC method for BEX, the comparison of the algorithm described here with those in Barsi and Maestrini [1974] and Jenkins [1982] in terms of MMULT and MADD is shown in Table 8.4. Clearly, this algorithm is superior to the previous algorithms by an order of magnitude. However, the algorithms in Barsi and Maestrini [1974] and Jenkins [1982] remain valid even for the case when the moduli and the code generator do not satisfy the condition stated in the Theorem 8.2.

In Chapter 11, we derive computationally efficient algorithms for simultaneous correction and detection of multiple errors and erasures in RNS-PC coding scheme. In the next section, we derive a superfast algorithm and describe its hardware implementation for single error correction in RNS.

Table 8.4: MMULT and MADD Required for Single Error Correction and Additive Overflow Detection in RNS-PC

♯ Error Correction	Previous Algorithms Barsi & Maestrini [1974], and Jenkins [1982]	New Algorithm	Modular Operations
1	$\frac{1}{2}n^3 + \frac{3}{2}n - 2$	$n^2 + 2n - 1$	MMULT
$d = 3$	$\frac{1}{2}n^3 + \frac{3}{2}n - 2$	$n^2 + 3n - 1$	MADD

8.4   A Superfast Algorithm for RNS

The algorithms developed in this chapter reduce the computational complexity of previously known algorithms by at least an order of magnitude. However, this superiority must be realized in hardware implementation as well.

It is clear that BEX is fundamental to error correction and detection in RNS. The BEX technique used in the decoding algorithm determines the computational and hardware implementation complexities of the various algorithms. The traditional BEX technique is based on the MRC algorithm which is inherently sequential in nature and needs k MMULT cycles and k MADD cycles to produce the residue in the extended moduli. An MMULT cycle is defined as the time taken to perform a modular multiplication. Similarly, a MADD cycle is defined as the time taken to perform a modular addition/subtraction. Recently, a CRT-I based fast BEX technique was proposed in Shenoy and Kumaresan [1989] which uses an additional redundant modulus. This method requires only one MMULT cycle and $\lceil \log_2(k+1) \rceil + 1$ MADD cycles and is superior to the MRC method from the point of view of hardware requirements. Here $\lceil a \rceil$ denotes the smallest integer greater than or equal to a.

In this section, a superfast algorithm for correcting single errors in the RNS is developed with a slight increase in redundancy. The hardware complexity of the superfast algorithm for single residue error correction is $O(k)$ while the hardware complexity of the previously known algorithms in Krishna et al. [1992] and Jenkins [1983] is $O(k^2)$. The latency of the superfast algorithm is 7 MMULT cycles and $\lceil \log_2 k \rceil + 6$ MADD cycles while the latency of the algorithm in Krishna et al. [1992] and Jenkins [1983] is $k+2$ MMULT cycles and $k+3$ MADD cycles, and $k^2 + 2k + 1$ MMULT cycles and $k^2 + 2k + 1$ MADD cycles, respectively.

Let $\underline{x} = [x_1 \, x_2 \, \ldots x_k]$ be the residue representation of an integer $X, 0 \leq X < M$, where $M = \prod_{i=1}^{k} m_i$. According to the CRT-I, given $[x_1 \, x_2 \, \ldots x_k]$, the corresponding integer X, can be computed using

$$X \equiv \sum_{i=1}^{k} x_i T_i M_i \pmod{M} \qquad (8.4.1)$$

$$\equiv \sum_{i=1}^{k} x_i' M_i \pmod{M}, \qquad (8.4.2)$$

where $M_i = M/m_i$, $T_i \equiv M_i^{-1} \pmod{m_i}$ is the multiplicative inverse of M_i modulo m_i, and

$$x_i' \equiv x_i T_i \pmod{m_i}. \qquad (8.4.3)$$

Equivalently, for some integer δ_x, $0 \le \delta_x < k$, (8.4.2) can alternatively be stated as

$$X = \sum_{i=1}^{k} x_i' M_i - \delta_x M. \qquad (8.4.4)$$

8.4.1 A Superfast Algorithm for Single Error Correction

In this section, a superfast algorithm for single-error correction in the RNS is derived based on the use of three redundant moduli, $\{m_{k+1}, m_{k+2}, m_{k+3}\}$, or the use of two redundant moduli, $\{m_{k+1}, m_{k+2}\}$, and a code generator A, where the moduli and the code generator satisfy the conditions

$$\begin{cases} m_{k+2} > m_{k+1} > m_k > \cdots > m_2 > m_1 \\ A, m_{k+3} \ge k. \end{cases} \qquad (8.4.5)$$

There are two encoding methods for this error control technique in the RNS. The first method is that the code-integer X, $X \in [0, M)$, is encoded to a $(k+3)$-tuple, $\underline{x} = [x_1\ x_2\ \cdots\ x_{k+2}\ x_{k+3}]$, with respect to the moduli $m_1, m_2, \ldots, m_{k+2}, m_{k+3}$. The second method is that the code-integer X in the range $[0, M)$ satisfies an additional constraint, $X \equiv 0 \pmod{A}$, $gcd(A, m_i) = 1$, $i = 1, 2, \ldots, k+2$. Clearly, the above constraint implies that every code-integer is a multiple of A. In this case, a code-integer is encoded to a $(k+2)$-tuple, $\underline{x} = [x_1\ x_2\ \cdots\ x_{k+2}]$, with respect to the moduli $m_1, m_2, \ldots, m_{k+2}$. The code obtained from the first encoding method is an RRNS code while the code obtained from the second encoding method is an RRNS-PC. In the following analysis, the decoding algorithm for the RRNS-PC can be derived by replacing m_{k+3} and y_{k+3} in the decoding algorithm for the RRNS by A and 0, respectively. Since the analysis of the decoding algorithm for the RRNS-PC is similar to the analysis of the decoding algorithm for the RRNS code, we only consider the decoding algorithm for RRNS code throughout this section.

Consider a received residue vector $\underline{y}$ corresponding to a code vector $\underline{x}$, $Y \longleftrightarrow \underline{y} = [y_1\ y_2\ \cdots\ y_{k+2}\ y_{k+3}]$ to be tested for errors occurred during transmission and/or computation. If no errors occur then $\underline{y} = \underline{x}$. If an error occurs in the ith residue digit, x_i, of the residue vector $\underline{x}$, then the received residue vector can be represented as in (7.2.18). According to the information part $[y_1\ y_2\ \cdots\ y_k]$ of $\underline{y}$, the corresponding altered information integer $\overline{Y} \longleftrightarrow [y_1\ y_2\ \cdots\ y_k]$ can be computed as follows:

$$\overline{Y} = \sum_{i=1}^{k} y_i' M_i - \delta_y M \qquad (8.4.6)$$

$$= \quad \tilde{Y} - \delta_y M, \tag{8.4.7}$$

where $y_i' \equiv y_i T_i \pmod{m_i}$, $0 \le \delta_y < k$, and

$$\tilde{Y} = \sum_{i=1}^{k} y_i' M_i. \tag{8.4.8}$$

The first step of the superfast algorithm for single error correction proposed here is to compute the quantities D_1, D_2 and D_3 using the congruences

$$D_r \equiv \tilde{Y} - y_{k+r} \pmod{m_{k+r}}, \ r = 1, 2, 3. \tag{8.4.9}$$

Next, the test quantities, called the modified syndromes, δ_1, δ_2 and δ_3 are computed as

$$\delta_r \equiv D_r M^{-1} \pmod{m_{k+r}}, \ r = 1, 2, 3, \tag{8.4.10}$$

where M^{-1} is the multiplicative inverse of M modulo m_{k+r}. A fundamental property of the modified syndromes is stated in the following Theorem.

THEOREM 8.3 *Under the assumption that no more than one error occurs and the moduli of a $(k+3, k)$ RRNS satisfy the condition in (8.4.5), we have the following three assertions:*

Assertion 1 : *A residue vector y is a codevector iff $0 \le \delta_1 = \delta_2 = \delta_3 < k$.*

Assertion 2 : *One of the information residue digits in y is in error iff $\delta_1 \ne \delta_2$, $\delta_2 \ne \delta_3$ and $\delta_3 \ne \delta_1$.*

Assertion 3 : *The first parity residue digit of y is in error iff $\delta_1 \ne \delta_2$, $\delta_1 \ne \delta_3$ and $0 \le \delta_2 = \delta_3 < k$. Similar statements can be made for the second and the third parity residue digits respectively.*

PROOF. **Assertion 1:** If there is no error in y, that is, $y = x$, then from (8.4.4) and (8.4.9) we have

$$\begin{aligned} D_r &\equiv \sum_{i=1}^{k} x_i' M_i - x_{k+r} \pmod{m_{k+r}} \\ &\equiv X + \delta_x M - x_{k+r} \pmod{m_{k+r}} \\ &\equiv \delta_x M \pmod{m_{k+r}}. \end{aligned}$$

Therefore, $\delta_x \equiv D_r M^{-1} \pmod{m_{k+r}}$, that is, $0 \le \delta_1 = \delta_2 = \delta_3 = \delta_x < k$. This proves the necessity. The sufficiency is proved by contradiction. We consider two situations, (i) one error in the information residue digits; (ii) one error in the parity residue digits.

Situation (i): Assume that one error is introduced in the pth information residue digit, then from (8.4.7) we have

$$\begin{aligned}
\tilde{Y} &= \overline{Y} + \delta_y M \\
&= X + E - aM + \delta_y M \\
&= X + e'_p M_p + (\delta_y - a)M \\
&= X + e'_p M_p + \delta'_y M, \quad (8.4.11)
\end{aligned}$$

where $E = e'_p M_p$, $e'_p \equiv e_p T_p \pmod{m_p}$, $0 < e_p, e'_p < m_p$, e_p being the error value. Also, $\delta'_y = \delta_y - a$. Here $a = 0$ if $X + E < M$, and $a = 1$ if $M \leq X + E < 2M$. Therefore,

$$\begin{aligned}
D_r &\equiv e'_p M_p + \delta'_y M \pmod{m_{k+r}} \\
&\equiv (e'_p + \delta'_y m_p)M_p \pmod{m_{k+r}}, \quad r = 1,2,3, \quad (8.4.12)
\end{aligned}$$

and

$$D_r M^{-1} m_p \equiv e'_p + \delta'_y m_p \pmod{m_{k+r}}, \quad r = 1,2,3. \quad (8.4.13)$$

If $0 \leq \delta_1 = \delta_2 = \delta_3 = \delta < k$, that is, $\delta \equiv D_r M^{-1} \pmod{m_{k+r}}$ for $r = 1,2,3$, then $e'_p + \delta'_y m_p - \delta m_p \equiv 0 \pmod{m_{k+1}m_{k+2}m_{k+3}}$. This is not possible as $|e'_p + \delta'_y m_p - \delta m_p| < km_p < m_{k+1}m_{k+2}m_{k+3}$.

Situation (ii): Assume that one error is present in the first parity residue digit (the case for one error in the second or third parity residue digit can be analyzed in a similar manner), that is, $y_i = x_i$, $i = 1,2,\ldots,k,k+2,k+3$, $y_{k+1} \equiv x_{k+1} + e_{k+1} \pmod{m_{k+1}}$, where e_{k+1} is the error value, $0 < e_{k+1} < m_{k+1}$. Then we have

$$\begin{cases}
D_1 \equiv \delta_x M - e_{k+1} \pmod{m_{k+1}} \\
D_2 \equiv \delta_x M \pmod{m_{k+2}} \\
D_3 \equiv \delta_x M \pmod{m_{k+3}}
\end{cases} \quad (8.4.14)$$

and

$$\begin{cases}
D_1 M^{-1} \equiv \delta_x - e_{k+1}M^{-1} \pmod{m_{k+1}} \\
D_2 M^{-1} \equiv \delta_x \pmod{m_{k+2}} \\
D_3 M^{-1} \equiv \delta_x \pmod{m_{k+3}}.
\end{cases} \quad (8.4.15)$$

If $0 \leq \delta_1 = \delta_2 = \delta_3 = \delta < k$, then $e_{k+1} = 0$. This contradicts the original assumption. This proves Assertion 1.

Assertion 2: The necessity is proved by contradiction. As in the proof of the sufficiency of Assertion 1, we assume that one error is introduced in the pth information residue digit. From (8.4.11) and (8.4.13), we have $\tilde{Y} = X + e'_p M_p + \delta'_y M$ and $D_r M^{-1} m_p \equiv e'_p + \delta'_y m_p \pmod{m_{k+r}}$, $r = 1,2,3$.

If any two of δ_1, δ_2 and δ_3 are equal, say $0 \leq \delta_1 = \delta_2 = \delta < k$ (other cases can be analyzed in a similar manner), then $e'_p + \delta'_y m_p - \delta m_p \equiv 0$ (mod $m_{k+1} m_{k+2}$). This is not possible as $|e'_p + \delta'_y m_p - \delta m_p| < k m_p < m_{k+1} m_{k+2}$. This proves the necessity. The sufficiency is proved in the following.

It follows from Assertion 1 that if $\delta_1 \neq \delta_2$, $\delta_2 \neq \delta_3$, and $\delta_3 \neq \delta_1$, then there must be one residue digit in error. If one error takes place in the first parity residue digit (the case for the second or third parity residue digit in error can be analyzed in a similar manner), then from (8.4.15) we know that $0 \leq \delta_2 = \delta_3 = \delta_x < k$. This contradicts the original assumption that $\delta_2 \neq \delta_3$. Therefore, one error must occur in the information digits. This proves Assertion 2.

Assertion 3: It follows from assertions 1 and 2 that one of the parity residue digits is in error *iff* one of the following three conditions is satisfied.

Condition 1: $\delta_1 \neq \delta_2$, $\delta_1 \neq \delta_3$ and $0 \leq \delta_2 = \delta_3 < k$.

Condition 2: $\delta_2 \neq \delta_1$, $\delta_2 \neq \delta_3$ and $0 \leq \delta_1 = \delta_3 < k$.

Condition 3: $\delta_3 \neq \delta_1$, $\delta_3 \neq \delta_2$ and $0 \leq \delta_1 = \delta_2 < k$.

Assuming that one error is introduced in the first parity residue digit (the case for the second or third parity residue digit in error can be analyzed in a similar manner), then from (8.4.14) and (8.4.15), $0 \leq \delta_2 = \delta_3 = \delta_x < k$ and $\delta_x \neq \delta_1$, where $\delta_1 \equiv \delta_x - e_{k+1} M^{-1}$ (mod m_{k+1}). This proves the necessity. The sufficiency is proved in the following.

If Condition 1 is satisfied (other conditions can be analyzed in a similar manner) and one error takes place in the second or third parity residue digit (say the third parity residue digit), then

$$\begin{cases} D_1 \equiv \delta_x M & (\text{mod } m_{k+1}) \\ D_2 \equiv \delta_x M & (\text{mod } m_{k+2}) \\ D_3 \equiv \delta_x M - e_{k+3} & (\text{mod } m_{k+3}), \end{cases}$$

$$\begin{cases} D_1 M^{-1} \equiv \delta_x & (\text{mod } m_{k+1}) \\ D_2 M^{-1} \equiv \delta_x & (\text{mod } m_{k+2}) \\ D_3 M^{-1} \equiv \delta_x - e_{k+3} M^{-1} & (\text{mod } m_{k+3}), \end{cases}$$

and $\delta_1 = \delta_2 = \delta_x$, where e_{k+3} is the error value. This contradicts the original assumption that $\delta_1 \neq \delta_2$. Therefore, an error must be present in the first parity residue digit. This proves Assertion 3.

From this theorem, we observe that the errors in the information digits and the errors in the parity digits affect the modified syndromes differently, and therefore, must be treated differently. If one of the conditions in Assertion 3 is satisfied, then the error-correcting algorithm declares that the

r_1th parity digit is in error and the estimate of the erroneous digit is given by

$$\hat{x}_{k+r_1} = \sum_{i=1}^{k} y_i' M_i - \delta_{r_2} M \quad (\text{mod } m_{k+r_1}) \qquad (8.4.16)$$

if $\delta_{r_1} \neq \delta_{r_2}$, $\delta_{r_1} \neq \delta_{r_3}$ and $0 \leq \delta_{r_2} = \delta_{r_3} < k$, $1 \leq r_1, r_2, r_3 \leq 3$. If Assertion 2 occurs, a procedure to determine the error location and the error value for single errors is described in the following section.

8.5 A Procedure for Single Error Correction

Assume that only the pth information digit is in error, then from (8.4.12) and (8.4.13), we have

$$D_r \equiv (e_p' + \delta_y' m_p) M_p \quad (\text{mod } m_{k+r}), \quad r = 1, 2, 3. \qquad (8.5.1)$$

This leads to

$$\delta_r m_p \equiv D_r M^{-1} m_p \equiv e_p' + \delta_y' m_p \quad (\text{mod } m_{k+r}), \quad r = 1, 2, 3, \qquad (8.5.2)$$

for $0 \leq \delta_y' < k$; and

$$- \delta_r m_p \equiv -(e_p' + \delta_y' m_p) \quad (\text{mod } m_{k+r}), \quad r = 1, 2, 3, \qquad (8.5.3)$$

for $\delta_y' = -1$. Given δ_1, δ_2 and δ_3, based on (8.5.1-8.5.3), a procedure to determine p (error location) and e_p (error value) can be outlined as follows:

Step A: For $j = 1, 2, \ldots, k$, solve the congruences

$$\begin{cases} e_j^{(1)} + \delta_y^{(1)} m_j = E_j' \equiv E_j^{(1)} \quad (\text{mod } m_{k+1}) \\ e_j^{(1)} + \delta_y^{(1)} m_j = E_j' \equiv E_j^{(3)} \quad (\text{mod } m_{k+3}), \end{cases} \qquad (8.5.4)$$

$$\begin{cases} e_j^{(2)} + \delta_y^{(2)} m_j = E_j'' \equiv E_j^{(2)} \quad (\text{mod } m_{k+2}) \\ e_j^{(2)} + \delta_y^{(2)} m_j = E_j'' \equiv E_j^{(3)} \quad (\text{mod } m_{k+3}), \end{cases} \qquad (8.5.5)$$

where

$$E_j^{(r)} \equiv \delta_r m_j \quad (\text{mod } m_{k+r}), \quad r = 1, 2, 3, \qquad (8.5.6)$$

to obtain the values E_j' and E_j''. It is possible that $\delta_y' = -1$. Therefore, we also compute the quantities $\check{E}_j'$ and $\check{E}_j''$ from E_j' and E_j'' using

$$\check{E}_j' = m_{k+1} m_{k+3} - E_j', \qquad (8.5.7)$$

$$\check{E}_j'' = m_{k+2} m_{k+3} - E_j''. \qquad (8.5.8)$$

Step B: Check if the solutions to (8.5.4) and (8.5.5) as computed in Step A are consistent, that is, one of the following two conditions is satisfied:

Condition (a) $0 \leq E_j' = E_j'' < km_j$ (or $0 \leq e_j^{(1)} = e_j^{(2)} < m_j$ and $0 \leq \delta_y^{(1)} = \delta_y^{(2)} < k$).

Condition (b) $0 \leq \check{E}_j' = \check{E}_j'' < m_j$ (or $0 \leq e_j^{(1)} = e_j^{(2)} < m_j$ and $\delta_y^{(1)} = \delta_y^{(2)} = -1$).

Let us say that either Condition (a) or Condition (b) is satisfied for $j = l$. Then, the lth information digit is declared to be in error, the value of the error being

$$e_l \equiv e_l^{(1)} M_l \quad (\text{mod } m_l), \tag{8.5.9}$$

where $e_j^{(1)}$ is obtained from the congruence, $e_j^{(1)} \equiv E_j' \pmod{m_j}$. The estimate of the lth residue digit is obtained by

$$\hat{x}_l \equiv y_l - e_l \quad (\text{mod } m_l). \tag{8.5.10}$$

If the corresponding value of $\delta_y' = \delta_y^{(1)} = \delta_y^{(2)}$ is required for further analysis, then it may be computed using

$$\delta_y' = \delta_y^{(1)} \equiv (E_j' - e_j^{(1)})m_j^{-1} \quad (\text{mod } m_{k+1}m_{k+3}),$$

if $E_j' = E_j''$ is a consistent solution; and

$$-\delta_y' = -\delta_y^{(1)} \equiv (\check{E}_j' + e_j^{(1)})m_j^{-1} \quad (\text{mod } m_{k+1}m_{k+3}),$$

if $\check{E}_j' = \check{E}_j''$ is a consistent solution. However, the value of δ_y' is not required for the error correcting algorithm. Therefore, the above equations are ignored in the hardware realization.

It is clear from (8.5.2, 8.5.4, 8.5.5) that if one error takes place in the pth information digit such that $0 \leq \delta_y' < k$, then for $j = p$, $0 \leq E_j' = E_j'' = (e_p' + \delta_y' m_p) < km_j$ (or $0 \leq e_j^{(1)} = e_j^{(2)} = e_p' < m_j$, $0 \leq \delta_y^{(1)} = \delta_y^{(2)} = \delta_y' < k$) and $l = p$. Similarly, if one error takes place in the pth information digit such that $\delta_y' = -1$, then for $j = p$, $0 \leq \check{E}_j' = \check{E}_j'' = -(e_p' + \delta_y' m_p) < m_j$ (or $0 \leq e_j^{(1)} = e_j^{(2)} = e_p' < m_j$, $\delta_y^{(1)} = \delta_y^{(2)} = \delta_y' = -1$) and $l = p$ from (8.5.3-8.5.5). However, it remains to be shown that if one error takes place in the pth information digit, then for $j \neq p$, $E_j' \neq E_j''$ and $\check{E}_j' \neq \check{E}_j''$, $0 \leq E_j', E_j'' < km_j$, $0 \leq \check{E}_j', \check{E}_j'' < m_j$, and $-1 \leq \delta_y^{(1)}, \delta_y^{(2)} < k$. Under the same constraint as in (8.4.5) on the form of the moduli, this result is established in the following theorem.

THEOREM 8.4 *If the moduli satisfy the constraints in* (8.4.5), *then for* $j \neq p$, $e_j^{(1)} \neq e_j^{(2)}$, $0 < e_j^{(1)}, e_j^{(2)} < m_j$ *and* $-1 \leq \delta_j^{(1)}, \delta_j^{(2)} < k$, *where the* pth *information digit is received in error.*

PROOF. First, we prove that for $-1 \leq \delta_y^{(1)}, \delta_y^{(2)} < k$, if $E_j' = E_j'' < km_j$ (the case for $\check{E}_j' = \check{E}_j'' < m_j$ can be analyzed in a similar manner), then $\delta_y^{(1)} = \delta_y^{(2)}$. It is clear from (8.5.4, 8.5.5) that if $E_j' = E_j'' < km_j$ and $\delta_y^{(1)} \neq \delta_y^{(2)}$, then $e_j^{(1)} = e_j^{(2)}$ and $(\delta_y^{(2)} - \delta_y^{(1)})m_j \equiv 0 \pmod{m_{k+1}m_{k+2}m_{k+3}}$. This is not possible as $|(\delta_y^{(2)} - \delta_y^{(1)})m_j| \leq km_j < m_{k+1}m_{k+2}m_{k+3}$. This proves the above statement. Now, for $j \neq p$, if $0 \leq E_j' = E_j'' < km_j$ (or $0 \leq \check{E}_j' = \check{E}_j'' < m_j$), and $-1 \leq \delta_y^{(1)} = \delta_y^{(2)} = \delta_y'' < k$, that is, $0 \leq e_j^{(1)} = e_j^{(2)} < m_j$, then

$$e_j^{(1)} + \delta_y'' m_j \equiv \delta_r m_j \pmod{m_{k+r}}, \quad r = 1, 2, 3; \tag{8.5.11}$$

or

$$e_j^{(1)} M_j + \delta_y'' M \equiv D_r \pmod{m_{k+r}}, \quad r = 1, 2, 3. \tag{8.5.12}$$

Comparing (8.5.12) to (8.5.1) we get

$$e_p' M_p + \delta_y' M - e_j^{(1)} M_j - \delta_y'' M \equiv 0 \pmod{m_{k+1}m_{k+2}m_{k+3}},$$

or

$$(e_p' m_j + \delta_y' m_j m_p - e_j^{(1)} m_p - \delta_y'' m_j m_p)\frac{M}{m_j m_p} \equiv 0 \pmod{m_{k+1}m_{k+2}m_{k+3}}.$$

This is not possible as

$$|e_p' m_j + \delta_y' m_j m_p - e_j^{(1)} m_p - \delta_y'' m_j m_p| < km_j m_p < m_{k+1}m_{k+2}m_{k+3}.$$

This proves the theorem.

Based on the concepts developed above and the assumption that the moduli satisfy the constraints stated in (8.4.5), the superfast algorithm to correct a single error is given in Table 8.5. In the following, we present several examples to illustrate the usage of the algorithm.

EXAMPLE 8.7 Consider a (9,6) RRNS code based on the moduli $m_1 = 23$, $m_2 = 25$, $m_3 = 27$, $m_4 = 29$, $m_5 = 31$, $m_6 = 32$, $m_7 = 37$, $m_8 = 41$ and $m_9 = 7$, where m_7, m_8 and m_9 are the redundant moduli. Clearly,

Table 8.5: The Superfast Algorithm for Correcting Single Error With 3 Redundant Moduli

Step 1:	According to the received vector, compute D_1, D_2, D_3, $\delta_1, \delta_2, \delta_3$ using (8.4.8-8.4.10).
Step 2:	(1). If $0 \leq \delta_1 = \delta_2 = \delta_3 < k$, then declare no error. Stop.
	(2). If $\delta_1 \neq \delta_2$, $\delta_1 \neq \delta_3$, and $0 \leq \delta_2 = \delta_3 < k$,
	then declare the first parity digit in error.
	Correct it using (8.4.16). Stop.
	(3). If $\delta_2 \neq \delta_1$, $\delta_2 \neq \delta_3$, and $0 \leq \delta_1 = \delta_3 < k$,
	then declare the second parity digit in error.
	Correct it using (8.4.16). Stop.
	(4). If $\delta_3 \neq \delta_1$, $\delta_3 \neq \delta_2$, and $0 \leq \delta_1 = \delta_2 < k$,
	then declare the third parity digit in error.
	Correct it using (8.4.16). Stop.
	(5). If $\delta_1 \neq \delta_2 \neq \delta_3$, then go to next step.
Step 3:	Let $j = 1$.
Step 4:	Perform single-error consistency-checking for the modulus m_j:
	(1). Compute $E_j^{(1)}, E_j^{(2)}, E_j^{(3)}, E_j', E_j'', \breve{E}_j', \breve{E}_j''$.
	(2). Check the consistency of the solutions,
	(a) $0 \leq E_j' = E_j'' < km_j$ or (b) $0 \leq \breve{E}_j' = \breve{E}_j'' < m_j$.
	If it is consistent, then go to Step 6.
	If it is inconsistent, then go to next step.
Step 5:	Let $j = j + 1$. Go to Step 4 for $j \leq k$.
	For $j = k + 1$, go to Step 7.
Step 6:	Declare one error in the jth position.
	Correct it using (8.5.9, 8.5.10) and stop.
Step 7:	Declare more than one error detected and stop.

$m_8 m_7 > m_6 m_5$ and $m_9 > k = 6$, therefore, the constraint in (8.4.5) is satisfied. $M = \prod_{i=1}^{6} m_i = 446,623,200$. Let $X = 391,804,672 \longleftrightarrow \underline{x} = $ [17 22 4 27 12 0 17 21 0]. Using the relation $x_i' \equiv x_i T_i \pmod{m_i}$, we have $[x_1' \; x_2' \; \ldots \; x_6'] = $ [22 24 26 28 1 0]. In this case,

$$X = \sum_{i=1}^{6} x_i' M_i - \delta_x M, \; \delta_x = 3.$$

Assume that no error occurs and $Y \longleftrightarrow \underline{y} = \underline{x}$. Following the decoding algorithm, we compute $D_1 \equiv 33 \pmod{37}$, $D_2 \equiv 14 \pmod{41}$ and $D_3 \equiv 6 \pmod 7$, and then compute $\delta_1 \equiv 3 \pmod{37}$, $\delta_2 \equiv 3 \pmod{41}$, and $\delta_3 \equiv 3 \pmod 7$. Since $0 \le \delta_1 = \delta_2 = \delta_3 < k$, we declare that no errors have occurred.

EXAMPLE 8.8 In the $(9,6)$ RRNS of Example 8.7, let $X = 391,804,672 \longleftrightarrow \underline{x} = $ [17 22 4 27 12 0 17 21 0]. Assume that one error takes place in the first parity digit, that is, $Y \longleftrightarrow \underline{y} = $ [17 22 4 27 12 0 15 21 0]. Following the decoding algorithm, we compute $D_1 \equiv 35 \pmod{37}$, $D_2 \equiv 14 \pmod{41}$ and $D_3 \equiv 6 \pmod 7$, and then compute $\delta_1 \equiv 20 \pmod{37}$, $\delta_2 \equiv 3 \pmod{41}$, and $\delta_3 \equiv 3 \pmod 7$. Since $\delta_1 \ne \delta_2$, $\delta_1 \ne \delta_3$, and $0 \le \delta_2 = \delta_3 = 3 < k$, we declare an error in the first parity digit, the estimate of the parity digit being

$$\widehat{x}_{k+1} \equiv \sum_{i=1}^{6} y_i' M_i - \delta_2 M \pmod{m_{k+1}} \equiv 17.$$

EXAMPLE 8.9 In the $(9,6)$ RRNS of Example 8.7, let $X = 391,804, 672 \longleftrightarrow \underline{x} = $ [17 22 4 27 12 0 17 21 0]. Assume that one error takes place in the third information digit, that is, $Y \longleftrightarrow \underline{y} = $ [17 22 0 27 12 0 17 21 0]. Following the decoding algorithm, we compute $D_1 \equiv 32 \pmod{37}$, $D_2 \equiv 9 \pmod{41}$ and $D_3 \equiv 2 \pmod 7$, and then compute $\delta_1 \equiv 13 \pmod{37}$, $\delta_2 \equiv 40 \pmod{41}$ and $\delta_3 \equiv 1 \pmod 7$. Since $\delta_1 \ne \delta_2$, $\delta_2 \ne \delta_3$ and $\delta_3 \ne \delta_1$, we declare that at least one error has occurred in the information digits. For $j = 1, 2, \ldots, 6$, we compute $E_j^{(1)}$, $E_j^{(2)}$, $E_j^{(3)}$, E_j', E_j'', $\check{E}_j'$ and $\check{E}_j''$. For $j = 3$, we get $E_3^{(1)} \equiv 18 \pmod{37}$, $E_3^{(2)} \equiv 14 \pmod{41}$, $E_3^{(3)} \equiv 6 \pmod 7$, $E_3' = 55$ and $E_3'' = 55$. We note here that E_j' and E_j'', $j = 1, 2, \ldots, 6$, are computed by solving the congruences in (8.5.4) and (8.5.5) respectively. Since $0 < E_3' = E_3'' < km_3$, we declare one error in the third information digit. The corresponding values $e_3^{(1)} = e_3^{(2)}$ and $\delta_y^{(1)} = \delta_y^{(2)}$ may be computed using $e_3^{(1)} = e_3^{(2)} \equiv E_3' \pmod{m_3} \equiv 1$ and $\delta_3^{(1)} = \delta_3^{(2)} \equiv (E_3' - e_3^{(1)})m_3^{-1} \pmod{m_{k+1}m_{k+3}} \equiv 2$. The estimate of the third information digit is given by $\widehat{x}_3 \equiv y_3 - e_3^{(1)} M_3 \pmod{m_3} \equiv 4$.

EXAMPLE 8.10 In the $(9, 6)$ RRNS of Example 8.7, let $X = 446{,}078{,}590$ $\longleftrightarrow \underline{x} = [7\ 15\ 7\ 10\ 29\ 30\ 4\ 25\ 6]$. Using the relation $x_i' \equiv x_i T_i \pmod{m_i}$, we have $[x_1'\ x_2'\ \ldots\ x_6'] = [5\ 5\ 5\ 5\ 5\ 2]$. In this case,

$$X = \sum_{i=1}^{6} x_i' M_i - \delta_x M,\ \delta_x = 0.$$

Assume that one error takes place in the sixth information digit, that is, $Y \longleftrightarrow \underline{y} = [7\ 15\ 7\ 10\ 29\ 15\ 4\ 25\ 6]$. Following the decoding algorithm, we compute $D_1 \equiv 17 \pmod{37}$, $D_2 \equiv 40 \pmod{41}$ and $D_3 \equiv 3 \pmod 7$, and then compute $\delta_1 \equiv 15 \pmod{37}$, $\delta_2 \equiv 32 \pmod{41}$ and $\delta_3 \equiv 5 \pmod 7$. Since $\delta_1 \neq \delta_2$, $\delta_2 \neq \delta_3$ and $\delta_3 \neq \delta_1$, we declare that at least one error has occurred in the information digits. For $j = 1, 2, \ldots, 6$, we compute $E_j^{(1)}$, $E_j^{(2)}$, $E_j^{(3)}$, E_j', E_j'', $\check{E}_j'$ and $\check{E}_j''$. For $j = 6$, $E_6^{(1)} \equiv 36 \pmod{37}$, $E_6^{(2)} \equiv 40 \pmod{41}$, $E_6^{(3)} \equiv 6 \pmod 7$, $E_6' = 258$, $E_6'' = 258$, $\check{E}_6'' = m_{k+1} m_{k+3} - E_6' = 1$ and $\check{E}_6'' = m_{k+2} m_{k+3} - E_6'' = 1$. Since $0 < \check{E}_6' = \check{E}_6'' < m_6 = 32$, we declare an error in the sixth information digit. The corresponding values $e_6^{(1)} = e_6^{(2)}$ and $-\delta_y^{(1)} = -\delta_y^{(2)}$ may be computed using $e_6^{(1)} = e_6^{(2)} \equiv m_6 - (\check{E}_6' \bmod m_6) \equiv 31$ and $-\delta_6^{(1)} = -\delta_6^{(2)} \equiv (\check{E}_6' + e_6^{(1)}) m_6^{-1} \pmod{m_{k+1} m_{k+3}} \equiv 1$. The estimate of the sixth information digit is given by $\widehat{x}_6 \equiv y_6 - e_6^{(1)} M_6 \pmod{m_6} \equiv 30$.

8.6 A Hardware Design for the Algorithms

In this section, we design architectures for the hardware implementation of two algorithms for single error correction in RRNS, one given in Table 8.1 and the second given in Table 8.5.

8.6.1 A Hardware Design for the Fast Algorithm

Figures 8.3 and 8.4 show an architecture for hardware implementation of the fast algorithm described in Table 8.1. Six ($k = 6$) nonredundant moduli m_1, m_2, ..., m_6, are used to describe the hardware architecture. This can be easily extended for other values of the number of nonredundant moduli. There are six steps for the fast algorithm as given in Table 8.1. First, the computation of syndromes is performed using the MRC based BEX method whose hardware architecture is shown in Figure 8.3. The right-hand triangular part of Figure 8.3 is used for computing the mixed radix digits a_1, a_2, ..., a_6, based on the MRC algorithm. The computation of $\bar{y}_{k+1}$ and $\bar{y}_{k+2}$ as shown in the two left-hand columns of Figure 8.3 is overlapped with the computation of the mixed radix digits, by making use of the mixed radix digits as soon as they are available. Each building block in the successive computational stages consists of an MMULT and an MADD, except the blocks in the last two stages. The one next to the last stage performs modulo subtraction to obtain syndrome digits $\Delta_r \equiv \bar{y}_{k+r} - y_{k+r} \pmod{m_{k+r}}$. The last stage performs comparison (COMP) with a known constant 0 to check how many syndromes are zero (Step 2 of Table 8.1). The output of COMP indicates the status of equal or unequal, that is, the result of comparing the two inputs. The computational block consisting of an MMULT and an MADD can be implemented by using a lookup table stored in the read-only memories (ROM) or by an adder and small-sized ROM. In general, the former is preferable for small moduli, and the latter is preferable for large moduli. The above process for obtaining the syndromes requires k MMULT cycles and $k+1$ MADD cycles. All of the consistency-checking operations (Step 4 of Table 8.1) for the moduli m_j, $j = 1, 2, \ldots, k$, are independent of each other, and therefore, can be performed in parallel. The hardware realization of the consistency-checking operations uses k identical blocks as shown in Figure 8.4. The two parity channels are also independent, and therefore, can be implemented in parallel using two identical blocks. The computational blocks perform either MMULT or MADD.

The COMP shown in Figure 8.4 is used for checking whether $e_j^{(1,1)} = e_j^{(1,2)} < m_j$ or $e_j^{(2,1)} = e_j^{(2,2)} < m_j$. Based on the consistency-checking scheme, the outputs of COMP are used as entries to the control logical circuit (CONTROL LOGIC) to generate the control variables (CONTROL) for the 2:1 multiplexers (MUX). The 2:1 MUX consists of gate arrays which control the direction of data flow. The COMP and CONTROL LOGIC can be implemented using gate arrays as well. The above consistency-checking process requires 2 MMULT cycles and 2 MADD cycles. Note that the time taken by MUX and COMP is negligible as compared to the time taken by the MMULT and MADD in the computational blocks. Also note that in the pipelined architecture a number of latches may have to be provided between two successive computational stages.

8.6.2 A Hardware Design for the Superfast Algorithm

Figures 8.5 and 8.6 show an architecture for the hardware implementation of the superfast algorithm developed in this chapter. There are seven steps for the superfast algorithm as given in Table 8.5. First, the modified syndromes

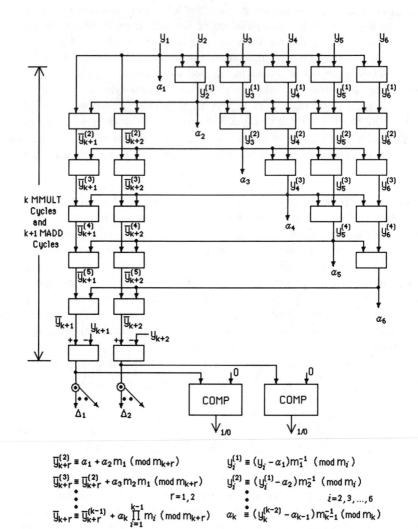

$$\overline{y}_{k+r}^{(2)} \equiv \alpha_1 + \alpha_2 m_1 \ (\text{mod } m_{k+r}) \qquad y_i^{(1)} \equiv (y_i - \alpha_1) m_1^{-1} \ (\text{mod } m_i)$$

$$\overline{y}_{k+r}^{(3)} \equiv \overline{y}_{k+r}^{(2)} + \alpha_3 m_2 m_1 \ (\text{mod } m_{k+r}) \qquad y_i^{(2)} \equiv (y_i^{(1)} - \alpha_2) m_2^{-1} \ (\text{mod } m_i)$$

$$\vdots \qquad\qquad r = 1, 2 \qquad\qquad \vdots$$

$$\qquad\qquad\qquad i = 2, 3, \dots, 6$$

$$\overline{y}_{k+r} \equiv \overline{y}_{k+r}^{(k-1)} + \alpha_k \prod_{i=1}^{k-1} m_i \ (\text{mod } m_{k+r}) \qquad \alpha_k \equiv (y_k^{(k-2)} - \alpha_{k-1}) m_{k-1}^{-1} \ (\text{mod } m_k)$$

Figure 8.3: A Hardware Implementation of the Fast Algorithm for Single Error Correction (Part I)

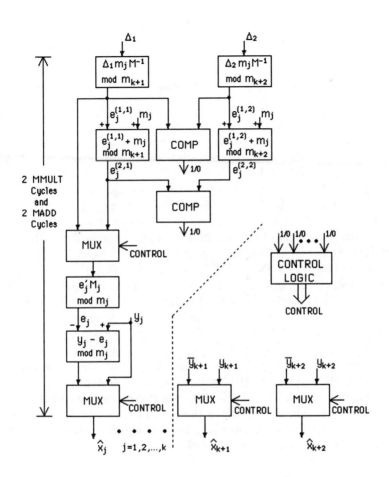

Figure 8.4: A Hardware Implementation of the Fast Algorithm for Single Error Correction (Part II)

δ_1, δ_2 and δ_3 are computed using

$$D_r \equiv \sum_{i=1}^{k} y_i' M_i - y_{k+r} \pmod{m_{k+r}}, \quad r = 1, 2, 3,$$

and

$$\delta_r \equiv D_r M^{-1} \pmod{m_{k+r}}, \quad r = 1, 2, 3.$$

It is clear that the steps in computations for δ_1, δ_2 and δ_3 are similar and independent of each other. Therefore, δ_1, δ_2 and δ_3 can be computed in parallel using three identical architectures as shown in Figure 8.5. Each building block in the successive computational stages consists of an MADD, except the first and the last two stages. Each block in the first stage consists of an MMULT. The one next to the last stage performs MMULT to obtain syndrome digits δ_1, δ_2 and δ_3. The last stage performs comparison to check how many syndromes are consistent (Step 2 of Table 8.5). The above process for obtaining the syndromes requires 2 MMULT cycles and $\lceil \log_2 k \rceil + 1$ MADD cycles. Second, all of the consistency-checking operations (Step 4 of Table 8.5) for the moduli m_j, $j = 1, 2, \ldots, k$, are independent of each other, and therefore, can be performed in parallel by using k identical blocks whose architectures are shown in Figure 8.6. The three parity channels are processed in parallel and have identical hardware realization. The COMP shown in Figure 8.6 is used for checking whether $0 \leq E_j' = E_j'' < km_j$ or $0 \leq \breve{E}_j' = \breve{E}_j'' < m_j$. The outputs of COMP are used as entries to the CONTROL LOGIC to generate the control variables for the 2:1 MUX. The 2:1 MUX is used for controlling the direction of data flow. Finally, as mentioned in the previous subsection, a number of latches may have to be provided between the successive computational stages in the pipelined architecture. The above consistency-checking process requires 5 MMULT cycles and 5 MADD cycles.

8.6.3 Comparison and Discussion

Under the assumption that the magnitude of the various moduli have the same order for the three algorithms to be compared, that is, similar hardware components are used, the comparison in terms of computational speed (latency in cycles), hardware requirements and computational complexity is shown in Table 8.6. From the computational speed point of view, the fast algorithm is superior to the algorithm in Jenkins [1983] by an order of magnitude. The fast algorithm requires $k + 2$ MMULT cycles and $k + 3$ MADD cycles while the Jenkins' algorithm requires $k^2 + 2k + 1$ MMULT cycles and $k^2 + 2k + 1$ MADD cycles. From Table 8.6, we note that the superfast

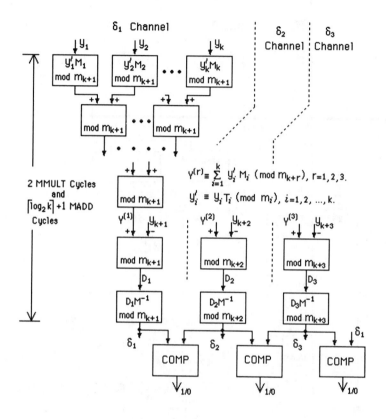

Figure 8.5: A Hardware Implementation of the Superfast Algorithm for Single Error Correction (Part I)

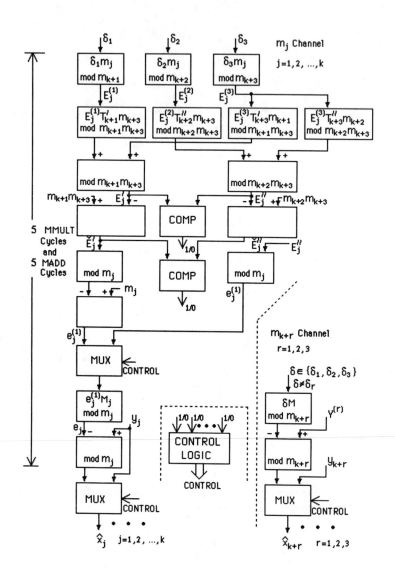

Figure 8.6: A Hardware Implementation of the Superfast Algorithm for Single Error Correction (Part II)

Table 8.6: The Latency, Hardware Requirements and Computational Complexity for Single Error Correcting Algorithms

	Algorithm in [Jenkins, 1983]	Fast Algorithm	Superfast Algorithm
Latency:			
MMULT cycles	$k^2 + 2k + 1$	$k + 2$	7
MADD cycles	$k^2 + 2k + 1$	$k + 3$	$\lceil \log_2 k \rceil + 6$
Hardware:			
MMULT	$\frac{3}{2}k^2 + \frac{5}{2}k$	$\frac{1}{2}k^2 + \frac{9}{2}k$	$13k + 6$
MADD	$\frac{3}{2}k^2 + \frac{5}{2}k$	$\frac{1}{2}k^2 + \frac{9}{2}k - 2$	$9k + 3$
MUX	$\frac{1}{2}k^2 + \frac{5}{2}k + 3$	$2k + 2$	$2k + 3$
COMP	2	$2k + 2$	$4k + 3$
Computational Complexity:			
MMULT	$\frac{1}{2}k^3 + \frac{3}{2}k^2 + 2k$	$\frac{1}{2}k^2 + \frac{7}{2}k - 1$	$10k + 5$
MADD	$\frac{1}{2}k^3 + \frac{3}{2}k^2 + 2k$	$\frac{1}{2}k^2 + \frac{7}{2}k + 1$	$7k + 2$

algorithm is even superior to the fast algorithm from both the speed and the hardware requirement point of view. The superfast algorithm requires 7 MMULT cycles and $\lceil \log_2 k \rceil + 6$ MADD cycles. The hardware requirement of the superfast algorithm is $13k + 6$ MMULT and $9k + 3$ MADD while the hardware requirement of the fast algorithm is $\frac{1}{2}k^2 + \frac{9}{2}k$ MMULT and $\frac{1}{2}k^2 + \frac{9}{2}k - 2$ MADD.

NOTES
The MRC has been extensively studied in the context of single error correcting RRNS in Barsi and Maestrini [1973] and Jenkins and Altman [1988]. It also forms the basis of the fast algorithms for single error correction in RRNS and RNS-PC. The superfast algorithm for single error correction is based on the direct computation of the modified syndromes from the

expression for the CRT-I reconstruction.

One of the reviewers of our papers brought Mandelbaum [1976], [1978], and [1984] to our attention that also deal with algorithms for error correction in RRNS. Subsequently, we discovered Barsi and Maestrini [1978]. The approach employed in these papers is completely different from our methodology. The algorithms developed here employ modulo operations exclusively, thereby avoiding the need to process large valued integers. The algorithms in these papers process large valued integers and employ continued fraction expansion along with Euclid's algorithm. A detailed and through comparison of these techniques with the techniques described here requires further analysis.

The modular algebraic structure of the RNS leads to modularity and parallelism in the hardware implementation for the RNS-based arithmetic processor, for example, see Garner [1959] and Szabo and Tanaka [1967]. Both modularity and parallelism are essential to fully utilize the very large scale integrated (VLSI) technology as pointed out in Bayoumi et al. [1984]. Several researchers have studied the application of RNS to digital signal processing systems which require large number of arithmetic operations with a view to designing high-speed fault-tolerant architectures. These include Soderstrand et al. [1986], Baranieka and Jullien [1980], Etzel and Jenkins [1980], Jenkins and Altman [1988], Jenkins [1983] and Huan et al. [1981].

The material presented in this chapter is derived from Krishna et al. [1992], Krishna and Sun [1993], and Sun et al. [1994].

8.7   Bibliography

[8.1] H.L. Garner, "The Residue Number System," *IRE Transactions Electronics Computers*, vol. EC-8, pp. 140-147, 1959.

[8.2] N.S. Szabo and R.I. Tanaka, *Residue Arithmetic and Its Applications to Computer Technology*, New York, McGraw-Hill, 1967.

[8.3] M.A. Bayoumi, G.A. Jullien, and W.C. Miller, "A VLSI Model for Residue Number System Architecture," *Integration, the VLSI journal*, vol. 2, pp. 191-211, 1984.

[8.4] M.A. Soderstrand, W.K. Jenkins, G.A. Jullien, and F.J. Taylor, *Modern Applications of Residue Number System Arithmetic to Digital Signal Processing*, New York, IEEE Press, 1986.

[8.5] A. Baraniecka and G.A. Jullien, "Residue Number System Implementations of Number Theoretic Transforms," *IEEE Transactions on*

Acoustics, Speech and Signal Processing, vol. ASSP-28, pp. 285-291, 1980.

[8.6] M.H. Etzel and W.K. Jenkins, "Redundant Residue Number Systems for Error Detection and Correction in Digital Filters," *IEEE Transactions on Acoustics, Speech and Signal Processing*, no. 5, pp. 538-544, 1980.

[8.7] W.K. Jenkins and E.J. Altman, "Self-Checking Properties of Residue Number Error Checkers Based on Mixed Radix Conversion," *IEEE Transactions on Circuits and Systems*, pp. 159-167, 1988.

[8.8] W.K. Jenkins, "The Design of Error Checkers for Self-Checking Residue Number Arithmetic," *IEEE Transactions on Computers*, vol. C-32, pp. 388-396, 1983.

[8.9] C.H. Huang, D.G. Peterson, H.E. Rauch, J.W. Teague, and D.F. Fraser, "Implementation of a Fast Digital Processor Using the Residue Number System," *IEEE Transactions on Circuits and Systems*, vol. CAS-28, pp. 32-38, 1981.

[8.10] S.S.-S. Yau and Y.-C. Liu, "Error Correction in Redundant Residue Number Systems," *IEEE Transactions on Computers*, vol. C-22, pp. 5-11, 1973.

[8.11] H. Krishna, K.-Y. Lin, and J.-D. Sun, "A Coding Theory Approach to Error Control in Redundant Residue Number Systems, Part I: Theory and Single Error Correction," *IEEE Transactions on Circuits and Systems*, vol. 39, pp. 8-17, 1992.

[8.12] H. Krishna and J.-D. Sun, "On Theory and Fast Algorithms for Error Correction in Residue Number System Product Codes," *IEEE Transactions on Computers*, vol. 42, pp. 840-852, 1993.

[8.13] J.-D. Sun, H. Krishna, and K.-Y. Lin, "A Superfast Algorithm for Single Error Correction in RRNS and Hardware Implementation," *Journal of VLSI Signal Processing*, vol. 6, pp. 259-269, 1993.

[8.14] A.P. Shenoy and R. Kumaresan, "Fast Base Extension Using a Redundant Modulus in RNS," *IEEE Transactions on Computers*, vol. C-38, pp. 292-297, 1989.

[8.15] D.M. Mandelbaum, "On a Class of Arithmetic Codes and a Decoding Algorithm," *IEEE Transactions on Information Theory*, vol. IT-22, pp. 85-88, 1976.

[**8.16**] D.M. Mandelbaum, "Further Results on Decoding Arithmetic Residue Codes," *IEEE Transactions on Information Theory*, vol. IT-24, pp. 643-644, 1978.

[**8.17**] D.M. Mandelbaum, "An Approach to an Arithmetic Analog of Berlekamp's Algorithm," *IEEE Transactions on Information Theory*, vol. IT-30, pp. 758-762, 1984.

[**8.18**] F. Barsi and P. Maestrini, "Improved Decoding Algorithm for Arithmetic Residue Codes," *IEEE Transactions on Information Theory*, vol. IT-24, pp. 640-643, 1978.

Chapter 9

Multiple Error Control in RRNS

In Chapter 7, a coding theory based mathematical framework for error control in the RNS was developed. Based on this framework, computationally efficient algorithms were described in Chapter 8 for correcting single residue errors in RRNS and RNS-PC. Also, a superfast algorithm along with its hardware implementation was described.

In this chapter, we extend the theory and present new algorithms for multiple error control in RRNS. The previously known algorithms are also extended for multiple error correction. The emphasis throughout this work is on deriving modulo arithmetic based computationally efficient algorithms for error control.

9.1 Errors and Consistency Checking

Consider an (n, k) RRNS based on the moduli $m_1, m_2, \ldots, m_k, m_{k+1}, \ldots, m_n$. The first k moduli form a set of information moduli, and their product represents the range, M, of the RNS,

$$M = \prod_{i=1}^{k} m_i. \tag{9.1.1}$$

The remaining $n - k$ moduli form the set of parity moduli that allows error detection and correction in the RRNS. Let M_R be the product of parity

227

moduli,

$$MR = \prod_{i=k+1}^{n} m_i. \tag{9.1.2}$$

Any integer X in the range $[0, M)$, called code-integer, is encoded to an n-tuple residue vector $\underline{x}$, called codevector,

$$X \longleftrightarrow \underline{x} = [x_1 \, x_2 \, \ldots \, x_n], \tag{9.1.3}$$

where

$$x_i \equiv X \pmod{m_i}, \, 0 \le x_i < m_i, \, i = 1, 2, \ldots, n. \tag{9.1.4}$$

According to Lemma 7.5, an (n, k) RRNS code having $d = \lambda + \beta + 1$ (or $n - k = \lambda + \beta$) is capable of correcting λ or fewer errors and simultaneously detecting β ($\beta > \lambda$) or fewer errors. For an (n, k) RRNS code, assume that there are c and g errors ($c + g \le \beta$) in the information and parity digits, respectively. Then the received residue vector can be represented as

$$Y \longleftrightarrow \underline{y} = [y_1 \, y_2 \, \ldots \, y_k \, y_{k+1} \, \ldots \, y_n]$$

$$\begin{aligned}
y_i &= x_i, \, 1 \le i \le n, \, i \ne i_\alpha, \, j_\nu, \\
&\quad \alpha = 1, 2, \ldots, c, \quad \nu = 1, 2, \ldots, g, \\
y_{i_\alpha} &\equiv x_{i_\alpha} + e_{i_\alpha} \pmod{m_{i_\alpha}}, \, 0 < e_{i_\alpha} < m_{i_\alpha}, \\
&\quad 1 \le i_\alpha \le k, \, \alpha = 1, 2, \ldots, c, \\
y_{j_\nu} &\equiv x_{j_\nu} + e_{j_\nu} \pmod{m_{j_\nu}}, \, 0 < e_{j_\nu} < m_{j_\nu}, \\
&\quad k + 1 \le j_\nu \le n, \, \nu = 1, 2, \ldots, g, \tag{9.1.5}
\end{aligned}$$

where $i_1, i_2, \ldots, i_c$ are the positions of errors in the information digits, the corresponding error values being $e_{i_1}, e_{i_2}, \ldots, e_{i_c}$, respectively; and $j_1, j_2, \ldots, j_g$ are the positions of errors in the parity, the corresponding error values being $e_{j_1}, e_{j_2}, \ldots, e_{j_g}$, respectively. The altered information number $\overline{Y}$ can be represented as

$$\begin{aligned}
\overline{Y} &\equiv X + E \pmod{M} \longleftrightarrow [y_1 \, y_2 \, \ldots \, y_k] \tag{9.1.6} \\
&\equiv [x_1 \, x_2 \, \ldots \, x_k] \\
&+ [0 \, \ldots \, 0 \, e_{i_1} \, 0 \, \ldots \, 0 \, e_{i_2} \, 0 \, \ldots \, 0 \, e_{i_c} \, 0 \, \ldots \, 0], \tag{9.1.7}
\end{aligned}$$

where

$$\begin{aligned}
X &\longleftrightarrow [x_1 \, x_2 \, \ldots \, x_k], \\
E &\longleftrightarrow [0 \, \ldots \, 0 \, e_{i_1} \, 0 \, \ldots \, 0 \, e_{i_2} \, 0 \, \ldots \, 0 \, e_{i_c} \, 0 \, \ldots \, 0].
\end{aligned}$$

Since $E \equiv 0 \pmod{m_i}$ for all $i \neq i_\alpha$, $\alpha = 1, 2, \ldots, c$, it is a multiple of all information moduli except $m_{i_1}, m_{i_2}, \ldots, m_{i_c}$. Therefore, E is an integer of the type,

$$
\begin{aligned}
E &= e' \frac{M}{m_{i_1} m_{i_2} \cdots m_{i_c}} \\
&\equiv e_{i_\alpha} \pmod{m_{i_\alpha}},
\end{aligned} \tag{9.1.8}
$$

where $0 < e' < m_{i_1} m_{i_2} \cdots m_{i_c}$. Based on $[y_1\ y_2\ \ldots\ y_k]$ and BEX as given in Table 7.1, we recompute the parity digits to get

$$
\overline{y}_{k+r} \equiv \overline{Y} \pmod{m_{k+r}}, \quad r = 1, 2, \ldots, n - k, \tag{9.1.9}
$$

and the syndromes

$$
\Delta_r \equiv \overline{y}_{k+r} - y_{k+r} \pmod{m_{k+r}}, \quad r = 1, 2, \ldots, n - k. \tag{9.1.10}
$$

Since both X and E in (9.1.6) are less than M, we consider the following two cases.

(i) $X + E < M$. In this case,

$$
\overline{Y} = X + E = X + e' \frac{M}{\prod_{\alpha=1}^{c} m_{i_\alpha}}, \tag{9.1.11}
$$

and

$$
\Delta_r \equiv e' \frac{M}{\prod_{\alpha=1}^{c} m_{i_\alpha}} \pmod{m_{k+r}}, \quad k + r \neq j_\nu, \tag{9.1.12}
$$

$$
\Delta_r \equiv e' \frac{M}{\prod_{\alpha=1}^{c} m_{i_\alpha}} - e_{k+r} \pmod{m_{k+r}}, \quad k + r = j_\nu, \tag{9.1.13}
$$

where $r = 1, 2, \ldots, n - k$, and $\nu = 1, 2, \ldots, g$.

(ii) $M \leq X + E < 2M$. In this case,

$$
\begin{aligned}
\overline{Y} = X + E - M &= X + e' \frac{M}{\prod_{\alpha=1}^{c} m_{i_\alpha}} - M \\
&= X + \left(e' - \prod_{\alpha=1}^{c} m_{i_\alpha} \right) \frac{M}{\prod_{\alpha=1}^{c} m_{i_\alpha}},
\end{aligned} \tag{9.1.14}
$$

and

$$
\Delta_r \equiv \left(e' - \prod_{\alpha=1}^{c} m_{i_\alpha} \right) \frac{M}{\prod_{\alpha=1}^{c} m_{i_\alpha}} \pmod{m_{k+r}}, \quad k + r \neq j_\nu, \tag{9.1.15}
$$

$$
\Delta_r \equiv \left(e' - \prod_{\alpha=1}^{c} m_{i_\alpha} \right) \frac{M}{\prod_{\alpha=1}^{c} m_{i_\alpha}} - e_{k+r} \pmod{m_{k+r}}, \quad k + r = j_\nu, \tag{9.1.16}
$$

where $r = 1, 2, \ldots, n - k$, and $\nu = 1, 2, \ldots, g$. A fundamental property of the syndrome digits is stated in the following theorem.

THEOREM 9.1 *Under the assumption that no more than β residues are in error for the RRNS with $d = \lambda + \beta + 1$ ($\lambda < \beta$), one of the following four cases occurs:*

Case 1: *If all the syndromes $\Delta_1, \ldots, \Delta_{n-k}$ are zero, then no residue is in error, and vice versa.*

For the following cases, let p be the number of non-zero syndromes.

Case 2: *If $p \leq \lambda$, then exactly p corresponding parity digits are in error, and all other residue digits are correct.*

Case 3: *If $\lambda + 1 \leq p \leq \beta$, then more than λ residue digits are in error.*

Case 4: *If $\beta + 1 \leq p \leq n - k$, then at least one of the information digits is in error.*

PROOF. The proofs of cases 1, 2 and 4 are based on the Theorem 7.3 and the observation that follows it. For Case 3, since k correct information residues uniquely determine the correct redundant residues, it is clear that if $\lambda + 1 \leq p \leq \beta$ and all the information residues are correct, then more than λ parity residues are in error. Now, suppose that c information residues are in error. We have $c \geq 1$, that is, at least one information residue is in error. Define $\Delta'_r \equiv \overline{y}_{k+r} - x_{k+r} \pmod{m_{k+r}}$. Then from (9.1.5-9.1.7), we have

$$X \longmapsto [x_1 \, x_2 \, \ldots \, x_k \, x_{k+1} \, \ldots \, x_n]$$

$$Y \longmapsto [y_1 \, y_2 \, \ldots \, y_k \, \overline{y}_{k+1} \, \ldots \, \overline{y}_n]$$

$$\overline{Y} - X \longmapsto [0 \, \ldots \, 0 \, e_{i_1} \, 0 \, \ldots \, 0 \, e_{i_2} \, 0 \, \ldots \, 0 \, e_{i_c} \, 0 \, \ldots \, 0 \, \Delta'_1 \, \Delta'_2 \, \ldots \, \Delta'_{n-k}].$$

It is obvious that if the received redundant residue digit y_{k+r} is erroneous, that is, $y_{k+r} \neq x_{k+r}$, then $\Delta'_r \neq \Delta_r$. For $\overline{Y} > X$ (the case for $\overline{Y} < X$ can be analyzed in an analogous manner), since $M > \overline{Y} - X$, by Lemma 7.5 we know that at least $d - c = \lambda + \beta + 1 - c$ of $\Delta'_1, \Delta'_2, \ldots, \Delta'_{n-k}$ are nonzero. Therefore, we have the following three assertions:

(1) If at most $\beta - c$ redundant residues are in error, then at least $\lambda + \beta + 1 - c - (\beta - c) = \lambda + 1$ syndromes are nonzero.

(2) If at most $\lambda + 1 - c$ redundant residues are in error, then at least β syndromes are nonzero.

(3) If at most $\lambda - c$ redundant residues are in error, then at least $\beta + 1$ syndromes are nonzero.

Now, for Case 3, if $\lambda + 1$ to β syndromes are nonzero and no more than λ residue digits are in error, then there is a contradiction with the assertion (3). So, this gives the proof for Case 3. This proves the theorem.

If Case 2 of Theorem 9.1 takes place, the new parity digit $\overline{y}_{k+r}$ is the correct value of the erroneous parity digit y_{k+r}. If Case 4 of Theorem 9.1 takes place, a procedure to determine error locations and error values for single/double errors is described in the following sections.

9.2 Single Error Correction: Continued

In this section, we continue our analysis for the case of single error correction and β $(\beta > 1)$ error detection in RRNS. By Lemma 7.5 an RRNS having $d = \lambda + \beta + 1$, $\lambda = 1$, and $\beta > \lambda$ can simultaneously correct single residue error and detect β $(\beta = d-2)$ errors. Assume that only the pth information digit is in error, then (9.1.11–9.1.16) become as follows.
(i) $X + E < M$. In this case,

$$\overline{Y} = X + e'\frac{M}{m_p}, \qquad (9.2.1)$$

and

$$\Delta_r \equiv e'\frac{M}{m_p} \pmod{m_{k+r}}, \ r = 1, 2, \ldots, n - k, \qquad (9.2.2)$$

where $0 < e' < m_p$.
(ii) $M \leq X + E < 2M$. In this case,

$$\overline{Y} = X + (e' - m_p)\frac{M}{m_p}, \qquad (9.2.3)$$

and

$$\Delta_r \equiv (e' - m_p)\frac{M}{m_p} \pmod{m_{k+r}}, \ r = 1, 2, \ldots, n - k, \qquad (9.2.4)$$

where $0 < e' < m_p$. Given $\Delta_1, \Delta_2, \ldots, \Delta_{n-k}$, based on (9.2.1–9.2.4) a procedure to correct single error and detect the presence of up to $d - 2$ errors can be outlined as follows:
For $j = 1, 2, \ldots, k$, solve the congruences

$$\Delta_r \equiv e_j^{(1,r)}\frac{M}{m_j} \pmod{m_{k+r}}, \ r = 1, 2, \ldots, n - k, \qquad (9.2.5)$$

and

$$\Delta_r \equiv (e_j^{(2,r)} - m_j)\frac{M}{m_j} \pmod{m_{k+r}}, \ r = 1, 2, \ldots, n-k, \qquad (9.2.6)$$

to obtain the values $e_j^{(1,r)}$ and $e_j^{(2,r)}$, $r = 1, 2, \ldots, n-k$. Compare to check if the solutions to (9.2.5) or (9.2.6) are consistent, that is, $e_j^{(1,1)} = e_j^{(1,2)} = \cdots = e_j^{(1,n-k)} < m_j$ or $e_j^{(2,1)} = e_j^{(2,2)} = \cdots = e_j^{(2,n-k)} < m_j$. If either of the two conditions is satisfied, for $j = l$, then the lth information digit is declared to be in error, the value of the error $e_l \equiv e\frac{M}{m_l} \pmod{m_l}$, where $e = e_j^{(1,1)}$ if $e_j^{(1,1)} = e_j^{(1,2)} = \cdots = e_j^{(1,n-k)}$; and $e = e_j^{(2,1)}$ if $e_j^{(2,1)} = e_j^{(2,2)} = \cdots = e_j^{(2,n-k)}$. The correct value of the lth residue digit is $(y_l - e_l) \bmod m_l$, and $\underline{y}$ is decoded to the codevector $\hat{\underline{x}}$, where

$$\hat{x}_j = y_j, \ j = 1, 2, \ldots, n, \ j \neq l, \qquad (9.2.7)$$
$$\hat{x}_l \equiv y_l - e_l \pmod{m_l}. \qquad (9.2.8)$$

It is clear from (9.2.1), (9.2.2) and (9.2.5) that if one error takes place in the pth information digit such that $X + E < M$, then for $j = p$, $e = e' = e_j^{(1,1)} = e_j^{(1,2)} = \cdots = e_j^{(1,n-k)} < m_j$ and $l = p$. Similarly, if one error takes place in the pth information digit such that $M \leq X + E < 2M$, then for $j = p$, $e = e' = e_j^{(2,1)} = e_j^{(2,2)} = \cdots = e_j^{(2,n-k)} < m_j$ and $l = p$ from (9.2.3), (9.2.4) and (9.2.6). However, it remains to be shown that if the following two cases occur: 1) only one error takes place in the pth information digit and $j \neq p$; and 2) more than one but less than $d - 1$ errors take place, then at least two of $\{e_j^{(1,r)}; \ r = 1, 2, \ldots, n-k\}$ and at least two of $\{e_j^{(2,r)}; \ r = 1, 2, \ldots, n-k\}$ are unequal. In other words, (9.2.5) and (9.2.6) do not give consistent solutions in these cases. Under an additional constraint on the form of the moduli, the result is established in the following theorem.

THEOREM 9.2 *If the moduli of an (n, k) RRNS code, are such that there do not exist integers n_j, n_C; $0 \leq n_j < m_j$, $0 \leq n_C < M_C = \prod_{\alpha=1}^{c} m_{i_\alpha}$, $1 \leq i_\alpha \leq k$, $1 \leq c \leq \beta$; that satisfy*

$$n_j M_C + n_C m_j = \prod_{i=1}^{c+1} m_{r_i}, \quad k < r_i \leq n, \ 1 \leq j \leq k, \qquad (9.2.9)$$

then for either of two cases, Case 1) only the pth information digit is received in error and $j \neq p$; and Case 2) more than one ($\leq \beta$) residue digits *are received in error, the solutions to (9.2.5) and (9.2.6) are inconsistent.*

PROOF. **Case 1):** For $X + E < M$ and $j \neq p$, if the solutions to (9.2.5) are consistent, that is, $e_j^{(1,1)} = e_j^{(1,2)} = \cdots = e_j^{(1,n-k)} = e < m_j$, then comparing (9.2.5) to (9.2.2) we obtain

$$\frac{M}{m_p m_j}(em_p - e'm_j) \equiv 0 \pmod{\prod_{r=1}^{n-k} m_{k+r}},$$

that is, $em_p - e'm_j$ is a multiple of $\prod_{r=1}^{n-k} m_{k+r}$. This is not possible as

$$|em_p - e'm_j| < \prod_{r=1}^{n-k} m_{k+r}.$$

Similarly, for $X + E < M$ and $j \neq p$, if $e_j^{(2,1)} = e_j^{(2,2)} = \cdots = e_j^{(2,n-k)} = e < m_j$, then comparing (9.2.6) to (9.2.2) we obtain

$$\frac{M}{m_p m_j}(e'm_j + (m_j - e)m_p) \equiv 0 \pmod{\prod_{r=1}^{n-k} m_{k+r}}.$$

This is not possible as

$$(e'm_j + (m_j - e)m_p) < 2m_p m_j < \prod_{r=1}^{n-k} m_{k+r}.$$

The case for $M \leq X + E < 2M$ can be analyzed in an analogous manner.

    Case 2): For $X + E < M$, if $e_j^{(1,1)} = e_j^{(1,2)} = \cdots = e_j^{(1,n-k)} = e$, then comparing (9.2.5) to (9.1.12) we obtain

$$\left(e \prod_{\alpha=1}^{c} m_{i_\alpha} - e'm_j\right)\frac{M}{m_j \prod_{\alpha=1}^{c} m_{i_\alpha}} \equiv 0 \pmod{\prod_{\substack{i=1 \\ r_i \neq j_\nu}}^{n-k-g} m_{r_i}},$$

where $k < r_i \leq n$, $\nu = 1, 2, \ldots, g$. This congruence can not hold as

$$\left|e \prod_{\alpha=1}^{c} m_{i_\alpha} - e'm_j\right| < m_j \prod_{\alpha=1}^{c} m_{i_\alpha} < \prod_{i=1}^{\beta+1-g} m_{r_i}.$$

For example, assume that the pth information digit and the first g parity digits are in error, and $\beta = c + g$, $c = 1$. Then the above congruence becomes

$$\frac{M}{m_j m_p}(em_p - e'm_j) \equiv 0 \pmod{m_{n-1} m_n}.$$

This is not possible as $|em_p - e'm_j| < m_{n-1}m_n$. In other words, at least two of $\{e_j^{(1,r)}; \; r = 1, \ldots, n-k\}$ are unequal. Similarly, for $X + E < M$, if $e_j^{(2,1)} = e_j^{(2,2)} = \cdots = e_j^{(2,n-k)} = e$, comparing (9.2.6) to (9.1.12) we obtain

$$(e'm_j + (m_j - e)) \prod_{\alpha=1}^{c} m_{i_\alpha}) \frac{M}{m_j \prod_{\alpha=1}^{c} m_{i_\alpha}} \equiv 0 \pmod{\prod_{\substack{i=1 \\ r_i \neq j_\nu}}^{n-k-g} m_{r_i}},$$

where $k < r_i \leq n, \; \nu = 1, 2, \ldots, g$. The above congruence can not hold as

$$(e'm_j + (m_j - e)) \prod_{\alpha=1}^{c} m_{i_\alpha}) < 2m_j \prod_{\alpha=1}^{c} m_{i_\alpha} < \prod_{i=1}^{\beta+1-g} m_{r_i}, \text{ for } c + g < \beta;$$

and (9.2.9) holds for $c + g = \beta$. The case for $M \leq X + E < 2M$ can also be analyzed in an analogous manner. This proves the theorem.

LEMMA 9.1 *If the moduli of RRNS satisfy the condition*

$$min\{m_{r_1} m_{r_2}\} > max\{2m_{i_1} m_{i_2} - m_{i_1} - m_{i_2}\}, \qquad (9.2.10)$$

where $k < r_1, r_2 \leq n$ and $1 \leq i_1, i_2 \leq k$, then for either of two cases, Case 1) only the pth information digit is received in error and $j \neq p$; and Case 2) more than one residue digit are received in error, the solutions to (9.2.5) and (9.2.6) are inconsistent.

PROOF. The condition is sufficient since, if the moduli satisfy (9.2.10), then (9.2.9) is satisfied trivially. This completes the proof.

It is interesting to note that this sufficient condition on the moduli for single error correction and multiple error detection is the same as the sufficient condition on the moduli for single error correction. Based on the concept developed above and under the assumption that the moduli m_i, $i = 1, 2, \ldots, n$, satisfy the condition in Theorem 9.2 or Lemma 9.1, the algorithm to correct a single error and simultaneously detect multiple errors in RRNS is given in Table 9.1. A flowchart for this algorithm is given in Figure 9.1.

EXAMPLE 9.1 Consider a $(10, 6)$ RRNS code based on the moduli $m_1 = 23$, $m_2 = 25$, $m_3 = 27$, $m_4 = 29$, $m_5 = 31$, $m_6 = 32$, $m_7 = 67$, $m_8 = 71$, $m_9 = 73$ and $m_{10} = 79$, where m_7, m_8, m_9 and m_{10} are the redundant moduli. Clearly, $m_7 m_8 > 2m_5 m_6 - m_5 - m_6$, therefore, the condition in Theorem 9.2 and Lemma 9.1 is satisfied. $M = \prod_{i=1}^{6} m_i = 446,623,200$. Let $X = 400,000,000$, then $\underline{x} = [8\,0\,22\,13\,25\,0\,17\,58\,4\,11]$. Since $0 \leq X < M$,

Table 9.1: An Algorithm for Single Error Correction and Multiple Error Detection in RRNS

Step 1:	According to the received vector, compute the syndromes using BEX as given in Table 7.1 and $\Delta_r \equiv \bar{y}_{k+r} - y_{k+r} \pmod{m_{k+r}}$, $r = 1, 2, \ldots, n-k$.
Step 2:	Check how many syndromes are zero: (1). If all the syndromes are zero, then no error occurs. Stop. (Theorem 9.1, Case 1) (2). If only one syndrome is non-zero, then the corresponding parity digit is declared to be in error. The estimate of erroneous parity digit is the corresponding $\bar{y}_{k+r}$. Stop. (Theorem 9.1, Case 2) (3). If $d-3$ to 1 syndromes are zero, then go to Step 6. (Theorem 9.1, Case 3) (4). If all the syndromes are non-zero, then go to Step 3.
Step 3:	Let $j = 1$.
Step 4:	Perform the single-error consistency-checking for the nonredundant modulus m_j: (1). Compute $e_j^{(1,1)}, e_j^{(1,2)}, \ldots, e_j^{(1,n-k)}$ using (9.2.5), and $e_j^{(2,1)}, e_j^{(2,2)}, \ldots, e_j^{(2,n-k)}$ using (9.2.6). (2). Check the consistency of the solutions, $e_j^{(1,1)} = e_j^{(1,2)} = \cdots = e_j^{(1,n-k)} < m_j$ or $e_j^{(2,1)} = e_j^{(2,2)} = \cdots = e_j^{(2,n-k)} < m_j$. If it is consistent, then go to Step 5. If it is not consistent, then $j = j + 1$. Go to Step 4 for $j \leq k$. For $j = k + 1$, go to Step 6.
Step 5:	Declare only one error in the jth position. Correct it using (9.2.8) and stop.
Step 6:	Declare more than one error detected and stop.

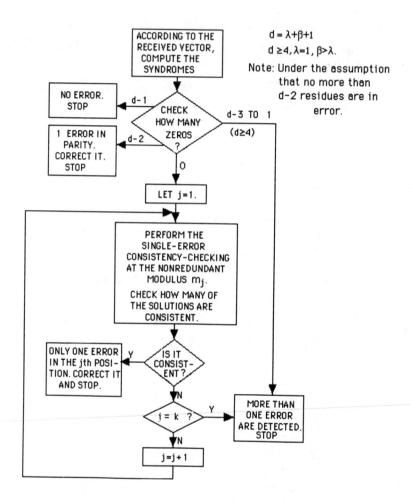

Figure 9.1: A Decoder Flowchart for Single Error Correction and Multiple Error Detection in RRNS

x is a codevector. Assume that one error takes place in the first residue digit, and the received vector is [0 0 22 13 25 0 17 58 4 11]. Based on the information part [0 0 22 13 25 0] and BEX, we compute the syndrome digits $\Delta_1 \equiv 14 \pmod{67}$, $\Delta_2 \equiv 48 \pmod{71}$, $\Delta_3 \equiv 50 \pmod{73}$, $\Delta_4 \equiv 17 \pmod{79}$. Following the decoding algorithm, we check the consistency for $j = 1, 2, \ldots, k$. Based on $e_j^{(1,r)}$ and $e_j^{(2,r)}$, $r = 1, 2, 3, 4$, it is seen that when $j = 1$, $e_1^{(2,1)} = e_1^{(2,2)} = e_1^{(2,3)} = e_1^{(2,4)} = e = 14$ is a consistent solution to (9.2.6). So, an error is declared in the first residue digit, the value of the error being $e_1 \equiv e\frac{M}{m_1} \pmod{m_1} = 15$. From (9.2.8), the estimated value of the erroneous digit is given by $\hat{x}_1 = 8$.

Now, assume that two errors take place in the first and third residue digits, and the received vector is [0 0 23 13 25 0 17 58 4 11]. Based on the information part [0 0 23 13 25 0] and the BEX, we compute the syndrome digits $\Delta_1 \equiv 23 \pmod{67}$, $\Delta_2 \equiv 50 \pmod{71}$, $\Delta_3 \equiv 63 \pmod{73}$ and $\Delta_4 \equiv 65 \pmod{79}$. Following the decoding algorithm, we check the consistency for $i = 1, 2, \ldots, 6$, and find that there is no consistent solution. Therefore, more than one error is detected.

9.3 Double Error Correction

In this section, we derive algorithms for double error correction and β ($\beta > 2$) error detection in RRNS. If Case 4 of Theorem 9.1 occurs and exactly two residue digits are received in error, there are two possible cases, **Case (a)** both errors in information part; **Case (b)** one error in information part and the other in parity part. By Lemma 7.5, the RRNS with $d = \lambda + \beta + 1$, $\lambda = 2$, and $\beta > \lambda$ can simultaneously correct two errors and detect β errors. Note that for correcting two errors only ($d = 5$), the analysis of the error correcting procedure is the same as in this section.

Consider **Case (a)**. If the pth and qth information digits are received in error, (9.1.11–9.1.16) become
(i) $X + E < M$. In this situation,

$$\overline{Y} = X + E = X + e'\frac{M}{m_p m_q}, \tag{9.3.1}$$

and

$$\Delta_r \equiv e'\frac{M}{m_p m_q} \pmod{m_{k+r}}, \quad r = 1, 2, \ldots, n - k, \tag{9.3.2}$$

where $0 < e' < m_p m_q$.

(ii) $M \leq X + E < 2M$. In this situation,

$$\overline{Y} = X + e'\frac{M}{m_p m_q} - M = X + (e' - m_p m_q)\frac{M}{m_p m_q}, \qquad (9.3.3)$$

$$\Delta_r \equiv (e' - m_p m_q)\frac{M}{m_p m_q} \pmod{m_{k+r}}, \ r = 1, 2, \ldots, n-k, \qquad (9.3.4)$$

where $0 < e' < m_p m_q$. Similar to the analysis in Section 9.2, given $\Delta_1, \Delta_2, \ldots, \Delta_{n-k}$, based on (9.3.1–9.3.4) a procedure to determine error locations and error values can be outlined as follows:

For $i = 1, 2, \ldots, k-1$, $j = i+1, i+2, \ldots, k$, solve the congruences

$$\Delta_r \equiv e^{(1,r)}\frac{M}{m_i m_j} \pmod{m_{k+r}}, \ r = 1, 2, \ldots, n-k, \qquad (9.3.5)$$

and

$$\Delta_r \equiv (e^{(2,r)} - m_i m_j)\frac{M}{m_i m_j} \pmod{m_{k+r}}, \ r = 1, 2, \ldots, n-k, \qquad (9.3.6)$$

to get the values

$$e^{(1,r)} \equiv \Delta_r \left(\frac{M}{m_i m_j}\right)^{-1} \pmod{m_{k+r}}, \qquad (9.3.7)$$

$$e^{(2,r)} \equiv \left(\Delta_r \left(\frac{M}{m_i m_j}\right)^{-1} + m_i m_j\right) \pmod{m_{k+r}}, \qquad (9.3.8)$$

where $r = 1, 2, 3, 4$. Based on the combination of any two digits of $\{e^{(1,r)}; r = 1, 2, \ldots, n-k\}$, the combination of any two digits of $e^{(2,r)}; r = 1, 2, \ldots, n-k\}$, say $\{e^{(1,1)}, e^{(1,2)}\}$ and $\{e^{(2,1)}, e^{(2,2)}\}$, and the CRT-I, we write

$$e_{ij}^{(1)} \equiv \sum_{r=1}^{2} a^{(1,r)}\frac{m_{k+1}m_{k+2}}{m_{k+r}} \pmod{m_{k+1}m_{k+2}}, \qquad (9.3.9)$$

$$e_{ij}^{(2)} \equiv \sum_{r=1}^{2} a^{(2,r)}\frac{m_{k+1}m_{k+2}}{m_{k+r}} \pmod{m_{k+1}m_{k+2}}, \qquad (9.3.10)$$

where

$$a^{(1,r)}\frac{m_{k+1}m_{k+2}}{m_{k+r}} \equiv e^{(1,r)} \pmod{m_{k+r}}, \ r = 1, 2 \qquad (9.3.11)$$

$$a^{(2,r)}\frac{m_{k+1}m_{k+2}}{m_{k+r}} \equiv e^{(2,r)} \pmod{m_{k+r}}, \ r = 1, 2. \qquad (9.3.12)$$

Then check if one of the following two conditions is satisfied: (i) $e_{ij}^{(1)} <$ $m_i m_j$ and (9.3.5) holds when $e_{ij}^{(1)}$ is substituted for $e^{(1,r)}$; if so then the solutions to (9.3.5) are consistent, and the consistent solution is $e_{ij}^{(1)}$; (ii) $e_{ij}^{(2)} < m_i m_j$ and (9.3.6) holds when $e_{ij}^{(2)}$ is substituted for $e^{(2,r)}$; if so then the solutions to (9.3.6) are consistent, and the consistent solution is $e_{ij}^{(2)}$. Note that $\{e^{(1,3)}, e^{(1,4)}\}$ and $\{e^{(2,3)}, e^{(2,4)}\}$ will be used to obtain the consistent solution for Case (b). Let us say either of the two conditions is satisfied for $i = f$ and $j = h$. Then the fth and hth information digits are declared to be in error, the value of errors being

$$e_f \equiv e \frac{M}{m_f m_h} \pmod{m_f}, \qquad (9.3.13)$$

$$e_h \equiv e \frac{M}{m_f m_h} \pmod{m_h}, \qquad (9.3.14)$$

where $e = e_{ij}^{(1)}$ if $e_{ij}^{(1)}$ is the consistent solution to (9.3.5); $e = e_{ij}^{(2)}$ if $e_{ij}^{(2)}$ is the consistent solution to (9.3.6). Note that if $e \equiv 0 \pmod{m_f}$ or $e \equiv 0 \pmod{m_h}$, then only the hth or fth information digit is in error. The correct values of the fth and hth residue digits are $(y_f - e_f) \bmod m_f$ and $(y_h - e_h) \bmod m_h$, respectively, and $\underline{y}$ is decoded to the codevector $\hat{\underline{x}}$, where

$$\hat{x}_j = y_j, \; j = 1, 2, \ldots, n, \; j \neq f, h, \qquad (9.3.15)$$
$$\hat{x}_f \equiv y_f - e_f \pmod{m_f}, \qquad (9.3.16)$$
$$\hat{x}_h \equiv y_h - e_h \pmod{m_h}. \qquad (9.3.17)$$

It is clear from (9.3.1), (9.3.2) and (9.3.5) that if two errors take place in the pth and qth information digits such that $X + E < M$, then for $i = p$ and $j = q$, $e = e' = e_{ij}^{(1)}$, $f = p$ and $h = q$. Similarly, if two errors take place in the pth and qth information digits such that $M \leq X + E < 2M$, then for $i = p$ and $j = q$, $e = e' = e_{ij}^{(2)}$, $f = p$ and $h = q$ from (9.3.3), (9.3.4) and (9.3.6). However, it remains to be shown that the solutions to either of (9.3.5) and (9.3.6) can not be consistent, if any one of the following cases occurs: 1) only one error takes place in the pth ($p \neq i, j$) information digit, 2) exactly two errors take place in the pth and qth information digits, where $p \neq i, j$ and/or $q \neq i, j$, 3) exactly two errors occur: one in the information part and the other in the parity part, 4) more than two ($\leq \beta$) errors occur. Under an additional constraint on the form of the moduli, this result is established in the following theorem.

THEOREM 9.3 *If the moduli of an (n, k) RRNS code are such that there do not exist integers n_{ij}, n_C; $0 \leq n_{ij} < m_i m_j$, $0 \leq n_C < M_C = \prod_{\alpha=1}^{c} m_{i_\alpha}$, $1 \leq i_\alpha \leq k$, $1 \leq c \leq \beta$; that satisfy*

$$n_{ij} M_C + n_C m_i m_j = \prod_{i=1}^{c+2} m_{r_i}, \quad k < r_i \leq n, 1 \leq i, j \leq k, \qquad (9.3.18)$$

then for any one of the following three cases: 1) only one error takes place in the pth $(p \neq i, j)$ information digit, 2) exactly two errors take place in the pth and qth digits, $p \neq i, j$ and/or $q \neq i, j$, 3) more than two $(\leq \beta)$ errors occur, the solutions to (9.3.5) and (9.3.6) are inconsistent.

The proof of this theorem is similar to the proof of Theorem 9.2, and therefore, is omitted here. The sufficient condition for Theorem 9.3 can be shown to be exactly the same as the sufficient condition stated in Lemma 9.1.

Now, consider **Case (b)**. Assume that one error takes place in the pth information digit and the other takes place in the $(k + u)$th digit (the uth parity digit), then (9.1.11–9.1.16) become:

(i) For $X + E < M$, $r = 1, 2, \ldots, n - k$ and $1 \leq u \leq n - k$,

$$\overline{Y} = X + E = X + e' \frac{M}{m_p}, \qquad (9.3.19)$$

and

$$\Delta_r \equiv e' \frac{M}{m_p} \pmod{m_{k+r}}, \ r \neq u$$

$$\equiv e' m_q \frac{M}{m_q m_p} \pmod{m_{k+r}}, \ r \neq u, \qquad (9.3.20)$$

$$\Delta_u \equiv e' \frac{M}{m_p} - e_{k+u} \pmod{m_{k+u}}, \qquad (9.3.21)$$

where $0 < e' < m_p$.

(ii) For $M \leq X + E < 2M$, $r = 1, 2, \ldots, n - k$ and $1 \leq u \leq n - k$,

$$\overline{Y} = X + E - M = X + (e' - m_p) \frac{M}{m_p}, \qquad (9.3.22)$$

and

$$\Delta_r \equiv (e' - m_p) \frac{M}{m_p} \pmod{m_{k+r}}, \ r \neq u$$

$$\equiv (e'm_q - m_p m_q)\frac{M}{m_q m_p} \quad (\text{mod } m_{k+r}),\ r \neq u, \quad (9.3.23)$$

$$\Delta_u \equiv (e' - m_p)\frac{M}{m_p} - e_{k+u} \quad (\text{mod } m_{k+u}), \quad (9.3.24)$$

where $0 < e' < m_p$ and m_q can be any one of the information moduli. Based on (9.3.19–9.3.24), the procedure to determine error locations and error values is the same as that for Case (a). The only exception is when no consistent solution is found in Case (a). In that case, we check if any $n-k-1$ congruences in either (9.3.5) or (9.3.6) have a consistent solution, e, which satisfies $e < m_i m_j$ and $e \equiv 0 \pmod{m_i}$ or $\pmod{m_j}$. Note that this consistent solution may be obtained from the combination $\{e^{(1,3)}, e^{(1,4)}\}$ or $\{e^{(2,3)}, e^{(2,4)}\}$, since one error may occur in the $(k+1)$th or $(k+2)$th digit. Let us say the above condition is satisfied for $i = f$ and $j = h$, and the only inconsistent congruence corresponds to the $(k+s)$th residue digit. Then the fth (if $e \equiv 0 \pmod{m_h}$) and $(k+s)$th residue digits are declared to be in error. The correct value of the fth and hth residue digit can be obtained from (9.3.13) and (9.3.14). The correct value of the $(k+s)$th residue digit is given by:

$$\hat{x}_{k+s} \equiv y_{k+s} + \Delta_s - e\frac{M}{m_f m_h} \quad (\text{mod } m_{k+s}), \quad (9.3.25)$$

if $n - k - 1$ congruences in (9.3.5) have a consistent solution e; and

$$\hat{x}_{k+s} \equiv y_{k+s} + \Delta_s + (m_f m_h - e)\frac{M}{m_f m_h} \quad (\text{mod } m_{k+s}), \quad (9.3.26)$$

if $n - k - 1$ congruences in (9.3.6) have a consistent solution e. It is clear from (9.3.19–9.3.21) and (9.3.5) that if one error takes place in the pth information digit and the other takes place in the $(k+u)$th digit such that $X + E < M$, then for i or $j = p$, say $i = p$, $e = e^{(1,r)} = e'm_j \equiv 0 \pmod{m_j}$, $f = p$, and $s = u$, $r = 1, 2, \ldots, n - k$, and $r \neq u$. Similarly, from (9.3.22–9.3.24) and (9.3.6) if one error takes place in the pth information digit and the other takes place in the $(k+u)$th digit such that $M \leq X + E < 2M$, then for i or $j = p$, say $i = p$, $e = e^{(2,r)} = e'm_j \equiv 0 \pmod{m_j}$, $f = p$, and $s = u$, $r = 1, 2, \ldots, n - k$, and $r \neq u$. However, it remains to be shown that neither (9.3.5) nor (9.3.6) has exactly $n - k - 1$ consistent solutions which are equivalent to zero mod m_i or mod m_j, if any one of the following cases occurs: (i) only one error takes place in the pth $(p \neq i, j)$ information digit, (ii) exactly two errors take place in the pth and qth information digits $(p \neq i, j,$ and/or $q \neq i, j)$, (iii) exactly two errors take place in the pth and $(k+u)$th digits $(p \neq i, j)$, (iv) more than two $(\leq \beta)$ errors take place. Under the same constraint on the form of the moduli as stated in Theorem 9.3, the above result is established in the following theorem.

THEOREM 9.4 *Consider an (n, k) RRNS code for which the moduli are such that there do not exist integers n_{ij}, n_C; $0 \leq n_{ij} < m_i m_j$, $0 \leq n_C < M_C = \prod_{\alpha=1}^{c} m_{i_\alpha}$, $1 \leq i_\alpha \leq k$, $1 \leq c \leq \beta$; that satisfy*

$$n_{ij} M_C + n_C m_i m_j = \prod_{i=1}^{c+2} m_{r_i}\,, \quad k < r_i \leq n, \quad 1 \leq i, j \leq k. \qquad (9.3.27)$$

When (9.3.5) or (9.3.6) has exactly $n - k - 1$ consistent solutions which are equivalent to zero mod m_j or mod m_i, the ith or jth information digit and the parity digit corresponding to the only inconsistent solution are received in error.

The proof of this theorem is similar to the proof of Theorem 9.2, and therefore, is omitted here.

Based on the concept developed above and the assumption that the moduli of MDS-RRNS satisfy the necessary and sufficient condition in theorems 9.3 or 9.4, or the sufficient condition in Lemma 9.1, the algorithm to correct double errors and simultaneously detect multiple errors is given in Table 9.2. A flowchart for this algorithm is given in Figure 9.2.

EXAMPLE 9.2 Consider the same RRNS as in Example 9.1 based on the moduli $m_1 = 23$, $m_2 = 25$, $m_3 = 27$, $m_4 = 29$, $m_5 = 31$, $m_6 = 32$, $m_7 = 67$, $m_8 = 71$, $m_9 = 73$ and $m_{10} = 79$, where m_7, m_8, m_9 and m_{10} are the redundant moduli. Clearly, $m_7 m_8 > 2m_5 m_6 - m_5 - m_6$, and therefore, the condition in theorems 9.3 and 9.4 is satisfied. $M = \prod_{i=1}^{6} m_i = 446,623,200$. Let $X = 400,000,000$, then $\underline{x} = [8\ 0\ 22\ 13\ 25\ 0\ 17\ 58\ 4\ 11]$. Since $0 \leq X < M$, $\underline{x}$ is a codevector. Assume that two errors take place in the first and third residue digits, and the received vector is $[0\ 0\ 23\ 13\ 25\ 0\ 17\ 58\ 4\ 11]$. Based on the information part $[0\ 0\ 23\ 13\ 25\ 0]$ and BEX, we compute the syndrome digits $\Delta_1 \equiv 23 \pmod{67}$, $\Delta_2 \equiv 50 \pmod{71}$, $\Delta_3 \equiv 63 \pmod{73}$ and $\Delta_4 \equiv 65 \pmod{79}$. Following the decoding algorithm, we check the consistency for $i = 1, 2, \ldots, 5$, $j = i + 1, i + 2, \ldots, 6$, based on computing $\{e^{(1,1)}, e^{(1,2)}, e^{(1,3)}, e^{(1,4)}\}$ and $\{e^{(2,1)}, e^{(2,2)}, e^{(2,3)}, e^{(2,4)}\}$ from (9.3.7) and (9.3.8), and $\{e_{ij}^{(1)}, e_{ij}^{(2)}\}$ from (9.3.9) and (9.3.10). It shows that when $i = 1$ and $j = 3$, $e_{13}^{(2)} = 217 < m_1 m_3 = 621$ is a consistent solution to (9.3.6), and $e_{13}^{(2)} \not\equiv 0 \pmod{23}$ or $\pmod{27}$. So, declare two errors in the first and third residue digits. From (9.3.13–9.3.17), the estimated values of the

Table 9.2: An Algorithm for Double Error Correction and Multiple Error Detection in RRNS

Step 1:	According to the received vector, compute the syndromes using BEX as given in Table 7.1 and $\Delta_r \equiv \overline{y}_{k+r} - y_{k+r} \pmod{m_{k+r}}, \; r = 1, 2, \ldots, n - k.$
Step 2:	Check how many syndromes are zero: (1). If all the syndromes are zero, then no error occurs. Stop. (2). If one or two syndromes are non-zero, then the parity digits are declared to be in error. The estimate of erroneous parity digits are the corresponding $\overline{y}_{k+r}$. Stop. (3). If $d - 4$ to 2 syndromes are zero, then go to Step 6. (4). If none or one syndrome is zero, then go to Step 3.
Step 3:	For $i = 1, 2, \ldots, k - 1$ and $j = i + 1, i + 2, \ldots, k$, perform double-error consistency-checking for the combination of the nonredundant moduli m_i and m_j. (1). Compute $e_{ij}^{(1)}, e_{ij}^{(2)}$ using (9.3.5)–(9.3.6). (2). Check the consistency of the solutions, $\quad e_{ij}^{(1)} < m_i m_j$ and (9.3.5) holds for $e^{(1,r)} = e_{ij}^{(1)} = e$ or $\quad e_{ij}^{(2)} < m_i m_j$ and (9.3.6) holds for $e^{(2,r)} = e_{ij}^{(2)} = e$. If it is consistent, then go to Step 4. Otherwise, go to Step 5.
Step 4:	Declare an error in the ith and jth positions. Correct them using (9.3.13)–(9.3.17) and stop.
Step 5:	If exactly $d - 2$ solutions are consistent and the consistent solution $e \equiv 0 \mod m_i$ (or $e \equiv 0 \mod m_j$), then declare two errors, one in the jth (or ith) position and the second in the parity for which the solution is inconsistent. Correct them using (9.3.16, 9.3.17, 9.3.25, 9.3.26) and stop. Otherwise, go to Step 6.
Step 6:	Declare more than two error detected and stop.

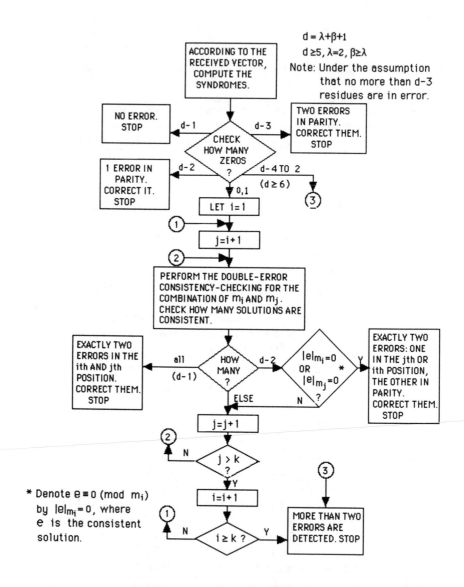

Figure 9.2: A Decoder Flowchart for Double Error Correction and Multiple Error Detection in RRNS

erroneous digits are

$$\widehat{x}_1 \equiv y_1 - e_{13}^{(2)} \frac{M}{m_1 m_3} \pmod{m_1} \equiv 8,$$

and

$$\widehat{x}_3 \equiv y_3 - e_{13}^{(2)} \frac{M}{m_1 m_3} \pmod{m_3} \equiv 22.$$

Now, assume that two errors take place in the first and $(k+4)$th residue digits, and the received vector is [0 0 22 13 25 0 17 58 4 12]. Based on the information part [0 0 22 13 25 0] and BEX, we compute the syndrome digits $\Delta_1 \equiv 14 \pmod{67}$, $\Delta_2 \equiv 48 \pmod{71}$, $\Delta_3 \equiv 50 \pmod{73}$ and $\Delta_4 \equiv 16 \pmod{79}$. Following the decoding algorithm, we check the consistency for $i = 1, 2, \ldots, 5$, $j = i+1, i+2, \ldots, 6$, based on computing $\{e^{(1,1)}, e^{(1,2)}, e^{(1,3)}, e^{(1,4)}\}$ and $\{e^{(2,1)}, e^{(2,2)}, e^{(2,3)}, e^{(2,4)}\}$ from (9.3.7) and (9.3.8), and $\{e_{ij}^{(1)}, e_{ij}^{(2)}\}$ from (9.3.9) and (9.3.10). It shows that when $i = 1$ and $j = 2$, $e_{12}^{(2)} = 350 \equiv 0 \pmod{m_2} < m_1 m_2$ is a consistent solution to the first three congruences of (9.3.6), that is, the only inconsistent congruence corresponds to the modulus m_{10}. So, declare that two errors occurred in the first and tenth residue digits. The estimated values of erroneous digits are computed using (9.3.16) and (9.3.26). They are

$$\widehat{x}_1 \equiv y_1 - e_{12}^{(2)} \frac{M}{m_1 m_2} \pmod{m_1} \equiv 8,$$

$$\widehat{x}_{10} \equiv y_{10} + \Delta_4 + (m_1 m_2 - e_{12}^{(2)}) \frac{M}{m_1 m_2} \pmod{m_{10}} \equiv 11.$$

Note that i is always equal to 1 if two errors occur in a way that one error is in information part and one error is in parity part.

9.4 Single-Burst Error Correction

A single-burst residue error vector of length b is defined as an RNS vector whose non-zero residue digits are confined to b consecutive digits, the first and last of which must be non-zero. In this case the decoding procedure can be summarized as — decode y to a codevector $\widehat{x}$ for which the length of burst-error vector $\widehat{e}$, $\widehat{e} = y - \widehat{x}$, is minimum. We will use this decoding procedure in our subsequent analysis. The proof of Theorem 9.5 is similar to the proof of Lemma 7.3.

THEOREM 9.5 *For detecting all single-burst errors of length b or less, an RRNS code must have $d \geq b + 1$.*

THEOREM 9.6 *For correcting all single-burst errors of length b or less, an RRNS code must have $d \geq 2b + 1$.*

PROOF. Let $\underline{x}_i$ be a codevector other than $\underline{x}$ in Ω. The Hamming distance among $\underline{x}$, $\underline{x}_i$, and $\underline{y}$ satisfies the triangular inequality

$$d(\underline{y}, \underline{x}) + d(\underline{y}, \underline{x}_i) \geq d(\underline{x}, \underline{x}_i).$$

Since $d(\underline{x}, \underline{x}_i) \geq d$ and $d(\underline{y}, \underline{x}) = wt(\underline{e}) \leq b$ $(\underline{e} = \underline{y} - \underline{x})$,

$$d(\underline{y}, \underline{x}_i) \geq d - b.$$

If $d \geq 2b + 1$, then $d(\underline{y}, \underline{x}_i) \geq b + 1$, thereby, implying that the length of burst-error vector $\underline{e}$ is smaller than the length of any other burst-error vector $\underline{e}_i$, $\underline{e}_i = \underline{y} - \underline{x}_i$. On the other hand, if the codevector $\underline{x}$ is such that $wt(\underline{x}) = d$, and the nonzero digits of $\underline{x}$ are confined to d consecutive places (examples of two such codevectors are codevectors corresponding to the integers $X_1 = \prod_{i=1}^{k-1} m_i$ and $X_2 = 2\prod_{i=1}^{k-1} m_i$ respectively), then for $d \leq 2b$, we can show that there exists at least one codevector $\underline{x}_i$ such that $d(\underline{y}, \underline{x}) = wt(\underline{e}) \geq wt(\underline{e}_i)$, $\underline{e}_i = \underline{y} - \underline{x}_i$. In this case, incorrect decoding will take place. This proves the theorem.

The proof of Theorem 9.7 is similar to the proof of Theorem 9.6.

THEOREM 9.7 *For correcting all single-burst errors of length b or less and simultaneously detecting all burst errors of length b' $(b' > b)$ or less, an RRNS code must have $d \geq b + b' + 1$.*

A fundamental property of the syndrome digits is stated in the following theorem.

THEOREM 9.8 *For an MDS-RRNS (n, k) code, $(d = n - k + 1 = b + b' + 1)$, under the assumption that no more than one single-burst residue error of length $\leq b'$ occurs, one of the following four cases occurs:*

Case 1: *If all the syndromes $\Delta_1, \Delta_2, \ldots, \Delta_{n-k}$ are zero, then no residue is in error, and vice versa.*

For the following cases, let p be the number of non-zero syndromes.

Case 2: *If $p \leq b$ syndromes are non-zero, then exactly p corresponding parity digits are in error, and all other residue digits are correct.*

Case 3: *If $b + 1 \leq p \leq b'$ syndromes are non-zero, then more than b residue digits are in error.*

Case 4: *If $b' + 1 \leq p \leq n - k$ syndromes are non-zero, then at least one of the information digits is in error, and all the last $b + 1$ redundant residue digits are correct.*

The proof of this theorem is similar to the proof of Theorem 9.1. If Case 4 of Theorem 9.8 happens, and exactly one single-burst residue error of length $\leq b$ occurs, there are two possible situations, that is, **Situation (a)** all the errors are in the information part; **Situation (b)** the errors are in both information and parity parts.

Consider **Situation (a)**. Assume that a single-burst error of length c ($1 \leq c \leq b$) takes place between the ith and $(i+c-1)$th information digits, that is, in the position interval $[i, i + c - 1]$. Then (9.1.11–9.1.16) become, (i) $X + E < M$. In this case,

$$\overline{Y} = X + E = X + e' \frac{M}{\prod_{\alpha=1}^{c} m_{i+\alpha-1}}, \qquad (9.4.1)$$

and

$$\Delta_r \equiv e' \frac{M}{\prod_{\alpha=1}^{c} m_{i+\alpha-1}} \pmod{m_{k+r}}, \; r = 1, 2, \ldots, n-k, \qquad (9.4.2)$$

where $0 < e' < \prod_{\alpha=1}^{c} m_{i+\alpha-1}$.
(ii) $M \leq X + E < 2M$. In this case,

$$\overline{Y} = X + E - M = X + (e' - \prod_{\alpha=1}^{c} m_{i+\alpha-1}) \frac{M}{\prod_{\alpha=1}^{c} m_{i+\alpha-1}}, \qquad (9.4.3)$$

and

$$\Delta_r \equiv (e' - \prod_{\alpha=1}^{c} m_{i+\alpha-1}) \frac{M}{\prod_{\alpha=1}^{c} m_{i+\alpha-1}} \pmod{m_{k+r}}$$
$$r = 1, 2, \ldots, n-k, \qquad (9.4.4)$$

where $0 < e' < \prod_{\alpha=1}^{c} m_{i+\alpha-1}$. Given $\Delta_1, \Delta_2, \ldots, \Delta_{n-k}$, based on (9.4.1–9.4.4) a procedure to determine error locations and error values can be outlined as follows:

For $j = 1, 2, \ldots, k - b + 1$, solve the congruences

$$\Delta_r \equiv e_j^{(1,r)} \frac{M}{\prod_{\alpha=1}^{b} m_{j+\alpha-1}} \pmod{m_{k+r}}$$
$$r = 1, 2, \ldots, n-k, \qquad (9.4.5)$$

$$\Delta_r \equiv (e_j^{(2,r)}) - \prod_{\alpha=1}^{b} m_{j+\alpha-1}) \frac{M}{\prod_{\alpha=1}^{b} m_{j+\alpha-1}} \quad (\bmod \; m_{k+r}),$$

$$r = 1, 2, \ldots, n - k, \qquad (9.4.6)$$

to get the values $e_j^{(1,r)}$ and $e_j^{(2,r)}$. Using the last b digits of $\{e_j^{(1,r)}; \; r = 1, 2, \ldots, n - k\}$, the last b digits of $\{e_j^{(2,r)}; \; r = 1, 2, \ldots, n - k\}$ and the CRT-I, we compute the following values:

$$e_j^{(1)} \equiv \sum_{r=n-k-b+1}^{n-k} e_j^{(1,r)} T_r' M_r' \quad (\bmod \; M_R'), \qquad (9.4.7)$$

$$e_j^{(2)} \equiv \sum_{r=n-k-b+1}^{n-k} e_j^{(2,r)} T_r' M_r' \quad (\bmod \; M_R'), \qquad (9.4.8)$$

where

$$M_R' = \prod_{i=n-b+1}^{n} m_i, \qquad (9.4.9)$$

$$M_r' = \frac{M_R'}{m_{k+r}}, \qquad (9.4.10)$$

$$T_r' M_r' \equiv 1 \quad (\bmod \; m_{k+r}), \qquad (9.4.11)$$

and then check if $e_j^{(1)}$ is a consistent solution to (9.4.5) or $e_j^{(2)}$ is a consistent solution to (9.4.6).

Let us say either of the two conditions is satisfied for $j = p$. Then declare that a single-burst error of length $\leq b$ has occurred in the position interval $[p, p + b - 1]$. The values of the errors are

$$e_q \equiv e \frac{M}{\prod_{\alpha=1}^{b} m_{p+\alpha-1}} \quad (\bmod \; m_q), \; q = p, p+1, \ldots, p+b-1, \qquad (9.4.12)$$

where $e = e_j^{(1)}$ if $e_j^{(1)}$ is the consistent solution to (9.4.5); $e = e_j^{(2)}$ if $e_j^{(2)}$ is the consistent solution to (9.4.6). The correct values of the erroneous residue digits are $(y_q - e_q) \bmod m_q$, $q = p, p+1, \ldots, p+b-1$. Then $\underline{y}$ is decoded to the codevector $\underline{\hat{x}}$, where

$$\begin{aligned} \hat{x}_q &= y_q, \; q = 1, 2, \ldots, n, \; q \neq p, p+1, \ldots, p+b-1, \\ \hat{x}_q &\equiv y_q - e_q \; (\bmod \; m_q), \; q = p, p+1, \ldots, p+b-1. \end{aligned} \qquad (9.4.13)$$

Note that $e_q \equiv 0 \; (\bmod \; m_q)$ implies no error occurs in the qth digit.

It is clear from (9.4.1), (9.4.2) and (9.4.5) that if a single-burst error of length c ($c \leq b$) takes place in the position interval $[i, i+c-1]$ and $X + E < M$, then for $[j, j+b-1]$ including $[i, i+c-1]$, $e'm = e_j^{(1)}$. Here m is the product of the moduli corresponding to the position interval $[p, p+b-1]$ exclusive of $[i, i+c-1]$. Note that if $c = b$, then $e' = e_j^{(1)}$ and $p = i$. Similarly, if a single-burst error of length c ($c \leq b$) takes place in the position interval $[i, i+c-1]$ and $M \leq X + E < 2M$, then for $[j, j+b-1]$ including $[i, i+c-1]$, $e'm = e_j^{(2)}$ from (9.4.3), (9.4.4) and (9.4.6). For this case, if $c = b$, then $e' = e_j^{(2)}$ and $p = i$. However, it remains to be shown that neither of (9.4.5) and (9.4.6) has a consistent solution, if a single-burst error of length c ($c \leq b'$) takes place in the position interval $[i, i+c-1]$ which is not included in $[j, j+b-1]$. Under an additional constraint on the form of the moduli, the result is established in the following theorem.

THEOREM 9.9 *If the moduli of an (n,k) RRNS code are such that there do not exist integers n_B, n_C; $0 \leq n_B < M_B = \prod_{\alpha=1}^{b} m_{j_\alpha}$, $0 \leq n_C < M_C = \prod_{\alpha=1}^{c} m_{i_\alpha}$, $1 \leq j_\alpha, i_\alpha < k$, $1 \leq c \leq b'$; that satisfy*

$$n_B M_C + n_C M_B = \prod_{i=1}^{c+b} m_{r_i}, \quad k < r_i \leq n, \qquad (9.4.14)$$

then for $[i, i+c-1]$ not included in $[j, j+b-1]$, neither (9.4.5) nor (9.4.6) has a consistent solution. It is assumed that a single-burst error of length c ($1 \leq c \leq b'$) occurs in the position interval $[i, i+c-1]$.

PROOF. Consider the extreme case, that is, $[i, i+c-1]$ and $[j, j+b-1]$ do not overlap and $c = b'$. For $X + E < M$, if (9.4.5) has a consistent solution $e_j^{(1)}$, then comparing (9.4.5) to (9.4.2) we get

$$\left(e' \prod_{\alpha=1}^{b} m_{j+\alpha-1} - e_j^{(1)} \prod_{\alpha=1}^{b'} m_{i+\alpha-1}\right) \frac{M}{\prod_{\alpha=1}^{b} m_{j+\alpha-1} \prod_{\alpha=1}^{b'} m_{i+\alpha-1}}$$
$$\equiv 0 \pmod{\prod_{r=1}^{n-k} m_{k+r}},$$

where $e' < \prod_{\alpha=1}^{b'} m_{i+\alpha-1}$. This can not hold due to

$$\left(e' \prod_{\alpha=1}^{b} m_{j+\alpha-1} - e_j^{(1)} \prod_{\alpha=1}^{b'} m_{i+\alpha-1}\right) < \prod_{\alpha=1}^{b} m_{j+\alpha-1} \prod_{\alpha=1}^{b'} m_{i+\alpha-1}$$

$$< \prod_{r=1}^{n-k} m_{k+r}.$$

Similarly, for $X + E < M$, if (9.4.6) has a consistent solution $e_j^{(2)}$, then comparing (9.4.6) to (9.4.2) we get

$$(e' \prod_{\alpha=1}^{b} m_{j+\alpha-1} + (\prod_{\alpha=1}^{b} m_{j+\alpha-1} - e_j^{(2)}) \prod_{\alpha=1}^{b'} m_{i+\alpha-1}).$$

$$\frac{M}{\prod_{\alpha=1}^{b} m_{j+\alpha-1} \prod_{\alpha=1}^{b'} m_{i+\alpha-1}} \equiv 0 \pmod{\prod_{r=1}^{n-k} m_{k+r}},$$

where $e' < \prod_{\alpha=1}^{b'} m_{i+\alpha-1}$. This can not hold due to (9.4.14). The case for $M \leq X + E < 2M$ can also be analyzed in an analogous manner. This proves the theorem.

A sufficient condition for Theorem 9.9 can be shown to be exactly the same as the sufficient condition stated in Lemma 9.1.

Now, consider **Situation (b)**. In Situation (a), when $j = k - b + 1$ and no consistent solution is found, it implies that a burst error of length $> b$ occurs or a burst error of length $\leq b$ occurs in both the information and parity digits. Assume that a single-burst error of length c $(1 \leq c \leq b)$ takes place in the position interval $[k - c_1 + 1, k + c_2]$, where $c_1, c_2 \geq 1$ and $c_1 + c_2 \leq b$. Then (9.1.11–9.1.16) become
(i) $X + E < M$. In this case,

$$\overline{Y} = X + E = X + e' \frac{M}{\prod_{i=k-c_1+1}^{k} m_i}, \tag{9.4.15}$$

and

$$\Delta_r \equiv e' \frac{M}{\prod_{i=k-c_1+1}^{k} m_i} - e_{k+r} \pmod{m_{k+r}}$$

$$r = 1, 2, \ldots, c_2, \tag{9.4.16}$$

$$\Delta_r \equiv e' \frac{M}{\prod_{i=k-c_1+1}^{k} m_i} \pmod{m_{k+r}}$$

$$r = c_2 + 1, \ldots, n - k, \tag{9.4.17}$$

where $e' < \prod_{i=k-c_1+1}^{k} m_i$.

(ii) $M \leq X + E < 2M$. In this case,

$$\overline{Y} = X + E - M = X + (e' - \prod_{i=k-c_1+1}^{k} m_i) \frac{M}{\prod_{i=k-c_1+1}^{k} m_i}, \qquad (9.4.18)$$

$$\Delta_r \equiv (e' - \prod_{i=k-c_1+1}^{k} m_i) \frac{M}{\prod_{i=k-c_1+1}^{k} m_i} - e_{k+r} \pmod{m_{k+r}},$$
$$r = 1, 2, \ldots, c_2, \quad (9.4.19)$$

$$\Delta_r \equiv (e' - \prod_{i=k-c_1+1}^{k} m_i) \frac{M}{\prod_{i=k-c_1+1}^{k} m_i} \pmod{m_{k+r}},$$
$$r = c_2 + 1, \ldots, n - k, \qquad (9.4.20)$$

where $e' < \prod_{i=k-c_1+1}^{k} m_i$. A procedure to determine error locations and error values is described in the following.

For $j = k - b + 1$, check if there exists the largest number c_3 in the range $[1, b)$ such that either of the following two conditions is satisfied:
Condition (1): The last $(b' + c_3)$ consecutive congruences of (9.4.5) have a consistent solution,

$$\begin{aligned}
e_j^{(1)} &\equiv e_j^{(1,b-c_3+1)} \pmod{m_{k+b-c_3+1}} \\
&\equiv e_j^{(1,b-c_3+2)} \pmod{m_{k+b-c_3+2}} \\
&\;\;\vdots \\
&\equiv e_j^{(1,n-k)} \pmod{m_n} \\
&\not\equiv e_j^{(1,b-c_3)} \pmod{m_{k+b-c_3}},
\end{aligned}$$

and

$$\prod_{\alpha=1}^{b} m_{k-b+\alpha} > e_j^{(1)} \equiv 0 \pmod{\prod_{\alpha=1}^{b-c_4} m_{k-b+\alpha}}, \; 1 \leq c_4 \leq c_3.$$

Condition (2): The last $(b' + c_3)$ consecutive congruences of (9.4.6) have a consistent solution, that is,

$$\begin{aligned}
e_j^{(2)} &\equiv e_j^{(2,b-c_3+1)} \pmod{m_{k+b-c_3+1}} \\
&\equiv e_j^{(2,b-c_3+2)} \pmod{m_{k+b-c_3+2}}
\end{aligned}$$

$$\vdots$$

$$\equiv e_j^{(2,n-k)} \pmod{m_n}$$

$$\not\equiv e_j^{(2,b-c_3)} \pmod{m_{k+b-c_3}},$$

and

$$\prod_{\alpha=1}^{b} m_{k-b+\alpha} > e_j^{(2)} \equiv 0 \pmod{\prod_{\alpha=1}^{b-c_4} m_{k-b+\alpha}}, \ 1 \le c_4 \le c_3.$$

Let us say there exists a number c_3 for which one of the above two conditions is satisfied. A single-burst error of length $\le b$ is declared to have taken place in the position interval $[k - c_3 + 1, k + b - c_3]$. The estimated values of the errors and erroneous information digits can be obtained by (9.4.12) and (9.4.13) for $p = k - b + 1$. The inconsistent congruences correspond to the erroneous parity digits, y_{k+s}, $s = 1, 2, \ldots, b - c_3$, the estimated values of erroneous parity digits being:

$$\widehat{x}_{k+s} \equiv y_{k+s} + \Delta_s - e_j^{(1)} \frac{M}{\prod_{\alpha=1}^{b} m_{k-b+\alpha}} \pmod{m_{k+s}}, \qquad (9.4.21)$$

if $e_j^{(1)}$ is the consistent solution to (9.4.5); and

$$\widehat{x}_{k+s} \equiv y_{k+s} + \Delta_s + (\prod_{\alpha=1}^{b} m_{k-b+\alpha} - e_j^{(2)}) \frac{M}{\prod_{\alpha=1}^{b} m_{k-b+\alpha}} \pmod{m_{k+s}},$$

$$(9.4.22)$$

if $e_j^{(2)}$ is the consistent solution to (9.4.6). It is clear from (9.4.15–9.4.17) and (9.4.5) that if a single–burst error of length c ($1 \le c \le b$) takes place in the position interval $[k - c_1 + 1, k + c_2]$ such that $X + E < M$, $c_1, c_2 \ge 1$ and $c_1 + c_2 = c$, then for $j = k - b + 1$, (9.4.6) has $b' + b - c_2$ solutions which are consistent and equivalent to zero mod $m_{k-b+\alpha}$. In other words, $e_j^{(1,r)} = e_j^{(1)} = e' \prod_{\alpha=1}^{b-c_1} m_{k-b+\alpha}$, $\alpha = 1, 2, \ldots, b - c_1$, $r = c_2 + 1, c_2 + 2, \ldots, n - k$, $b - c_2 = c_3$ and $c_1 = c_4$. Similarly, from (9.4.18–9.4.20) and (9.4.6), if a single–burst error of length c ($1 \le c \le b$) takes place in the position interval $[k - c_1 + 1, k + c_2]$ such that $M \le X + E < 2M$, $c_1, c_2 \ge 1$ and $c_1 + c_2 = c$, then for $j = k - b + 1$, (9.4.6) has $b' + b - c_2$ solutions which are consistent and equivalent to zero mod $m_{k-b+\alpha}$. In other words, $e_j^{(2,r)} = e_j^{(2)} = e' \prod_{\alpha=1}^{b-c_1} m_{k-b+\alpha}$, $\alpha = 1, 2, \ldots, b - c_1$, $r = c_2 + 1, c_2 + 2, \ldots, n - k$, $b - c_2 = c_3$ and $c_1 = c_4$. However, it remains to be shown that no c_3 and c_4 exist to satisfy Condition (1) and Condition (2) if a single–burst error of length $> b$ (but $\le b'$) occurs or a single–burst error of length $\le b$ occurs which is not in the position interval $[k - b + 2, k + b - 1]$. The result is established in the following theorem.

THEOREM 9.10 *Consider an (n, k) RRNS code for which the moduli are such that there do not exist integers n_B, n_C; $0 \leq n_B < M_B = \prod_{\alpha=1}^{b} m_{j_\alpha}$, $0 \leq n_C < M_C = \prod_{\alpha=1}^{c} m_{i_\alpha}$, $1 \leq j_\alpha, i_\alpha < k$, $1 \leq c \leq b'$; that satisfy*

$$n_B M_C + n_C M_B = \prod_{i=1}^{c+b} m_{r_i}, \quad k < r_i \leq n. \tag{9.4.23}$$

When a single–burst error of length $> b$ (but $\leq b'$) occurs or a single–burst error of length $\leq b$ occurs outside of the position interval $[k - b + 2, k + b - 1]$, c_3 and c_4 do not exist to satisfy conditions (1) and (2).

The proof of this theorem is similar to the proof of Theorem 9.9. Based on the concepts developed above and under the assumption that the moduli satisfy the condition in Theorem 9.9 or the sufficient condition in Lemma 9.1, the algorithm to correct single-burst error of length b or less and simultaneously detect the presence of single-burst error of length b' ($b' > b$) in RRNS is given in Table 9.3. A flowchart for this algorithm is given in Figure 9.3.

9.5   General Fault Detection and Correction

In this section, we give an alternate proof of conditions which allow detection of up to β faults and correction of up to λ faults in a maximally redundant RRNS. We follow this with the general procedure for finding and correcting these faults.

Let $i_1, \ldots, i_c$ be the actual locations of the faults in the information digits, and let $j_1, \ldots, j_g$ be the locations of the faults in the parity digits, where $c + g \leq \lambda$. Let $\overline{Y}$ be the reconstruction of the value modulo M using only the information digits, but let us define the error E slightly differently than in Section 9.1. In particular, define:

$$E = \overline{Y} - X.$$

The value E can take one of M possible values. Without computing $\overline{Y}$ or X directly, however, we can definitely state that E is confined to the range:

$$-M < E < M.$$

Now let $e_i \equiv E \pmod{m_i}$. Clearly all the digits e_i are zero except for digits $i = i_1, \ldots, i_c$. We can therefore state that E can be written in the

---

Table 9.3: An Algorithm for Correcting Single Burst Error of Length b and Detecting Burst-Error of Length b' ($b' > b$) in RRNS

Step 1:	According to the received vector, compute the syndromes using BEX as given in Table 7.1 and $\Delta_r \equiv \bar{y}_{k+r} - y_{k+r} \pmod{m_{k+r}}, \ r = 1, 2, \ldots, n - k$.
Step 2:	Check how many syndromes are zero:
	(1). If all the syndromes are zero, then no error occurs. Stop.
	(2). If p ($1 \leq p \leq b$) syndromes are non-zero, then the corresponding parity digits are declared to be in error. The estimate of erroneous parity digits are the corresponding $\bar{y}_{k+r}$. Stop.
	(3). If $b + 1$ to b' syndromes are non-zero, then go to Step 6.
	(4). If $b' + 1$ to all syndromes are non-zero, then go to Step 3.
Step 3:	For $j = 1, 2, \ldots, k - b + 1$, perform b-error consistency-checking for the combination of the nonredundant moduli $m_j, m_{j+1}, \ldots, m_{j+b-1}$.
	(1). Compute $e_j^{(1)}, e_j^{(2)}$ using (9.4.7, 9.4.8).
	(2). Check the consistency of the solutions, that is, (9.4.5) holds for $e_j^{(1,r)} = e_j^{(1)}$ or (9.4.6) holds for $e_j^{(2,r)} = e_j^{(2)}$.
	If it is consistent, then go to Step 4.
	Otherwise, go to Step 5.
Step 4:	Declare a single burst-error in the position interval $[j, j + b - 1]$.
	Correct it using (9.4.12, 9.4.13) and stop.
Step 5:	If $j = k - b + 1$ and the last $b' + c$ consecutive congruences of (9.4.5) or (9.4.6) have a consistent solution which is equivalent to zero mod $\prod_{\alpha=1}^{b-c} m_{k-b+\alpha}$, then declare a single burst-error in the position interval $[k - c + 1, k + b - c]$. Correct them using (9.4.12, 9.4.13, 9.4.21, 9.4.22) and stop. Otherwise, go to Step 6.
Step 6:	Declare that a burst-error of length $> b$ is detected. Stop.

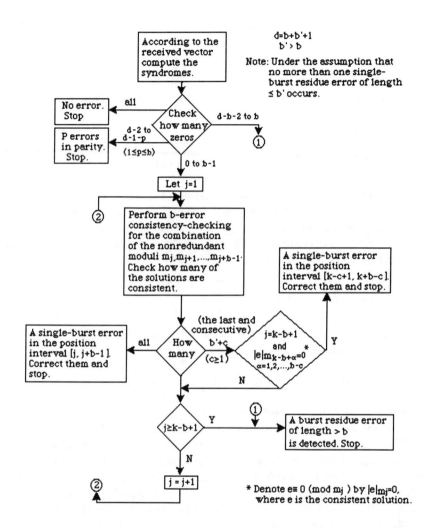

Figure 9.3: A Decoder Flowchart for Single Burst Error Correction in RRNS

form:

$$E = e \frac{M}{\prod_{\alpha=1}^{c} m_{i_\alpha}},$$

where

$$-\prod_{\alpha=1}^{c} m_{i_\alpha} < e < \prod_{\alpha=1}^{c} m_{i_\alpha}. \qquad (9.5.1)$$

Let us define syndromes in the same way as before. Then it is easy to show that

$$\Delta_r \equiv \begin{cases} e \dfrac{M}{\prod_{\alpha=1}^{c} m_{i_\alpha}} \pmod{m_r} & \text{for } r \neq j_1, \ldots, j_g \\ \text{arbitrary} & \text{otherwise.} \end{cases} \qquad (9.5.2)$$

Let Φ and Φ' be any (possibly overlapping) sets of indices in the range $1, \ldots, k$. Suppose Φ has c elements and Φ' has c' elements. Let $\Phi^{1,2} = \Phi \cap \Phi'$ be the intersection, with $c'' \leq min\{c, c'\}$ elements. Let $\Phi^1 = \Phi - \Phi^{1,2}$ be the elements of Φ that are not in Φ', and let $\Phi^2 = \Phi' - \Phi^{1,2}$ be the elements of Φ' that are not in Φ. Now we assume that:

$$2 \prod_{i \in \Phi^{1,2}} m_i \prod_{j \in \Phi^1} m_j \prod_{k \in \Phi^2} m_k - \prod_{j \in \Phi^1} m_j - \prod_{k \in \Phi^2} m_k < \prod_{r \in \Psi} m_r, \qquad (9.5.3)$$

where Ψ is any subset of at least $c + c'$ of the redundant moduli $k+1, \ldots, n$. Note that this assumption is trivially satisfied if:

$$2 m_i m_j < m_r m_s \qquad (9.5.4)$$

for any $1 \leq i, j \leq k$ and $k + 1 \leq r, s \leq n$.

Now we derive the main result. Assume (9.5.3) holds and that the RRNS is maximally redundant. Now hypothesize that the syndromes we observe might have been caused by some alternative "different" combination of up to β digit failures in information digit positions $i'_1, \ldots, i'_{c'}$ and parity digit positions $j'_1, \ldots, j'_{g'}$, where $c' + g' \leq \beta$, and $\lambda + \beta = n - k$ ("Different" implies that at least one error information digit in the alternative hypothesis is assumed to have a different value than in the correct set of error information digits). We show by contradiction that the syndromes of this alternative hypothesis cannot equal those from the correct hypothesis. Under the alternative assumption, we hypothesize an error $-M < E' < M$, and can write this error in the form:

$$E' = e' \frac{M}{\prod_{\alpha=1}^{c'} m_{i'_\alpha}},$$

where

$$-\prod_{\alpha=1}^{c'} m_{i'_\alpha} < e' < \prod_{\alpha=1}^{c'} m_{i'_\alpha}. \tag{9.5.5}$$

Let $\Phi = \{i_1, \ldots, i_c\}$ be the set of information digits that actually failed. Let $\Phi' = \{i'_1, \ldots, i'_{c'}\}$ be the set of information digits assumed to have failed in the alternative hypothesis. Let $\Psi = \{j_1, \ldots, j_g\}$ be the set of parity digits that have actually failed. Let $\Psi' = \{j'_1, \ldots, j'_{g'}\}$ be the set of parity digits assumed to have failed in the alternative hypothesis. Let $e'_i \equiv E' \pmod{m_i}$ be the error digits in the alternative hypothesis, with $e'_i \neq 0$ for all $i \in \Phi'$. Note that this implies that $e' \pmod{m_i} \not\equiv 0$ for $i \in \Phi'$. Define:

$$\begin{aligned}
\Phi^{1,2} &= \Phi \cap \Phi' \\
\Phi^1 &= \Phi - \Phi^{1,2} \\
\Phi^2 &= \Phi' - \Phi^{1,2} \\
\Psi^{1,2} &= \Psi \cap \Psi' \\
\Psi^1 &= \Psi - \Psi^{1,2} \\
\Psi^2 &= \Psi' - \Psi^{1,2}.
\end{aligned}$$

Note that

$$\begin{aligned}
\Phi &= \Phi^{1,2} + \Phi^1 \\
\Phi' &= \Phi^{1,2} + \Phi^2 \\
\Psi &= \Psi^{1,2} + \Psi^1 \\
\Psi' &= \Psi^{1,2} + \Psi^2.
\end{aligned}$$

Now if the alternative hypothesis could indeed explain the observed syndromes, then we would have

$$\Delta_r \equiv e' \frac{M}{\prod_{\alpha=1}^{c'} m_{i'_\alpha}} \pmod{m_r} \quad \text{for } r \neq j'_1, \ldots, j'_{g'}.$$

Equating this with the actual value of the syndromes in (9.5.2) in terms of e, and using our index set definitions, we get:

$$\left(e \frac{M}{\prod_{i \in \Phi} m_i} - e' \frac{M}{\prod_{i \in \Phi'} m_i} \right) \pmod{m_r} \equiv 0 \quad \text{for } r \notin \Psi, \Psi'$$

which implies that

$$\left[\left(e \prod_{i \in \Phi^2} m_i - e' \prod_{i \in \Phi^1} m_i \right) \frac{M}{\prod_{i \in \Phi^{1,2}} m_i \prod_{i \in \Phi^1} m_i \prod_{i \in \Phi^2} m_i} \right]$$

$$\left(\text{mod} \prod_{r \in \{k+1,\ldots,n\} - \Psi^{1,2} - \Psi^1 - \Psi^2} m_r\right) \equiv 0.$$

But since the moduli are relatively prime, this implies that

$$\left(e \prod_{i \in \Phi^2} m_i - e' \prod_{i \in \Phi^1} m_i\right) = \delta \prod_{r \in \{k+1,\ldots,n\} - \Psi^{1,2} - \Psi^1 - \Psi^2} m_r$$

for some integer δ. Multiply both sides by the product of the redundant moduli in index sets $\Psi^{1,2}$, Ψ^1 and Ψ^2 to get

$$\left(e \prod_{i \in \Phi^2} m_i - e' \prod_{i \in \Phi^1} m_i\right) \prod_{r \in \Psi^{1,2} + \Psi^1 + \Psi^2} m_r = \delta M_R. \tag{9.5.6}$$

Now

$$\left| e \prod_{i \in \Phi^2} m_i - e' \prod_{i \in \Phi^1} m_i \right| \leq \left(\prod_{i \in \Phi^1 + \Phi^{1,2}} m_i - 1\right) \prod_{i \in \Phi^2} m_i$$

$$+ \left(\prod_{i \in \Phi^2 + \Phi^{1,2}} m_i - 1\right) \prod_{i \in \Phi^1} m_i = 2 \prod_{i \in \Phi^1 + \Phi^2 + \Phi^{1,2}} m_i - \prod_{i \in \Phi^2} m_i - \prod_{i \in \Phi^1} m_i.$$

Suppose set $\Phi^{1,2}$ has c'' elements, where $c'' \leq min(c, c')$, and suppose set $\Psi^{1,2}$ has g'' elements, where $g'' \leq min(g, g')$. By our assumption (9.5.3), the expression above is guaranteed to be less than the product of any set of $c + c'$ or more redundant moduli. Note that the set $\Psi^{1,2} + \Psi^1 + \Psi^2$ has $g + g' - g''$ redundant moduli in it, which leaves $\lambda + \beta - (g + g' - g'') \geq c + c' + g''$ redundant moduli which are not in that set. Choose any $c + c'$ of these, and call this set $\bar{\Psi}$. Then

$$\left| e \prod_{i \in \Phi^2} m_i - e' \prod_{i \in \Phi^1} m_i \right| \prod_{r \in \Psi^{1,2} + \Psi^1 + \Psi^2} m_r < \prod_{r \in \bar{\Psi} + \Psi^{1,2} + \Psi^1 + \Psi^2} m_r$$

$$\leq M_R.$$

Combining this with (9.5.6) shows that the constant $\delta = 0$, and we have

$$\left(e \prod_{i \in \Phi^2} m_i - e' \prod_{i \in \Phi^1} m_i\right) = 0. \tag{9.5.7}$$

But this is only possible if

$$e = \bar{e} \prod_{i \in \Phi^1} m_i \qquad e' = \bar{e} \prod_{i \in \Phi^2} m_i$$

which would imply that

$$E = E' = \bar{e} \frac{M}{\prod_{i \in \Phi^{1,2}} m_i}.$$

Also this implies that $e_i = 0$ for all $i \in \Phi^1$, and that $e'_i = 0$ for all $i \in \Phi^2$.

Now, we originally assumed that there were non-zero errors in all the index positions in the set Φ, including those in the set Φ^1. We must conclude that Φ^1 must be a null set, and so $\Phi = \Phi^{1,2}$. Our alternative hypothesis assumed that there were non-zero errors in all the index positions in the set Φ', including those in the set Φ^2. We must conclude that if the syndromes match, then Φ^2 is a null set, and so $\Phi' = \Phi^{1,2}$. Putting this together, $\Phi = \Phi'$, and so if the syndromes match then the alternative hypothesis must have correctly identified the faulty information digits. Also note that (9.5.7) implies that if Φ^1 and Φ^2 are null, then $e = e'$ and $E = E'$. Thus the error information digits e_i and e'_i must also match, and the alternative hypothesis exactly matches the actual errors. We conclude that with assumption (9.5.3) and with up to λ faults, the syndromes uniquely specify the erroneous information error digits.

A procedure to correct up to λ errors (at least one of these errors in an information digit) is as follows: Check the number of non-zero syndromes. If the number is greater than β, then consider all possible sets of up to λ faults, with at least one of these faults in an information digit. For each hypothesis, try to solve (9.5.2) for a consistent value e in the range (9.5.1). At most one of these combinations can yield a coherent solution e (either positive or negative) which would explain the non-faulty syndromes. If none yield a coherent solution e, there must be more than λ faults. Total complexity is the time required to compute BEX, plus the time to evaluate every possible hypothesis involving λ failures.

9.6   Extensions of Previous Algorithms

The previous algorithms for locating a single residue digit error are based on the properties of modulus projection and MRC. These can be found in Barsi and Maestrini [1973], Jenkins and Altman [1988] and Jenkins [1983].

In an (n, k) RRNS code, the first k mixed radix digits $a_1, a_2, \ldots, a_k$ are the nonredundant mixed radix digits, and the remaining $a_{k+1}, a_{k+2}, \ldots, a_n$ are the redundant mixed radix digits. It is obvious that if all the redundant mixed radix digits are zero, the number X is a legitimate number; otherwise, X is an illegitimate number. The modulus m_i-projection of X in an (n, k)

RRNS code, denoted by X_i, is defined as

$$X_i \equiv X \pmod{\frac{M M_R}{m_i}}. \tag{9.6.1}$$

Clearly, X_i can be represented as $[x_1\, x_2\, \ldots\, x_{i-1}\, x_{i+1}\, \ldots\, x_n]$ which is the residue representation of X in a reduced RRNS with the ith residue digit x_i deleted. The mixed radix representation of X_i is

$$X_i = \sum_{\substack{l=1 \\ l \neq i}}^{n} a_l \prod_{\substack{r=1 \\ r \neq i}}^{l-1} m_r. \tag{9.6.2}$$

The new first k mixed radix digits are called the nonredundant mixed radix digits and the rest are called the redundant mixed radix digits. For example, if $i \leq k$, $a_1, a_2, \ldots, a_{i-1}, a_{i+1}, \ldots, a_k, a_{k+1}$ are the nonredundant mixed radix digits, and $a_{k+2}, a_{k+3}, \ldots, a_n$ are the redundant mixed radix digits.

It is obvious that if X_i is a legitimate number, the redundant mixed radix digits are all zero. However, if the redundant mixed radix digits are all zero, mathematically X_i could still be an illegitimate number. In Etzel and Jenkins [1980], it was shown that an illegitimate projection resulting from a single error can not be smaller than the smallest non-zero number represented by (9.6.2) with all the nonredundant mixed radix digits zero. Therefore, the algorithm in Jenkins and Altman [1988] for locating single residue digit error consists in checking if the redundant mixed radix digits are all zero for each modulus m_i-projection, $i = 1, 2, \ldots, n$. In the following analysis, we will show, from the coding theory point of view, that their algorithm can be extended for detecting and correcting multiple residue digit errors.

The M_Λ-projection of X, denoted by X_Λ, is defined by

$$X_\Lambda \equiv X \pmod{\frac{M M_R}{M_\Lambda}}, \tag{9.6.3}$$

where $M_\Lambda = \prod_{\alpha=1}^{\lambda} m_{i_\alpha}$, $\Lambda = \{i_1, i_2, \ldots, i_c, \ldots, i_\lambda;\ i_1 < i_2 < \cdots < i_c < \cdots < i_\lambda\}$, and $\lambda \leq n - k = d - 1$. X_Λ can also be represented as a reduced residue representation of X with the residue digits $x_{i_1}, x_{i_2}, \ldots, x_{i_\lambda}$ deleted. Then, the mixed radix representation of X_Λ is

$$X_\Lambda = \sum_{\substack{l=1 \\ l \notin \Lambda}}^{n} a_l \prod_{\substack{r=1 \\ r \notin \Lambda}}^{l-1} m_r. \tag{9.6.4}$$

The legitimate and illegitimate range of the reduced RRNS are $[0, M')$ and $[M', \frac{MM_R}{M_\Lambda})$, respectively, where M' is the product of the first k moduli of the reduced RRNS,

$$M' = \prod_{\substack{r=1 \\ r \notin \Lambda}}^{k+c} m_r \ , \ \ i_c \le k+c, \ i_{c+1} > k+c, \ i_1 \le k$$

$$M' = M = \prod_{r=1}^{k} m_r \ , \ i_1 > k. \tag{9.6.5}$$

It is obvious that $M' \ge M$. M' is also the smallest nonzero number represented by (9.6.4) with all the nonredundant mixed radix digits zero. It follows from (9.6.3) that the M_Λ-projection of any legitimate number X in RRNS is still a legitimate number in $[0, M)$, that is, $X_\Lambda = X$. The proof of the following theorem has been given in [Barsi and Maestrini 1973].

THEOREM 9.11 *Let $\overline{X}$ be a number in RRNS and $\overline{X}_\Lambda$ be the M_Λ-projection of $\overline{X}$, where $M_\Lambda = \prod_{\alpha=1}^{\lambda} m_{i_\alpha}$. If $\overline{X}_\Lambda \ne \overline{X}$, the residue representation of $\overline{X}_\Lambda$ uniquely differs from $\overline{X}$ in one or more of the residue digits corresponding to the moduli $m_{i_1}, m_{i_2}, \ldots, m_{i_\lambda}$.*

Before considering the range of the M_Λ-projection $\overline{X}_\Lambda$ of any illegitimate number $\overline{X}$ in RRNS, we give the following theorems. Theorem 9.12 is based on the fact that all codevectors differ in at least d places in an RRNS with minimum distance d.

THEOREM 9.12 *If X is a legitimate number in the RRNS having minimum distance d, then any integer $\overline{X}$ differing from X in at least one and no more than $d-1$ residue digits is an illegitimate number.*

THEOREM 9.13 *Let $\overline{X}$ be an illegitimate number in the RRNS. If there exists a legitimate number X differing from $\overline{X}$ in the i_1th, i_2th, $\ldots$, i_λth residue digits, then the M_Λ-projection $\overline{X}_\Lambda$ is a legitimate number, where $M_\Lambda = \prod_{\alpha=1}^{\lambda} m_{i_\alpha}$.*

PROOF. Since $\overline{X}$ differs from X in the i_1th, i_2th, $\ldots$, i_λth residue digits, $\overline{X}$ can be expressed as

$$\overline{X} \equiv X + e' \frac{MM_R}{\prod_{\alpha=1}^{\lambda} m_{i_\alpha}} \pmod{MM_R},$$

$0 < e' < \prod_{\alpha=1}^{\lambda} m_{i_\alpha}$. By definition,

$$\overline{X}_\Lambda \equiv \overline{X} \pmod{\frac{MM_R}{M_\Lambda}}$$

$$\equiv X + e'\frac{MM_R}{\prod_{\alpha=1}^{\lambda} m_{i_\alpha}} \pmod{\frac{MM_R}{M_\Lambda}}$$

$$\equiv X \pmod{\frac{MM_R}{M_\Lambda}}$$

$$\equiv X_\Lambda < M.$$

Therefore, $\overline{X}_\Lambda$ is a legitimate number. This proves the theorem.

THEOREM 9.14 *Let $\overline{X}$ be an illegitimate number in the RRNS having minimum distance d. If the M_Λ-projection $\overline{X}_\Lambda$ ($M_\Lambda = \prod_{\alpha=1}^{\lambda} m_{i_\alpha}$, $1 \leq \lambda < d$) is a legitimate number, then there exists only one legitimate number X differing from $\overline{X}$ in one or more residue digits that correspond to the moduli m_{i_1}, m_{i_2}, ..., m_{i_λ}.*

PROOF. Since $\overline{X} \neq \overline{X}_\Lambda$, it is obvious from Theorem 9.11 that the legitimate number $X = \overline{X}_\Lambda$ is a solution to the problem. Now assume that there exists two different legitimate number X and X', both differing from $\overline{X}$ in one or more residue digits that correspond to the moduli m_{i_1}, m_{i_2}, ..., m_{i_λ}. Then,

$$X = X_\Lambda + e\frac{MM_R}{\prod_{\alpha=1}^{\lambda} m_{i_\alpha}},$$

$$X' = X'_\Lambda + e'\frac{MM_R}{\prod_{\alpha=1}^{\lambda} m_{i_\alpha}},$$

where $0 \leq e, e' < \prod_{\alpha=1}^{\lambda} m_{i_\alpha}$. Since $0 \leq X$, $X' < M$, $0 \leq X_\Lambda = X'_\Lambda < M$, and $\frac{MM_R}{\prod_{\alpha=1}^{\lambda} m_{i_\alpha}} \geq M$, it follows that $e = e' = 0$ and $X = X'$. This contradicts the original assumption $X \neq X'$. This proves the theorem.

THEOREM 9.15 *In an RRNS with minimum distance $d = \lambda + \beta + 1$, an illegitimate number $\overline{X}$ is originated from a legitimate number X affected by λ errors corresponding to the moduli m_{i_1}, m_{i_2}, ..., m_{i_λ}, iff the M_Λ-projection $\overline{X}_\Lambda$ is a legitimate number and the M_P-projection $\overline{X}_P$ is an illegitimate number, $M_\Lambda = \prod_{\alpha=1}^{\lambda} m_{i_\alpha}$, $M_P = \prod_{\alpha=1}^{p} m_{j_\alpha}$, $1 \leq p \leq \beta$, and $gcd(M_P, M_\Lambda) \neq \prod_{\alpha=1}^{\lambda} m_{i_\alpha} = M_\Lambda$.*

PROOF. By Theorem 9.13, the M_Λ-projection $\overline{X}_\Lambda$ is a legitimate number. Since $gcd(M_P, M_\Lambda) \neq \prod_{\alpha=1}^{\lambda} m_{i_\alpha}$, $\overline{X}_P$ can be treated as a number originating from X_P affected by λ or fewer errors corresponding to the moduli m_{i_1}, $m_{i_2}, \ldots, m_{i_\lambda}$, that is,

$$\overline{X}_P = X_P + e' \frac{M M_R}{\prod_{\alpha=1}^{p} m_{j_\alpha} \prod_{\alpha=1}^{\lambda} m_{i_\alpha}},$$

where $0 < e' \leq \prod_{\alpha=1}^{\lambda} m_{i_\alpha}$. By Theorem 7.1, $M_R \geq \prod_{\alpha=1}^{p} m_{j_\alpha} \prod_{\alpha=1}^{\lambda} m_{i_\alpha}$. It follows that $\overline{X}_P$ is an illegitimate number. This proves the necessity. The sufficiency is proved by contradiction. Since $\overline{X}_\Lambda$ is a legitimate number, by Theorem 9.14 there exists only one legitimate number X differing from $\overline{X}$ in one or more of residue digits corresponding to the moduli m_{i_1}, $m_{i_2}, \ldots,$ m_{i_λ}. However, if the legitimate number X differs from $\overline{X}$ in less than λ residue digits (say p ($p < \lambda$) residue digits corresponding to the moduli m_{i_1}, $m_{i_2}, \ldots, m_{i_p}$) then, by Theorem 9.13, the M_P-projection $\overline{X}_P$ is a legitimate number, where $M_P = \prod_{\alpha=1}^{p} m_{i_\alpha}$ and $gcd(M_P, M_\Lambda) \neq M_\Lambda$. This contradicts the original assumption. This proves the theorem.

The range of the illegitimate projection $\overline{X}_\Lambda$ of any illegitimate number $\overline{X}$ is given in the following theorem.

THEOREM 9.16 *Assume that no more than β residue errors occur for an RRNS with $d = \lambda + \beta + 1$, $\beta > \lambda$. If the M_Λ-projection $\overline{X}_\Lambda$ of any illegitimate number $\overline{X}$ is illegitimate, $\overline{X}_\Lambda$ will be in the range $[M', \frac{M M_R}{M_\Lambda})$, $M_\Lambda = \prod_{\alpha=1}^{\lambda} m_{i_\alpha}$, $1 \leq i_\alpha \leq n$, and M' is the lower bound of the illegitimate range of the reduced RRNS.*

PROOF. The illegitimate projection $\overline{X}_\Lambda$ can be treated as a number originating from X_Λ in the reduced RRNS affected by β or less errors corresponding to the moduli m_{j_1}, $m_{j_2}, \ldots, m_{j_\beta}$. From the definition of the reduced RRNS in (9.6.3), $\overline{X}_\Lambda$ can be expressed as follows:

$$
\begin{aligned}
\overline{X}_\Lambda &\equiv X_\Lambda + e' \frac{M M_R}{\prod_{\alpha=1}^{\lambda} m_{i_\alpha} \prod_{\alpha=1}^{\beta} m_{j_\alpha}} \quad \left(\text{mod } \frac{M M_R}{M_\Lambda}\right) \\
&= X_\Lambda + e' \left(\frac{M \prod_{\alpha=1}^{c} m_{k+\alpha}}{\prod_{\alpha=1}^{c} m_{i_\alpha}} \right) \frac{M_R}{\prod_{\alpha=1}^{c} m_{k+\alpha} \prod_{\alpha=c+1}^{\lambda} m_{i_\alpha} \prod_{\alpha=1}^{\beta} m_{j_\alpha}} \\
&= X_\Lambda + e' M' \frac{M_R}{\prod_{\alpha=1}^{c} m_{k+\alpha} \prod_{\alpha=c+1}^{\lambda} m_{i_\alpha} \prod_{\alpha=1}^{\beta} m_{j_\alpha}},
\end{aligned}
$$

where $e' < \prod_{\alpha=1}^{\beta} m_{j_\alpha}$, $i_c \le k + c$ and $i_{c+1} > k + c$. By Theorem 7.1,

$$M_R \ge \prod_{\alpha=1}^{c} m_{k+\alpha} \prod_{\alpha=c+1}^{\lambda} m_{i_\alpha} \prod_{\alpha=1}^{\beta} m_{j_\alpha}.$$

It follows that $\overline{X}_\Lambda \ge X_\Lambda + M' \ge M'$. This proves the theorem.

Therefore, by checking only the redundant mixed radix digits, it can be determined whether or not the projection is legitimate. If all the redundant mixed radix digits are zero, the projection is legitimate; otherwise, it is illegitimate. Based on the above analysis, a procedure for correcting double errors and detecting multiple errors can be outlined as follows:

Step 1: Based on the received residue vector, compute the mixed radix digits and check if all the redundant mixed radix digits are zero. If yes, then declare no error and stop. Otherwise, go to next step.

Step 2: Compute the mixed radix digits of the $m_i m_j$-projection for $i = 1, 2, \ldots, n-1$, $j = i + 1, 2, \ldots, n$. Check if all the redundant mixed radix digits are zero. If yes, then declare the ith and jth residue digits in error, correct them by BEX, and stop. Otherwise, go to next step.

Step 3: Declare more than two errors detected. Stop.

9.7   Computational Complexity Analysis

The present-day practical algorithms for residue error correction and detection mostly focus on single-error correction because of the considerations of memory space and computational inefficiency for multiple error correction. The algorithms developed above do not require large memory space as required by table lookup techniques. They are superior to the algorithms in Yau and Lin [1973] and Jenkins and Altman [1988] from a computational efficiency point of view. The comparison of our algorithm with those in Yau and Lin [1973] and Jenkins and Altman [1988] in terms of the requirement of MADD and MMULT is shown in Table 9.4. We should point out that for double error correction the expression in column 2 of Table 9.4 is obtained by extending the algorithm in [Jenkins and Altman 1988] as described in the previous section. A detailed computational complexity analysis for double error correction is given in Appendix B. For single burst error correction, our method only needs to find $n - k$ syndromes, whereas the method in Yau and Liu [1973] has to find n syndromes. The comparison is shown in Table 9.5. However, our algorithm needs some comparison to check solution consistency.

Table 9.4: MMULT and MADD Required for Single/Double Errors Correction in RRNS

♯ Error Correction	Yau and Liu Algorithm [1973]	Jenkins and Altman Algorithm [1988]	New Algorithm
1 $d = 3$	$\frac{1}{4}k^3 + \frac{5}{4}k^2 + \frac{1}{2}k - 2$ MMULT $\frac{1}{4}k^3 + \frac{5}{4}k^2 + \frac{3}{2}k$ MADD	$\frac{1}{2}k^3 + \frac{3}{2}k^2 + 2k$ MMULT $\frac{1}{2}k^3 + \frac{3}{2}k^2 + 2k$ MADD	$\frac{1}{2}k^2 + \frac{7}{2}k - 1$ MMULT $\frac{1}{2}k^2 + \frac{7}{2}k + 1$ MADD
2 $d = 5$	Not Applicable	$\frac{1}{4}k^4 + \frac{10}{4}k^3 + \frac{35}{4}k^2 + \frac{58}{4}k$ MMULT $\frac{1}{4}k^4 + \frac{10}{4}k^3 + \frac{35}{4}k^2 + \frac{58}{4}k$ MADD	$\frac{42}{4}k^2 - \frac{26}{4}k$ -2 MMULT $\frac{26}{4}k^2 - \frac{10}{4}k$ +2 MADD

Table 9.5: MMULT and MADD Required for Single Burst Error Correction in RRNS

	Yau and Liu Algorithm [1973]	New Algorithm	Modular Operations
$d = n - k + 1$ $= 2b + 1$	$\geq \frac{1}{b+1}[\frac{1}{2}k^3 + \frac{6b-1}{2}k^2$ $\quad +(4b^2 - 3b)k - 4b^2]$ $\geq \frac{1}{b+1}[\frac{1}{2}k^3 + \frac{6b-1}{2}k^2$ $\quad +(4b^2 - b)k]$	$\frac{1}{2}k^2 + \frac{14b-1}{2}k$ $\quad -(5b^2 - 4b)$ $\frac{1}{2}k^2 + \frac{12b-5}{2}k$ $\quad -(4b^2 - 7b + 2)$	MMULT MADD
$k = 8$ $b = 4$	≥ 268 ≥ 294	188 166	MMULT MADD
$k = 12$ $b = 4$	≥ 616 ≥ 648	338 292	MMULT MADD
$k = 16$ $b = 8$	≥ 1280 ≥ 1336	728 654	MMULT MADD

NOTES

In this chapter, the theory was extended and new algorithms for multiple error control in RRNS were presented. The focus in this work is on modulo arithmetic based computationally efficient algorithms for error control. As we noted earlier also, most of the previous work has been oriented towards single error control. The exceptions to this are the papers Mandelbaum [1976], [1978] and [1984] which describe decoding algorithms for multiple error correction based on continued fraction expansion and Euclid's algorithm. Since the algorithms in these papers need to process large valued integers and use an iterative process, they appear to be more suitable for a general purpose computer.

Finally, this chapter is based on the paper Sun and Krishna [1992]. Section 9.5 is taken from the review of one of the reviewers. Of course, he was anonymous at the time. He agreed to come forward upon mutual agreement after the paper was accepted for publication. He is Bruce Musicus of Bolt, Beranek and Newman, Cambridge, MA 021138. Thanks Bruce. We were fortunate to have you as a reviewer.

9.8 Bibliography

[9.1] F. Barsi and P. Maestrini, "Error Correcting Properties of Redundant Residue Number Systems," *IEEE Transactions on Computers*, vol. C-25, pp. 307-315, 1973.

[9.2] W.K. Jenkins and E.J. Altman, "Self-Checking Properties of Residue Number Error Checkers Based on Mixed Radix Conversion," *IEEE Transactions on Circuits and Systems*, vol. CAS-35, pp. 159-167, 1988.

[9.3] W.K. Jenkins, "The Design of Error Checkers for Self-Checking Residue Number Arithmetic," *IEEE Transactions on Computers*, vol. C-32, pp. 388-396, 1983.

[9.4] M.H. Etzel and W.K. Jenkins, "Redundant Residue Number Systems for Error Detection and Correction in Digital Filters," *IEEE Transactions on Acoustics, Speech, and Signal Processing*, vol. ASSP-28, pp. 538-544, 1980.

[9.5] S.S.-S. Yau and Y.-C. Liu, "Error Correction in Redundant Residue Number Systems," *IEEE Transactions on Computers*, vol. C-27, pp. 5-11, 1973.

[9.6] J.-D. Sun and H. Krishna, "A Coding Theory Approach to Error Control in Redundant Residue Number Systems, Part II: Multiple Error Detection and Correction," *IEEE Transactions on Circuits and Systems*, vol. 39, pp. 18-34, 1992.

[9.7] D.M. Mandelbaum, "On a Class of Arithmetic Codes and a Decoding Algorithm," *IEEE Transactions on Information Theory*, vol. IT-22, pp. 85-88, 1976.

[9.8] D.M. Mandelbaum, "Further Results on Decoding Arithmetic Residue Codes," *IEEE Transactions on Information Theory*, vol. IT-24, pp. 643-644, 1978.

[9.9] D.M. Mandelbaum, "An Approach to an Arithmetic Analog of Berlekamp's Algorithm," *IEEE Transactions on Information Theory*, vol. IT-30, pp. 758-762, 1984.

Chapter 10

Erasure and Error Control in RRNS

In some situations such as presence of interference or a transient malfunction, it may be desirable for the decoder to correct erasures in addition to errors. In this chapter we show that an RRNS code can be used as an error control code with a receiver that is capable of declaring a residue digit erased when the receiver recognizes the presence of interference or a transient malfunction. We establish the relationship between minimum distance and erasure correcting capability of the RRNS. Based on the mathematical framework developed in this chapter for the RRNS, new algorithms are derived for simultaneously correcting multiple erasures and single errors and detecting multiple errors.

10.1 Erasures in RRNS

Consider an (n, k) RRNS based on the moduli m_1, m_2, ..., m_k, m_{k+1}, ..., m_n. The first k moduli form the set of information moduli, and their product represents the range, M, of the RNS,

$$M = \prod_{i=1}^{k} m_i.$$

$\qquad(10.1.1)$

The remaining $n - k$ moduli form the set of parity moduli that allows error control in the RRNS. Let M_R be the product of parity moduli,

$$M_R = \prod_{i=k+1}^{n} m_i.$$

$\qquad(10.1.2)$

Any integer X in the range $[0, M)$, called code-integer, is encoded to an n-tuple residue vector $\underline{x}$, called codevector, that is,

$$X \longleftrightarrow \underline{x} = [x_1 \; x_2 \; \ldots \; x_n] \qquad (10.1.3)$$

where

$$x_i \equiv X \pmod{m_i}, \; 0 \leq x_i < m_i, \; i = 1, 2, \ldots, n. \qquad (10.1.4)$$

DEFINITION 10.1 An erasure in an RRNS code is an erased residue digit whose location is known.

Since, to the decoder, an erasure is an error whose location is known, the theorems in Section 7.2 can be generalized to the following theorems for the erasure case.

THEOREM 10.1 *The erasure correcting capability, l_e, of an RRNS Ω is $d - 1$.*

PROOF. Since for an (n, k) RRNS code, k correct residues uniquely determine the information integer, if no more than $d - 1 = n - k$ residue digits are erased, then the remaining residues can correctly determine the information integer. This proves the theorem.

THEOREM 10.2 *An RRNS is capable of correcting ρ or fewer erasures and simultaneously detecting β or fewer errors if $d \geq \beta + \rho + 1$.*

PROOF. It follows from Theorem 10.1 that an RRNS with $d \geq \beta + \rho + 1 > \rho + 1$ can correct ρ or fewer erasures. Now, consider that up to β $(\beta \geq 1)$ errors occur. If the decoder makes a decoding error, then the received vector with ρ or fewer erasures is decoded to a codevector $\hat{\underline{x}}$. It is clear that $\hat{\underline{x}} \neq \underline{x}$, where $\underline{x}$ is the transmitted codevector. Also,

$$d(\hat{\underline{x}}, \underline{x}) \leq \rho + \beta < d.$$

However, by Definition 7.3, if $d(\hat{\underline{x}}, \underline{x}) < d$, then $\hat{\underline{x}}$ can not be a codevector. This contradicts the original assumption. This proves the theorem.

THEOREM 10.3 *An RRNS is capable of correcting λ or fewer errors and simultaneously correcting ρ or fewer erasures if $d \geq 2\lambda + \rho + 1$.*

PROOF. It follows from Theorem 10.1 that an RRNS with $d \geq 2\lambda + \rho + 1 \geq \rho + 1$ can correct ρ or fewer erasures, and from Lemma 7.4 that an RRNS with $d \geq 2\lambda + \rho + 1 \geq 2\lambda + 1$ can correct λ or fewer errors. This proves the theorem.

THEOREM 10.4 *An RRNS is capable of correcting λ or fewer errors, correcting ρ or fewer erasures, and simultaneously detecting β ($\beta > \lambda$) or fewer errors if $d \geq \lambda + \beta + \rho + 1$.*

PROOF. It follows from theorems 10.1 and 10.3 that an RRNS with $d \geq \lambda + \beta + \rho + 1 > 2\lambda + \rho + 1 \geq \rho + 1$ can simultaneously correct ρ or fewer erasures and λ or fewer errors. Now, assume that more than λ but less than $\beta + 1$ errors occur. If the RRNS makes a decoding error, that is the received vector with ρ or fewer erasures is decoded to $\hat{x}$ but $\hat{x} \neq x$, where x is the transmitted codevector, then

$$d(\hat{x}, x) \leq \lambda + \rho + \beta < d.$$

However, by Definition 7.3, if $d(\hat{x}, x) < d$, then $\hat{x}$ cannot be a codevector. This contradicts the original assumption. This proves the theorem.

10.2 Consistency Checking

Given an (n, k) RRNS code, if β_1 errors and ρ_1 erasures occurred in information digits, and β_2 errors and ρ_2 erasures occurred in parity digits, then the received residue representation can be expressed as

$$[y_1 \, y_2 \, \cdots \, *_{i_1} \, \cdots \, *_{i_2} \, \cdots \, *_{i_{\rho_1}} \, \cdots \, y_k \, y_{k+1} \, \cdots \, y_{n-\rho_2} \, *_{n-\rho_2+1} \, \cdots \, *_n],$$

where $*_i$ denotes the erasure in the ith position. For the convenience of description and without any loss of generality, we assume that the last ρ_2 residue digits are erased. *The decoder decodes the received vector with zeros substituted for all the erasures.* The altered residue vector can, therefore, be represented as follows:

$$y = [y_1 \, y_2 \, \cdots \, y_n] \tag{10.2.1}$$

$$\begin{aligned}
y_i &= x_i, \, 1 \leq i \leq n, \, i \neq i_\alpha \neq j_\gamma, \alpha = 1, 2, \ldots, \rho_1 + \rho_2, \\
&\qquad\qquad \gamma = 1, 2, \ldots, \beta_1 + \beta_2, \\
y_{j_\gamma} &\equiv x_{j_\gamma} + e_{j_\gamma} \, (\text{mod } m_{j_\gamma}), 0 < e_{j_\gamma} < m_{j_\gamma}, \, 1 \leq j_\gamma \leq n - \rho_2, \\
&\qquad\qquad \gamma = 1, 2, \ldots, \beta_1 + \beta_2, \\
y_{i_\alpha} &= 0 \equiv x_{i_\alpha} + e_{i_\alpha} \, (\text{mod } m_{i_\alpha}), 0 < e_{i_\alpha} < m_{i_\alpha}, \, 1 \leq i_\alpha \leq n, \\
&\qquad\qquad \alpha = 1, 2, \ldots, \rho_1 + \rho_2.
\end{aligned}$$

where $\rho_1 + \rho_2 \leq \rho$, $\beta_1 + \beta_2 \leq \beta$. Here, $j_1, j_2, \ldots, j_{\beta_1+\beta_2}$ are the positions of errors, the corresponding error values being $e_{j_1}, e_{j_2}, \ldots, e_{j_{\beta_1+\beta_2}}$, and $i_1, i_2, \ldots,$

$i_{\rho_1+\rho_2}$ are the positions of erasures, the corresponding erasure values being $-e_{i_1}, -e_{i_2}, \ldots, -e_{i_{\rho_1+\rho_2}}$. Now, the altered information integer can be represented as

$$\overline{Y} \equiv X + E \pmod{M} \tag{10.2.2}$$
$$\longleftrightarrow [y_1 \, y_2 \, \cdots \, y_k]$$
$$\equiv [x_1 \, x_2 \, \cdots \, x_k]$$
$$+ [0 \, \cdots \, 0 \, e_{i_1} \, 0 \, \cdots \, 0 \, \cdots \, 0 \, e_{i_2} \, 0 \, \cdots \, 0 \, e_{i_{\rho_1}} \, 0 \, \cdots \, 0]$$
$$+ [0 \, \cdots \, 0 \, e_{j_1} \, 0 \, \cdots \, 0 \, \cdots \, 0 \, e_{j_2} \, 0 \, \cdots \, 0 \, e_{j_{\beta_1}} \, 0 \, \cdots \, 0],$$

where

$$X \longleftrightarrow [x_1 \, x_2 \, \cdots \, x_k]$$
$$E \longleftrightarrow [0 \, \cdots \, 0 \, e_{i_1} \, 0 \, \cdots \, 0 \, \cdots \, 0 \, e_{i_2} \, 0 \, \cdots \, 0 \, e_{i_{\rho_1}} \, 0 \, \cdots \, 0]$$
$$+ [0 \, \cdots \, 0 \, e_{j_1} \, 0 \, \cdots \, 0 \, \cdots \, 0 \, e_{j_2} \, 0 \, \cdots \, 0 \, e_{j_{\beta_1}} \, 0 \, \cdots \, 0].$$

Since $E \equiv 0 \pmod{m_i}$ for all $i \neq i_\alpha$ and $i \neq j_\gamma$, $\alpha = 1, 2, \ldots, \rho_1$, $\gamma = 1, 2, \ldots, \beta_1$, it is a multiple of all information moduli except $m_{i_1}, m_{i_2}, \ldots, m_{i_{\rho_1}}, m_{j_1}, m_{j_2}, \ldots, m_{j_{\beta_1}}$. Therefore, E is an integer of the type,

$$E = e' \frac{M}{\prod_{\alpha=1}^{\rho_1} m_{i_\alpha} \prod_{\gamma=1}^{\beta_1} m_{j_\gamma}}$$
$$\equiv e_{i_\alpha} \pmod{m_{i_\alpha}} \tag{10.2.3}$$
$$\equiv e_{j_\gamma} \pmod{m_{j_\gamma}},$$

where $0 < e' < \prod_{\alpha=1}^{\rho_1} m_{i_\alpha} \prod_{\gamma=1}^{\beta_1} m_{j_\gamma}$, $\prod_{\alpha=1}^{0} m_{i_\alpha} = 1$ and $\prod_{\gamma=1}^{0} m_{j_\gamma} = 1$. In addition $e' = 0$ for $\rho_1 = \beta_1 = 0$.

Based on information part $[y_1 \, y_2 \, \cdots \, y_k]$ and BEX, the parity digits are recomputed to get

$$\overline{y}_{k+r} \equiv \overline{Y} \pmod{m_{k+r}}, \quad r = 1, 2, \ldots, n - k, \tag{10.2.4}$$

and the syndromes Δ_r

$$\Delta_r \equiv (\overline{y}_{k+r} - y_{k+r}) \pmod{m_{k+r}}, \quad r = 1, 2, \ldots, n - k. \tag{10.2.5}$$

Since both E and X in (10.2.2) are less than M, we consider the following two cases:
(i) $X + E < M$. In this case,

$$\overline{Y} = X + E = X + e' \frac{M}{\prod_{\alpha=1}^{\rho_1} m_{i_\alpha} \prod_{\gamma=1}^{\beta_1} m_{j_\gamma}}, \tag{10.2.6}$$

and

$$\Delta_r \equiv e' \frac{M}{\prod_{\alpha=1}^{\rho_1} m_{i_\alpha} \prod_{\gamma=1}^{\beta_1} m_{j_\gamma}} \pmod{m_{k+r}},$$
$$k + r \neq j_\gamma, k + r \neq i_\alpha, \tag{10.2.7}$$

$$\Delta_r \equiv e' \frac{M}{\prod_{\alpha=1}^{\rho_1} m_{i_\alpha} \prod_{\gamma=1}^{\beta_1} m_{j_\gamma}} - e_{k+r} \pmod{m_{k+r}},$$
$$k + r = j_\gamma, \tag{10.2.8}$$

$$\Delta_r \equiv e' \frac{M}{\prod_{\alpha=1}^{\rho_1} m_{i_\alpha} \prod_{\gamma=1}^{\beta_1} m_{j_\gamma}} - e_{k+r} \pmod{m_{k+r}},$$
$$k + r = i_\alpha, \tag{10.2.9}$$

$\alpha = \rho_1 + 1, \rho_1 + 2, \ldots, \rho_1 + \rho_2$, $\gamma = \beta_1 + 1, \beta_1 + 2, \ldots, \beta_1 + \beta_2$ and $r = 1, 2, \ldots, n - k$.
(ii) $M \leq X + E < 2M$. In this case,

$$\overline{Y} = X + E - M = X + e' \frac{M}{\prod_{\alpha=1}^{\rho_1} m_{i_\alpha} \prod_{\gamma=1}^{\beta_1} m_{j_\gamma}} - M$$
$$= X + (e' - \prod_{\alpha=1}^{\rho_1} m_{i_\alpha} \prod_{\gamma=1}^{\beta_1} m_{j_\gamma}) \frac{M}{\prod_{\alpha=1}^{\rho_1} m_{i_\alpha} \prod_{\gamma=1}^{\beta_1} m_{j_\gamma}}, \tag{10.2.10}$$

and

$$\Delta_r \equiv (e' - \prod_{\alpha=1}^{\rho_1} m_{i_\alpha} \prod_{\gamma=1}^{\beta_1} m_{j_\gamma}) \frac{M}{\prod_{\alpha=1}^{\rho_1} m_{i_\alpha} \prod_{\gamma=1}^{\beta_1} m_{j_\gamma}} \pmod{m_{k+r}}$$
$$k + r \neq j_\gamma, \ k + r \neq i_\alpha; \tag{10.2.11}$$

$$\Delta_r \equiv (e' - \prod_{\alpha=1}^{\rho_1} m_{i_\alpha} \prod_{\gamma=1}^{\beta_1} m_{j_\gamma}) \frac{M}{\prod_{\alpha=1}^{\rho_1} m_{i_\alpha} \prod_{\gamma=1}^{\beta_1} m_{j_\gamma}}$$
$$-e_{k+r} \pmod{m_{k+r}}, \quad k + r = j_\gamma, \tag{10.2.12}$$

$$\Delta_r \equiv (e' - \prod_{\alpha=1}^{\rho_1} m_{i_\alpha} \prod_{\gamma=1}^{\beta_1} m_{j_\gamma}) \frac{M}{\prod_{\alpha=1}^{\rho_1} m_{i_\alpha} \prod_{\gamma=1}^{\beta_1} m_{j_\gamma}}$$
$$-e_{k+r} \pmod{m_{k+r}}, \quad k + r = i_\alpha, \tag{10.2.13}$$

where $\alpha = \rho_1 + 1, \rho_1 + 2, \ldots, \rho_1 + \rho_2$, $\gamma = \rho_1 + 1, \rho_1 + 2, \ldots, \beta_1 + \beta_2$ and $r = 1, 2, \ldots, n - k$.
A fundamental property of the syndrome digits is stated in the following theorem. This theorem can be proved in a manner similar to the proof of

the Theorem 9.1.

THEOREM 10.5 *Under the assumption that no more than β errors and no more than ρ erasures occur for an RRNS with minimum distance $d = \rho + \beta + 1$, one of the following three cases occurs: (note that the syndromes corresponding to the erasures in parity are not included.)*

Case 1: *If all the syndromes $\Delta_1, \Delta_2, \ldots, \Delta_{n-k-\rho_2}$ are zero, then no residue is in error.*

For the following cases, let p be the number of non-zero syndromes.

Case 2: *If $1 \le p \le \beta + \rho - (\rho_1 + \rho_2)$ and $\beta \ge 1$, then at least one residue (not including the zeros assigned to erasures) is in error.*

Case 3: *If $\beta + \rho - (\rho_1 + \rho_2) + 1 \le p \le \beta + \rho - \rho_2$ and $\rho_1 \ge 1$, then at least one information residue (including the zeros assigned to erasures) is in error.*

If Case 1 of Theorem 10.5 occurs, the correct values of erasures in the information and parity are zero and $\overline{y}_{k+r}$ respectively. Note that for $\rho_1 = 0$, (no erasure is in the information part) either Case 1 or Case 2 occurs. If Case 3 of Theorem 10.5 occurs, a procedure to determine the erasure values is described in the following sections.

10.3 Multiple-Erasure Correction

Under the assumption that no residue digit is in error, according to Theorem 10.1, an (n, k) RRNS code with $d - 1 = n - k = \rho$ can correct ρ erasures. Now, assuming that no error occurs and at least one ($\rho_1 \ge 1$) erasure takes place in the information part, (10.2.6–10.2.13) become (i) $X + E < M$. In this case,

$$\overline{Y} = X + e' \frac{M}{\prod_{\alpha=1}^{\rho_1} m_{i_\alpha}}, \tag{10.3.1}$$

and

$$\Delta_r \equiv e' \frac{M}{\prod_{\alpha=1}^{\rho_1} m_{i_\alpha}} \pmod{m_{k+r}},$$
$$r = 1, 2, \ldots, n - k - \rho_2, \tag{10.3.2}$$

$$\Delta_r \equiv e' \frac{M}{\prod_{\alpha=1}^{\rho_1} m_{i_\alpha}} - e_{k+r} \pmod{m_{k+r}},$$
$$r = n - k - \rho_2 + 1, \ldots, n - k, \tag{10.3.3}$$

where $0 < e' < \prod_{\alpha=1}^{\rho_1} m_{i_\alpha}$.

(ii) $M \leq X + E < 2M$. In this case,

$$
\begin{aligned}
\overline{Y} &= X + e' \frac{M}{\prod_{\alpha=1}^{\rho_1} m_{i_\alpha}} - M \\
&= X + (e' - \prod_{\alpha=1}^{\rho_1} m_{i_\alpha}) \frac{M}{\prod_{\alpha=1}^{\rho_1} m_{i_\alpha}},
\end{aligned}
\tag{10.3.4}
$$

and

$$
\Delta_r \equiv (e' - \prod_{\alpha=1}^{\rho_1} m_{i_\alpha}) \frac{M}{\prod_{\alpha=1}^{\rho_1} m_{i_\alpha}} \quad (\bmod\ m_{k+r}),
$$
$$
r = 1, 2, \ldots, n - k - \rho_2,
\tag{10.3.5}
$$

$$
\Delta_r \equiv (e' - \prod_{\alpha=1}^{\rho_1} m_{i_\alpha}) \frac{M}{\prod_{\alpha=1}^{\rho_1} m_{i_\alpha}} - e_{k+r} \quad (\bmod\ m_{k+r}),
$$
$$
r = n - k - \rho_2 + 1, \ldots, n - k,
\tag{10.3.6}
$$

where $0 < e' < \prod_{\alpha=1}^{\rho_1} m_{i_\alpha}$.

Given Δ_r, $r = 1, 2, \ldots, n - k$, based on (10.3.1–10.3.6), a procedure to determine erasure value $\widehat{x}_{i_\alpha}$ can be outlined as follows:

Solve the congruences

$$
\Delta_r \equiv e^{(1)} \frac{M}{\prod_{\alpha=1}^{\rho_1} m_{i_\alpha}} \quad (\bmod\ m_{k+r}),
$$
$$
r = 1, 2, \ldots, n - k - \rho_2,
\tag{10.3.7}
$$

$$
\Delta_r \equiv (e^{(2)} - \prod_{\alpha=1}^{\rho_1} m_{i_\alpha}) \frac{M}{\prod_{\alpha=1}^{\rho_1} m_{i_\alpha}} \quad (\bmod\ m_{k+r}),
$$
$$
r = 1, 2, \ldots, n - k - \rho_2,
\tag{10.3.8}
$$

to obtain the values $e^{(1)}$ and $e^{(2)}$. Then check if $e^{(1)} < \prod_{\alpha=1}^{\rho_1} m_{i_\alpha}$ or $e^{(2)} < \prod_{\alpha=1}^{\rho_1} m_{i_\alpha}$. The solution $e^{(1)}$ is consistent if $e^{(1)} < \prod_{\alpha=1}^{\rho_1} m_{i_\alpha}$. Similarly, the solution $e^{(2)}$ is consistent if $e^{(2)} < \prod_{\alpha=1}^{\rho_1} m_{i_\alpha}$. Let us say either of the two conditions is satisfied. Then y is decoded to the codevector $\widehat{x}$,

$$
\widehat{x} = [\widehat{x}_1\ \widehat{x}_2\ \ldots\ \widehat{x}_n],
$$

$$
\widehat{x}_j = y_j, \quad j = 1, 2, \ldots, n - \rho_2, \quad j \neq i_\alpha, \alpha = 1, \ldots, \rho_1,
$$

and

$$
\widehat{x}_{i_\alpha} \equiv -e^{(1)} \frac{M}{\prod_{\alpha=1}^{\rho_1} m_{i_\alpha}} \quad (\bmod\ m_{i_\alpha}), \quad \alpha = 1, \ldots, \rho_1,
\tag{10.3.9}
$$

$$\widehat{x}_{k+r} \equiv \Delta_r - e^{(1)}\frac{M}{\prod_{\alpha=1}^{\rho_1} m_{i_\alpha}} \pmod{m_{k+r}},$$
$$k + r = n - \rho_2 + 1, \ldots, n, \tag{10.3.10}$$

if $e^{(1)}$ is a consistent solution to (10.3.7); and

$$\widehat{x}_{i_\alpha} \equiv -e^{(2)}\frac{M}{\prod_{\alpha=1}^{\rho_1} m_{i_\alpha}} \pmod{m_{i_\alpha}}, \quad \alpha = 1, \ldots, \rho_1, \tag{10.3.11}$$

$$\widehat{x}_{k+r} \equiv \Delta_r + \left(\prod_{\alpha=1}^{\rho_1} m_{i_\alpha} - e^{(2)}\right)\frac{M}{\prod_{\alpha=1}^{\rho_1} m_{i_\alpha}} \pmod{m_{k+r}},$$
$$k + r = n - \rho_2 + 1, \ldots, n, \tag{10.3.12}$$

if $e^{(2)}$ is a consistent solution to (10.3.8).

It is clear from (10.3.2) and (10.3.7) that if $X + E < M$, then e' is a solution to (10.3.7), and $0 < e^{(1)} = e' < \prod_{\alpha=1}^{\rho_1} m_{i_\alpha}$. Similarly, from (10.3.5) and (10.3.8), if $M \le X + E < 2M$, then e' is a solution to (10.3.8), and $0 < e^{(2)} = e' < \prod_{\alpha=1}^{\rho_1} m_{i_\alpha}$. However, it remains to be shown that if $0 < e^{(1)} < \prod_{\alpha=1}^{\rho_1} m_{i_\alpha}$, then $e^{(2)} > \prod_{\alpha=1}^{\rho_1} m_{i_\alpha}$; and if $0 < e^{(2)} < \prod_{\alpha=1}^{\rho_1} m_{i_\alpha}$, then $e^{(1)} > \prod_{\alpha=1}^{\rho_1} m_{i_\alpha}$. Under an additional constraint on the form of the moduli, this result is established in the following theorem.

THEOREM 10.6 *Assume that no residue digits is in error and no more than ρ erasures occur for an RRNS with $d - 1 = n - k = \rho$. If the moduli satisfy the constraint*

$$min\{m_{r_1} m_{r_2}\} > max\{2m_{i_1} m_{i_2} - m_{i_1} - m_{i_2}\} \quad \text{for } \rho > 1,$$
$$min\{m_{r_1}\} > max\{2m_{i_1} - 1\} \quad \text{for } \rho = 1,$$

$k < r_1, r_2 \le n$ *and* $1 \le i_1, i_2 \le k$, *then* $e^{(2)} > \prod_{\alpha-1}^{\rho_1} m_{i_\alpha}$ *if* $0 < e^{(1)} < \prod_{\alpha=1}^{\rho_1} m_{i_\alpha}$, *and* $e^{(1)} > \prod_{\alpha-1}^{\rho_1} m_{i_\alpha}$ *if* $0 < e^{(2)} < \prod_{\alpha=1}^{\rho_1} m_{i_\alpha}$.

PROOF. If $e^{(1)}$ and $e^{(2)}$ are the solutions to (10.3.7) and (10.3.8), respectively, where $0 < e^{(1)} < \prod_{\alpha=1}^{\rho_1} m_{i_\alpha}$ and $0 < e^{(2)} < \prod_{\alpha=1}^{\rho_1} m_{i_\alpha}$, then comparing these two equations we get

$$\left(e^{(1)} - e^{(2)} + \prod_{\alpha=1}^{\rho_1} m_{i_\alpha}\right)\frac{M}{\prod_{\alpha=1}^{\rho_1} m_{i_\alpha}} \equiv 0 \pmod{M'_R},$$

where $M'_R = \prod_{r=1}^{n-k-\rho_2} m_{k+r}$. This implies that $(e^{(1)} - e^{(2)} + \prod_{\alpha=1}^{\rho_1} m_{i_\alpha})$ is a multiple of M'_R. This is not possible as

$$0 < \left(e^{(1)} - e^{(2)} + \prod_{\alpha=1}^{\rho_1} m_{i_\alpha}\right) < 2\prod_{\alpha=1}^{\rho_1} m_{i_\alpha} < M'_R.$$

If $e^{(1)}$, $0 < e^{(1)} < \prod_{\alpha=1}^{\rho_1} m_{i_\alpha}$, is a solution to (10.3.7), then from (10.3.5) we get

$$e^{(2)} = (e^{(1)} + \prod_{\alpha=1}^{\rho_1} m_{i_\alpha}) > \prod_{\alpha=1}^{\rho_1} m_{i_\alpha}.$$

The case $e^{(1)} > \prod_{\alpha-1}^{\rho_1} m_{i_\alpha}$ for $0 < e^{(2)} < \prod_{\alpha=1}^{\rho_1} m_{i_\alpha}$ can also be analyzed in an analogous manner. This proves the theorem.

Based on the concepts developed above and under the assumption that the moduli m_i, $i = 1, 2, \ldots, n$, satisfy the condition in Theorem 10.6, the algorithm to correct multiple erasures is given in Table 10.1.

EXAMPLE 10.1 Consider a $(7,6)$ RRNS code based on the moduli $m_1 = 23$, $m_2 = 25$, $m_3 = 27$, $m_4 = 29$, $m_5 = 31$, $m_6 = 32$ and $m_7 = 67$, where m_7 is the redundant modulus. Clearly, $m_7 > 2m_i - 1$, $i = 1, 2, \ldots, 6$, and therefore, the condition in Theorem 10.6 is satisfied for single erasure correction. $M = \prod_{i=1}^{6} m_i = 446,623,200$. Let $X = 400,000,000$, then $\underline{x} = [8\ 0\ 22\ 13\ 25\ 0\ 17]$. Since $0 \le X < M$, $\underline{x}$ is a codevector. Assume that one erasure takes place in the first position and the received vector is $[\ast\ 0\ 22\ 13\ 25\ 0\ 17]$.

After substituting zero for the erasure $\ast$, the received vector becomes $[0\ 0\ 22\ 13\ 25\ 0\ 17]$. Based on the information part $[0\ 0\ 22\ 13\ 25\ 0]$ and BEX, we compute the syndrome digit $\Delta_1 \equiv 14 \pmod{67}$. Then from (10.3.7) and (10.3.8), we obtain

$$e^{(1)} \equiv \Delta_1 m_1 M^{-1} \pmod{m_7} \equiv 58,$$

$$e^{(2)} \equiv (\Delta_1 m_1 M^{-1} + m_1) \pmod{m_7} \equiv 14.$$

Since $e^{(1)} > m_1$ and $e^{(2)} < m_1$, $e^{(2)}$ is the consistent solution and the erasure value is

$$\hat{x}_1 \equiv -e^{(2)} \frac{M}{m_1} \pmod{m_1} \equiv 8.$$

EXAMPLE 10.2 Consider an $(8,6)$ RRNS based on the moduli $m_1 = 23$, $m_2 = 25$, $m_3 = 27$, $m_4 = 29$, $m_5 = 31$, $m_6 = 32$, $m_7 = 41$ and $m_8 = 47$, where m_7 and m_8 are redundant moduli. Clearly, $m_7 m_8 = 1927 > 2m_5 m_6 - m_5 - m_6 = 1921$, and therefore, the condition in Theorem 10.6 are satisfied for correcting double erasures. Let $X = 400,000,000 < M = \prod_{i=1}^{6} m_i = 446,623,200$. Therefore, $\underline{x} = [8\ 0\ 22\ 13\ 25\ 0\ 23\ 14]$ is a codevector. Assume that two erasures take place in the first and fourth residue digits, and the received vector is $[\ast\ 0\ 22\ \ast\ 25\ 0\ 23\ 14]$. After substituting zeros for the erasures, the received vector becomes $[0\ 0\ 22\ 0\ 25\ 0\ 23\ 14]$. Based on the information part $[0\ 0\ 22\ 0\ 25\ 0]$ and BEX, we compute the syndrome digits $\Delta_1 \equiv 22 \pmod{41}$, $\Delta_2 \equiv 35 \pmod{47}$. Following the decoding algorithm,

Table 10.1: An Algorithm for Multiple Erasures Correction in RRNS

Step 1:	According to the received vector with zeros substituted for all erasures, compute the syndromes using BEX as given in Table 7.1 and $\Delta_r \equiv \bar{y}_{k+r} - y_{k+r} \pmod{m_{k+r}}$, $r = 1, 2, \ldots, n - k$.
Step 2:	Check how many syndromes are zero (without including the syndromes corresponding to the erasures in the parity).
	(1). If all the syndromes are zero, then the values of erasures in the information are all zero, and $\bar{y}_{k+r}$ is the value of the rth parity residue. Stop.
	(2). If $\rho - (\rho_1 + \rho_2) + 1$ to all syndromes are non-zero, then go to Step 3.
Step 3:	Perform consistency-checking:
	(1). Compute $e^{(1)}$ and $e^{(2)}$ using (10.3.7) and (10.3.8), respectively.
	(2). Check the consistency of the solutions: $e^{(1)} < \prod_{\alpha=1}^{\rho_1} m_{i_\alpha}$ or $e^{(2)} < \prod_{\alpha=1}^{\rho_1} m_{i_\alpha}$ where m_{i_α} are the nonredundant moduli corresponding to the erasures in the information digits.
	If it is consistent, then correct the erasures using (10.3.9–10.3.12) and stop. Otherwise, go to next step.
Step 4:	Declare one or more errors detected and stop.

we check the consistency of the solutions. From (10.3.7) and (10.3.8) we obtain $e^{(1)} = 1850 > m_1 m_4$ and $e^{(2)} = 590 < m_1 m_4$. Since $e^{(2)}$ is a consistent solution, the erasure values are computed as

$$\widehat{x}_1 \equiv -e^{(2)} \frac{M}{m_1 m_4} \pmod{m_1} \equiv 8$$

and

$$\widehat{x}_4 \equiv -e^{(2)} \frac{M}{m_1 m_4} \pmod{m_4} \equiv 13.$$

EXAMPLE 10.3 Consider the same RRNS as in Example 10.1. Let $X = 400,000,000$, $\underline{x} = [8\ 0\ 22\ 13\ 25\ 0\ 17]$. Now, assume that one erasure and one error take place in the first and third residue digits, respectively, and the received vector is $[*\ 0\ 23\ 13\ 25\ 0\ 17]$. After substituting zero for the erasure $*$, the received vector becomes $[0\ 0\ 23\ 13\ 25\ 0\ 17]$. Based on the information part $[0\ 0\ 23\ 13\ 25\ 0]$ and BEX, we compute syndrome digit $\Delta_1 \equiv 23 \pmod{67}$. Then from (10.3.7) and (10.3.8) we obtain $e^{(1)} = 57 > m_1$ and $e^{(2)} = 13 < m_1$, and the estimated erasure value is given by $\widehat{x}_1 = 14$. It is obvious that the decoder makes a decoding error.

EXAMPLE 10.4 Consider the same RRNS as in Example 10.2. Let $X = 400,000,000$, $\underline{x} = [8\ 0\ 22\ 13\ 25\ 0\ 23\ 14]$. Now, assume that two erasures and one error take place in the first, fourth and seventh residue digits, respectively, the received vector being $[*\ 0\ 22\ *\ 25\ 0\ 16\ 14]$. After substituting zeros for the erasures, the received vector becomes $[0\ 0\ 22\ 0\ 25\ 0\ 16\ 14]$. Based on the information part $[0\ 0\ 22\ 0\ 25\ 0]$ and BEX, we compute the syndromes $\Delta_1 \equiv 29 \pmod{m_7}$ and $\Delta_2 \equiv 35 \pmod{m_8}$. Then, from (10.3.7) and (10.3.8) we obtain $e^{(1)} = 534 < m_1 m_4$ and $e^{(2)} = 1201 > m_1 m_4$. The erasure values are computed as

$$\widehat{x}_1 \equiv -e^{(1)} \frac{M}{m_1 m_4} \pmod{m_1} \equiv 18$$

and

$$\widehat{x}_4 \equiv -e^{(1)} \frac{M}{m_1 m_4} \pmod{m_4} \equiv 4.$$

It is obvious that the decoder makes a decoding error. The following section will deal with this problem.

10.4 Erasure Correction and Error Detection

It was established in Theorem 10.2 that an RRNS with minimum distance $d = \rho + \beta + 1$ can correct ρ or fewer erasures and simultaneously detect the presence of β or fewer errors. A procedure to determine the erasure values is the same as in the previous section. Consistency-checking is performed to test if the solution to the congruences in (10.3.7) or (10.3.8) is consistent. If either of the two conditions is satisfied, then y is decoded to the codevector $\hat{x}$ as given in (10.3.9–10.3.12). It is clear from (10.3.1, 10.3.2, 10.3.7) that if $X + E < M$, then $e^{(1)} = e'$ is a consistent solution to (10.3.7). Similarly, if $M \leq X + E < 2M$, then $e^{(2)} = e'$ is a consistent solution to (10.3.8). However, it remains to be shown that if at least one residue digit is in error, then neither $e^{(1)}$ nor $e^{(2)}$ will be a consistent solution. Under an additional constraint on the form of the moduli, this result is established in the following theorem.

THEOREM 10.7 *Assume that no more than β ($\beta \geq 1$) errors and no more than ρ ($\rho \geq 1$) erasures occur in an RRNS with minimum distance $d = \rho + \beta + 1$. If the moduli are such that there are do not exist integers $n_{\rho_1 \beta_1}$, n_{ρ_1}, $0 < n_{\rho_1 \beta_1} < \prod_{\alpha=1}^{\rho_1} m_{i_\alpha} \prod_{\gamma=1}^{\beta_1} m_{j_\gamma}$, $0 < n_{\rho_1} < \prod_{\alpha=1}^{\rho_1} m_{i_\alpha}$, $1 \leq i_\alpha, j_\gamma \leq k$, that satisfy the equation*

$$n_{\rho_1 \beta_1} + n_{\rho_1} \prod_{\gamma=1}^{\beta_1} m_{j_\gamma} = \prod_{i=1}^{\rho_1+\beta_1} m_{r_i}, \quad k < r_i \leq n, \tag{10.4.1}$$

$1 \leq \rho_1 \leq \rho$ and $1 \leq \beta_1 \leq \beta$, and at least one residue digit is received in error, then neither $e^{(1)}$ nor $e^{(2)}$ is a consistent solution.

PROOF. Assume that at least one residue digit is received in error. For $X + E < M$, if $e^{(1)}$ is the consistent solution to (10.3.7), then comparing (10.2.7) to (10.3.7) we get

$$(e' - e^{(1)}) \prod_{\gamma=1}^{\beta_1} m_{j_\gamma}) \frac{M}{\prod_{\alpha=1}^{\rho_1} m_{i_\alpha} \prod_{\gamma=1}^{\beta_1} m_{j_\gamma}} \equiv 0 \; (\mathrm{mod} \prod_{\substack{i=1 \\ k+r \neq j_\gamma}}^{n-k-\rho_2} m_{k+r}),$$

$$\gamma = \beta_1 + 1, \ldots, \beta_1 + \beta_2,$$

that is, $(e' - e^{(1)} \prod_{\gamma=1}^{\beta_1} m_{j_\gamma})$ is a multiple of

$$\prod_{\substack{i=1 \\ k+r \neq j_\gamma}}^{n-k-\rho_2} m_{k+r}, \; \gamma = \beta_1 + 1, \ldots, \beta_1 + \beta_2.$$

This is not possible as

$$|e' - e^{(1)} \prod_{\gamma=1}^{\beta_1} m_{j_\gamma}| < \prod_{\alpha=1}^{\rho_1} m_{i_\alpha} \prod_{\gamma=1}^{\beta_1} m_{j_\gamma} < \prod_{\substack{i=1 \\ k+r \neq j_\gamma}}^{n-k-\rho_2} m_{k+r},$$

$$\gamma = \beta_1 + 1, \ldots, \beta_1 + \beta_2.$$

Similarly, for $X + E < M$, if $e^{(2)}$ is the consistent solution to (10.3.8), then comparing (10.2.7) to (10.3.8), we get

$$(e' + (\prod_{\alpha=1}^{\rho_1} m_{i_\alpha} - e^{(2)}) \prod_{\gamma=1}^{\beta_1} m_{j_\gamma}) \frac{M}{\prod_{\alpha=1}^{\rho_1} m_{i_\alpha} \prod_{\gamma=1}^{\beta_1} m_{j_\gamma}} \equiv 0$$

$$(\bmod \prod_{\substack{i=1 \\ k+r \neq j_\gamma}}^{n-k-\rho_2} m_{k+r}), \ \gamma = \beta_1 + 1, \ldots, \beta_1 + \beta_2.$$

This can not hold as

$$(e' + (\prod_{\alpha=1}^{\rho_1} m_{i_\alpha} - e^{(2)}) \prod_{\gamma=1}^{\beta_1} m_{j_\gamma}) < 2 \prod_{\alpha=1}^{\rho_1} m_{i_\alpha} \prod_{\gamma=1}^{\beta_1} m_{j_\gamma} <$$

$$\prod_{\substack{i=1 \\ k+r \neq j_\gamma}}^{n-k-\rho_2} m_{k+r}, \ \gamma = \beta_1 + 1, \ldots, \beta_1 + \beta_2,$$

for $\rho_1 + \rho_2 < \rho$ and/or $\beta_1 + \beta_2 < \beta$, and (10.4.1) holds for $\rho_1 + \rho_2 = \rho$ and $\beta_1 + \beta_2 = \beta$. The case for $M \leq X + E < 2M$ can also be analyzed in an analogous manner. This proves the theorem.

LEMMA 10.1 *Assume that no more than β ($\beta \geq 1$) errors and no more than ρ ($\rho \geq 1$) erasures occur for an RRNS with minimum distance $d = \rho + \beta + 1$. If the moduli satisfy the condition*

$$min\{m_{r_1} m_{r_2}\} > max\{2m_{i_1} m_{i_2} - m_{i_1} - m_{i_2}\}, \tag{10.4.2}$$

$k < r_1, r_2 \leq n$, $1 \leq i_1, i_2 \leq k$, *and at least one residue digit is received in error, then neither $e^{(1)}$ nor $e^{(2)}$ is a consistent solution.*

PROOF. The condition is sufficient. If the moduli satisfy (10.4.2), then the condition in Theorem 10.7 is satisfied trivially. This completes the proof.

Based on the concept developed above and under the assumption that the moduli m_i, i=1, 2, ..., n, satisfy the conditions in Theorem 10.7 or Lemma 10.1, the algorithm to correct multiple erasures and simultaneously detect the presence of multiple errors is given in Table 10.2. A flowchart for this algorithm is given in Figure 10.1.

EXAMPLE 10.5 Consider an (8,6) RRNS based on the moduli $m_1 = 23$, $m_2 = 25$, $m_3 = 27$, $m_4 = 29$, $m_5 = 31$, $m_6 = 32$, $m_7 = 41$ and $m_8 = 47$, where m_7 and m_8 are redundant moduli. Clearly, $m_7 m_8 > 2m_5 m_6 - m_5 - m_6$, and therefore, the conditions in Lemma 10.1 and Theorem 10.7 are satisfied for simultaneously correcting single erasure and detecting single error. Let $X = 400,000,000 < M = \prod_{i=1}^{6} m_i = 446,623,200$. Therefore, $\underline{x} = [8\ 0\ 22\ 13\ 25\ 0\ 23\ 14]$ is a codevector. Assume that one erasure takes place in the first residue digit and the received vector is $[*\ 0\ 22\ 13\ 25\ 0\ 23\ 14]$. After substituting zero for the erasure $*$, the received vector becomes $[0\ 0\ 22\ 13\ 25\ 0\ 23\ 14]$. Based on the information part $[0\ 0\ 22\ 13\ 25\ 0]$ and BEX, we compute the syndrome digits $\Delta_1 \equiv 16\ (\text{mod } 41)$ and $\Delta_2 \equiv 46\ (\text{mod } 47)$. Following the decoding algorithm, we check the consistency of the solutions. From (10.3.7) and (10.3.8), we obtain $e^{(1)} = 1918 > m_1$ and $e^{(2)} = 14 < m_1$. Since $e^{(2)}$ is a consistent solution, the erasure value is

$$\hat{x}_1 \equiv -e^{(2)} \frac{M}{m_1} \quad (\text{mod } m_1) \equiv 8.$$

Now, assume the same situation as in Example 10.3, that is, one erasure and one error take place in the first and third residue digits, respectively. The received vector is $[*\ 0\ 23\ 13\ 25\ 0\ 23\ 14]$. After substituting zero for the erasure $*$, the received vector becomes $[0\ 0\ 23\ 13\ 25\ 0\ 23\ 14]$. Based on the information part $[0\ 0\ 23\ 13\ 25\ 0]$ and BEX, we compute the syndrome digits $\Delta_1 \equiv 32\ (\text{mod } 41)$ and $\Delta_2 \equiv 20\ (\text{mod } 47)$. Following the decoding algorithm, we check the consistency of the solutions. From (10.3.7) and (10.3.8), we obtain $e^{(1)} = 556 > m_1$ and $e^{(2)} = 579 > m_1$. Since there is no consistent solution, we declare one or more errors detected.

EXAMPLE 10.6 Consider a (9,6) RRNS code based on the moduli $m_1 = 23$, $m_2 = 25$, $m_3 = 27$, $m_4 = 29$, $m_5 = 31$, $m_6 = 32$, $m_7 = 41$, $m_8 = 47$ and $m_9 = 53$, where m_7, m_8 and m_9 are the redundant moduli. Clearly, $m_7 m_8 > 2m_5 m_6 - m_5 - m_6$, and therefore, the conditions in Theorem 10.7 and Lemma 10.1 are satisfied. Let $X = 400,000,000 < \prod_{i=1}^{6} m_i = 446,623,200$. Then $\underline{x} = [8\ 0\ 22\ 13\ 25\ 0\ 23\ 14\ 43]$ is a codevector. Assume that two erasures and one error take place in the first, fourth and seventh positions, respectively. Let the received vector be $[*\ 0\ 22\ *\ 25\ 0\ 16\ 14\ 43]$. After substituting zeros for the erasures, the received vector becomes $[0\ 0\ 22\ 0\ 25\ 0\ 16\ 14\ 43]$. Based on the information part $[0\ 0\ 22\ 0\ 25\ 0]$ and BEX, we compute the syndrome digits $\Delta_1 \equiv 29\ (\text{mod } 41)$, $\Delta_2 \equiv 35\ (\text{mod } 47)$ and $\Delta_3 \equiv 48\ (\text{mod } 53)$. Following the decoding algorithm, we check the consistency of the solutions. From (10.3.7) and (10.3.8), we obtain $e^{(1)} = 62,198 > m_1 m_4$ and $e^{(2)} = 62,865 > m_1 m_4$. Since there is no consistent solution, we declare one or more errors detected.

Table 10.2: An Algorithm for Multiple Erasure Correction and Multiple Error Detection

Step 1:	According to the received vector with zeros substituted for all the erasures, compute the syndromes using BEX as given in Table 7.1 and $\Delta_r \equiv \bar{y}_{k+r} - y_{k+r} \pmod{m_{k+r}}$, $r = 1, 2, \ldots, n - k$.
Step 2:	Check how many syndromes are zero (without including the syndromes corresponding to the erasures in parity). (1). If all the syndromes are zero, then the values of erasures in the information and parity are zero and $\bar{y}_{k+r}$, respectively. Stop. (2). If 1 to $\beta + \rho - (\rho_1 + \rho_2)$ $(\beta \geq 1)$ syndromes are non-zero, then go to Step 4. (3). If $\beta + \rho - (\rho_1 + \rho_2) + 1$ to $\beta + \rho - \rho_2$ $(\rho_1 \geq 1)$ syndromes are non-zero, then go to Step 3.
Step 3:	Perform consistency-checking: (1). Compute $e^{(1)}$ and $e^{(2)}$ using (10.3.7) and (10.3.8), respectively. (2). Check the consistency of the solutions: $e^{(1)} < \prod_{\alpha=1}^{\rho_1} m_{i_\alpha}$ or $e^{(2)} < \prod_{\alpha=1}^{\rho_1} m_{i_\alpha}$ where m_{i_α} are the nonredundant moduli corresponding to the erasures in the information digits. If it is consistent, then correct the erasures using (10.3.9–10.3.12) and stop. Otherwise, go to next step.
Step 4:	Declare that one or more errors are detected and stop.

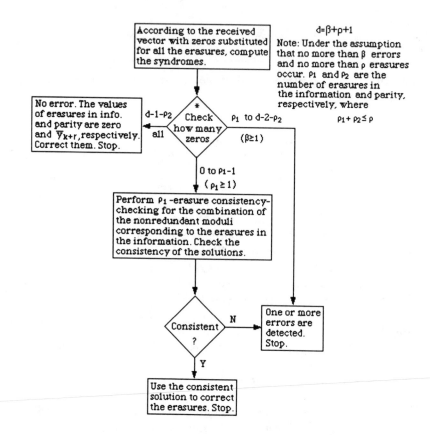

According to the received vector with zeros substituted for all the erasures, compute the syndromes.

$d=\beta+\rho+1$

Note: Under the assumption that no more than β errors and no more than ρ erasures occur. ρ_1 and ρ_2 are the number of erasures in the information and parity, respectively, where

$$\rho_1+\rho_2\le\rho$$

No error. The values of erasures in info. and parity are zero and $\overline{y}_{k+r}$, respectively. Correct them. Stop.

$d-1-\rho_2$ / all

Check how many zeros

ρ_1 to $d-2-\rho_2$

$(\beta\ge1)$

0 to ρ_1-1
$(\rho_1\ge1)$

Perform ρ_1-erasure consistency-checking for the combination of the nonredundant moduli corresponding to the erasures in the information. Check the consistency of the solutions.

Consistent ?

N → One or more errors are detected. Stop.

Y

Use the consistent solution to correct the erasures. Stop.

* Without including the syndromes corresponding to the erasures in the parity.

Figure 10.1: A Decoder Flowchart for Multiple Erasure Correction and Multiple Error Detection

10.5   Error and Erasure Correction

An RRNS with minimum distance $d = \lambda + \beta + \rho + 1$ $(\beta > \lambda)$ can correct λ or fewer errors, ρ or fewer erasures and simultaneously detect β or fewer errors. A fundamental property of the syndrome digits is stated in the following theorem. The proof of this theorem is similar to the proof of Theorem 9.1.

THEOREM 10.8 *Under the assumption that no more than β errors and ρ erasures occur, one of the following four cases occurs (the syndromes corresponding to the erasures in parity are not included):*

Case 1: *If all the syndromes $\Delta_1, \Delta_2, \ldots, \Delta_{n-k-\rho_2}$ are zero, then no residue digit is in error, and the values of erasures in the information and parity are zero and $\overline{y}_{k+r}$, respectively.*

For the following cases, let p be the number of non-zero syndromes.

Case 2: *If $1 \le p \le \lambda$, then exactly p corresponding parity digits are in error, and all other residue digits are correct. The values of erasures in the information are zero and the values of errors and erasures in the parity are $\overline{y}_{k+r}$.*

Case 3: *If $\lambda + 1 \le p \le \beta + \rho - (\rho_1 + \rho_2)$, then more than λ residue digits (not including zeros assigned to erasures) are in error.*

Case 4: *If $\beta + \rho - (\rho_1 + \rho_2) + 1 \le p \le n - k - \rho_2$, then at least one of the information digits (including zeros assigned to erasures) is in error.*

If Case 4 of Theorem 10.8 occurs, a procedure to determine error location and error and erasure values for single-error $(\lambda = 1)$ and multiple-erasure correction and multiple-error detection is described in the following. We consider the following two situations.
Situation (A): one error takes place in the information part.
Situation (B): one error takes place in the parity part.
Situation (A): Assuming that one error takes place in the pth information digit, (10.2.6–10.2.13) become
(i) $X + E < M$. In this case,

$$\overline{Y} = X + E = X + e' \frac{M}{m_p \prod_{\alpha=1}^{\rho_1} m_{i_\alpha}}, \tag{10.5.1}$$

and

$$\Delta_r \equiv e' \frac{M}{m_p \prod_{\alpha=1}^{\rho_1} m_{i_\alpha}} \pmod{m_{k+r}}$$

$$r = 1, 2, \ldots, n - k - \rho_2, \tag{10.5.2}$$

$$\Delta_r \equiv e' \frac{M}{m_p \prod_{\alpha=1}^{\rho_1} m_{i_\alpha}} - e_{k+r} \pmod{m_{k+r}}$$

$$k + r = n - \rho_2 + 1, \ldots, n, \tag{10.5.3}$$

where $0 < e' < m_p \prod_{\alpha=1}^{\rho_1} m_{i_\alpha}$ and $\prod_{\alpha=1}^{0} m_{i_\alpha} = 1$ for $\rho_1 = 0$.
(ii) $M \leq X + E < 2M$. In this case,

$$\begin{aligned} \overline{Y} &= X + E - M = X + e' \frac{M}{m_p \prod_{\alpha=1}^{\rho_1} m_{i_\alpha}} - M \\ &= X + (e' - m_p \prod_{\alpha=1}^{\rho_1} m_{i_\alpha}) \frac{M}{m_p \prod_{\alpha=1}^{\rho_1} m_{i_\alpha}}, \end{aligned} \tag{10.5.4}$$

and

$$\Delta_r \equiv (e' - m_p \prod_{\alpha=1}^{\rho_1} m_{i_\alpha}) \frac{M}{m_p \prod_{\alpha=1}^{\rho_1} m_{i_\alpha}} \pmod{m_{k+r}}$$

$$r = 1, 2, \ldots, n - k - \rho_2, \tag{10.5.5}$$

$$\Delta_r \equiv (e' - m_p \prod_{\alpha=1}^{\rho_1} m_{i_\alpha}) \frac{M}{m_p \prod_{\alpha=1}^{\rho_1} m_{i_\alpha}} - e_{k+r} \pmod{m_{k+r}}$$

$$k + r = n - \rho_2 + 1, \ldots, n, \tag{10.5.6}$$

where $0 < e' < m_p \prod_{\alpha=1}^{\rho_1} m_{i_\alpha}$ and $\prod_{\alpha=1}^{0} m_{i_\alpha} = 1$ for $\rho_1 = 0$.

Given Δ_r, $r = 1, 2, \ldots, n - k$, based on (10.5.1–10.5.6) a procedure to determine error location, error and erasure values can be outlined as follows: For $j = 1, 2, \ldots, k$, $j \neq i_\alpha$, $\alpha = 1, 2, \ldots, \rho_1$, solve the congruences

$$\Delta_r \equiv e_j^{(1)} \frac{M}{m_j \prod_{\alpha=1}^{\rho_1} m_{i_\alpha}} \pmod{m_{k+r}}$$

$$r = 1, 2, \ldots, n - k - \rho_2, \tag{10.5.7}$$

$$\Delta_r \equiv (e_j^{(2)} - m_j \prod_{\alpha=1}^{\rho_1} m_{i_\alpha}) \frac{M}{m_j \prod_{\alpha=1}^{\rho_1} m_{i_\alpha}} \pmod{m_{k+r}}$$

$$r = 1, 2, \ldots, n - k - \rho_2, \tag{10.5.8}$$

to get the values $e_j^{(1)}$ and $e_j^{(2)}$. Then check if $e_j^{(1)} < m_j \prod_{\alpha=1}^{\rho_1} m_{i_\alpha}$ or $e_j^{(2)} < m_j \prod_{\alpha=1}^{\rho_1} m_{i_\alpha}$. The solution $e_j^{(1)}$ is consistent if $e_j^{(1)} < m_j \prod_{\alpha=1}^{\rho_1} m_{i_\alpha}$. Similarly, the solution $e_j^{(2)}$ is consistent if $e_j^{(2)} < m_j \prod_{\alpha=1}^{\rho_1} m_{i_\alpha}$. Let us say either of the two conditions is satisfied for $j = l$. Then $\underline{y}$ is decoded to the codevector $\hat{\underline{x}}$,

$$\hat{x}_j = y_j, \quad j = 1, 2, \ldots, n - \rho_2, \quad j \neq i_\alpha, \ j \neq l, \alpha = 1, 2, \ldots, \rho_1,$$

$$\hat{x}_{i_\alpha} \equiv -e_l^{(1)} \frac{M}{m_l \prod_{\alpha=1}^{\rho_1} m_{i_\alpha}} \pmod{m_{i_\alpha}}, \quad \alpha = 1, 2, \ldots, \rho_1 \quad (10.5.9)$$

$$\hat{x}_l \equiv y_l - e_l^{(1)} \frac{M}{m_l \prod_{\alpha=1}^{\rho_1} m_{i_\alpha}} \pmod{m_l}, \quad (10.5.10)$$

and

$$\hat{x}_{k+r} \equiv \Delta_r - e_l^{(1)} \frac{M}{m_l \prod_{\alpha=1}^{\rho_1} m_{i_\alpha}} \pmod{m_{k+r}}$$
$$k + r = n - \rho_2 + 1, \ldots, n, \quad (10.5.11)$$

if $e_l^{(1)}$ is a consistent solution to (10.5.7); and

$$\hat{x}_{i_\alpha} \equiv -e_l^{(2)} \frac{M}{m_l \prod_{\alpha=1}^{\rho_1} m_{i_\alpha}} \pmod{m_{i_\alpha}}, \quad \alpha = 1, 2, \ldots, \rho_1 \quad (10.5.12)$$

$$\hat{x}_l \equiv y_l - e_l^{(2)} \frac{M}{m_l \prod_{\alpha=1}^{\rho_1} m_{i_\alpha}} \pmod{m_l}, \quad (10.5.13)$$

and

$$\hat{x}_{k+r} \equiv \Delta_r + \left(m_l \prod_{\alpha=1}^{\rho_1} m_{i_\alpha} - e_l^{(2)} \right) \frac{M}{m_l \prod_{\alpha=1}^{\rho_1} m_{i_\alpha}} \pmod{m_{k+r}},$$
$$k + r = n - \rho_2 + 1, \ldots, n, \quad (10.5.14)$$

if $e_l^{(2)}$ is a consistent solution to (10.5.8).

It is clear from (10.5.1–10.5.3, 10.5.7) that if one error takes place in the pth information digit such that $X + E < M$, then for $j = p$, $e_l^{(1)} = e'$ and $l = p$. Similarly, if one error takes place in the pth information digit such that $M \leq X + E < 2M$, then for $j = p$, $e_l^{(2)} = e'$ and $l = p$ from (10.5.4–10.5.6, 10.5.8). It remains to be shown that the solutions to either of (10.5.7) and (10.5.8) are inconsistent if only one error takes place in the pth information digit and $j \neq p$, or if more than one but up to β errors take place. Under an additional constraint on the form of the moduli, this result is established in the following theorem.

THEOREM 10.9 *Assume that up to β errors and ρ $(\rho \geq 1)$ erasures occur in an RRNS with minimum distance $d = \rho + \lambda + \beta + 1$ $(\lambda = 1, \beta > \lambda)$. If the moduli are such that there are do not exist integers $n_{\rho_1 \beta_1}$, $n_{j \rho_1}$, $0 < n_{\rho_1 \beta_1} < \prod_{\alpha=1}^{\rho_1} m_{i_\alpha} \prod_{\gamma=1}^{\beta_1} m_{j_\gamma}$, $0 < n_{j \rho_1} < m_j \prod_{\alpha=1}^{\rho_1} m_{i_\alpha}$, $1 \leq j, i_\alpha, j_\gamma \leq k$, that satisfy the equation*

$$n_{\rho_1 \beta_1} m_j + n_{j \rho_1} \prod_{\gamma=1}^{\beta_1} m_{j_\gamma} = \prod_{\substack{i=1 \\ k < r_i \leq n}}^{\rho_1 + \beta_1 + 1} m_{r_i}, \quad k < r_i \leq n, \quad (10.5.15)$$

$1 \leq \rho_1 \leq \rho$ *and* $1 \leq \beta_1 \leq \beta$, *then the solution to either* (10.5.7) *or* (10.5.8) *is inconsistent, when the following two cases occur, Case* (a): *Only the pth information digit is received in error and* $(j \neq p)$, *and Case* (b): *More than one but up to* β *residue digits are received in error.*

PROOF. Consider Case (a). For $X + E < M$ and $j \neq p$, if the solutions to (10.5.7) are consistent, then $e_j^{(1)} < m_j \prod_{\alpha=1}^{\rho_1} m_{i_\alpha}$. Comparing (10.5.7) to (10.5.2) we obtain

$$(e_j^{(1)} m_p - e' m_j) \frac{M}{m_p m_j \prod_{\alpha=1}^{\rho_1} m_{i_\alpha}} \equiv 0 \quad (\mathrm{mod} \ \prod_{r=1}^{n-k-\rho_2} m_{k+r}),$$

that is, $(e_j^{(1)} m_p - e' m_j)$ is a multiple of $\prod_{r=1}^{n-k-\rho_2} m_{k+r}$. This is not possible as

$$|e_j^{(1)} m_p - e' m_j| < m_p m_j \prod_{\alpha=1}^{\rho_1} m_{i_\alpha} < \prod_{r=1}^{n-k-\rho_2} m_{k+r}.$$

Similarly, for $X + E < M$ and $j \neq p$, if the solutions to (10.5.8) are consistent, then $e_j^{(2)} < m_j \prod_{\alpha=1}^{\rho_1} m_{i_\alpha}$. Comparing (10.5.8) to (10.5.2) we obtain

$$(e' m_j + (m_j \prod_{\alpha=1}^{\rho_1} m_{i_\alpha} - e_j^{(2)}) m_p) \frac{M}{m_p m_j \prod_{\alpha=1}^{\rho_1} m_{i_\alpha}}$$

$$\equiv 0 \quad (\mathrm{mod} \ \prod_{r=1}^{n-k-\rho_2} m_{k+r}).$$

This is not possible as

$$e' m_j + (m_j \prod_{\alpha=1}^{\rho_1} m_{i_\alpha} - e_j^{(2)}) m_p < 2 m_p m_j \prod_{\alpha=1}^{\rho_1} m_{i_\alpha} < \prod_{r=1}^{n-k-\rho_2} m_{k+r}.$$

The case for $M \leq X + E < 2M$ can be analyzed in an analogous manner.

Now, consider Case (b). For $X + E < M$, if (10.5.7) has a consistent solution $e_j^{(1)}$, then comparing (10.5.7) to (10.2.7) we obtain

$$(e_j^{(1)} \prod_{\gamma=1}^{\beta_1} m_{j_\gamma} - e' m_j) \frac{M}{m_j \prod_{\alpha=1}^{\rho_1} m_{i_\alpha} \prod_{\gamma=1}^{\beta_1} m_{j_\gamma}} \equiv$$

$$0 \quad (\mathrm{mod} \ \prod_{\substack{r=1 \\ k+r \neq j_\gamma}}^{n-k-\rho_2} m_{k+r}), \ \gamma = \beta_1 + 1, \ldots, \beta_1 + \beta_2.$$

This can not hold as

$$|e_j^{(1)} \prod_{\gamma=1}^{\beta_1} m_{j_\gamma} - e'm_j| < m_j \prod_{\alpha=1}^{\rho_1} m_{i_\alpha} \prod_{\gamma=1}^{\beta_1} m_{j_\gamma} < \prod_{\substack{r=1 \\ k+r \neq j_\gamma}}^{n-k-\rho_2} m_{k+r},$$

$$\gamma = \beta_1 + 1, \ldots, \beta_1 + \beta_2.$$

Similarly, for $X + E < M$, if (10.5.8) has a consistent solution $e_j^{(2)}$, then comparing (10.5.8) to (10.2.7) we obtain

$$(e'm_j + (m_j \prod_{\alpha=1}^{\rho_1} m_{i_\alpha} - e_j^{(2)}) \prod_{\gamma=1}^{\beta_1} m_{j_\gamma}) \frac{M}{m_j \prod_{\alpha=1}^{\rho_1} m_{i_\alpha} \prod_{\gamma=1}^{\beta_1} m_{j_\gamma}}$$

$$\equiv 0 (\bmod \prod_{\substack{r=1 \\ k+r \neq j_\gamma}}^{n-k-\rho_2} m_{k+r}), \ \gamma = \beta_1 + 1, \ldots, \beta_1 + \beta_2.$$

This can not hold as

$$e'm_j + (m_j \prod_{\alpha=1}^{\rho_1} m_{i_\alpha} - e_j^{(2)}) \prod_{\gamma=1}^{\beta_1} m_{j_\gamma} < 2m_j \prod_{\alpha=1}^{\rho_1} m_{i_\alpha} \prod_{\gamma=1}^{\beta_1} m_{j_\gamma}$$

$$< \prod_{\substack{r=1 \\ k+r \neq j_\gamma}}^{n-k-\rho_2} m_{k+r},$$

for $\rho_1 + \rho_2 < \rho$ and/or $\beta_1 + \beta_2 < \beta$; and (10.5.15) holds for $\rho_1 + \rho_2 = \rho$ and $\beta_1 + \beta_2 = \beta$. The case for $M \leq X + E < 2M$ can be analyzed in an analogous manner. This proves the theorem.

The sufficient condition for Theorem 10.9 can be shown to be exactly the same as the sufficient condition stated in Lemma 10.1. Let us now consider the second situation.

Situation (B): Assuming that one error takes place in the $(k+q)$th digit, that is, the qth parity digit, (10.2.6–10.2.13) become
(i) $X + E < M$. In this case,

$$\overline{Y} = X + E = X + e' \frac{M}{\prod_{\alpha=1}^{\rho_1} m_{i_\alpha}}, \qquad (10.5.16)$$

and

$$\Delta_r \equiv e' \frac{M}{\prod_{\alpha=1}^{\rho_1} m_{i_\alpha}} \equiv e'm_j \frac{M}{m_j \prod_{\alpha=1}^{\rho_1} m_{i_\alpha}} \ (\bmod \ m_{k+r})$$

$$r = 1, 2, \ldots, n - k - \rho_2, \ r \neq q, \tag{10.5.17}$$

$$\Delta_q \equiv e' \frac{M}{\prod_{\alpha=1}^{\rho_1} m_{i_\alpha}} - e_{k+q} \pmod{m_{k+q}}$$

$$1 \leq q \leq n - k - \rho_2, \tag{10.5.18}$$

$$\Delta_r \equiv e' \frac{M}{\prod_{\alpha=1}^{\rho_1} m_{i_\alpha}} - e_{k+r} \pmod{m_{k+r}}$$

$$k + r = n - \rho_2 + 1, \ldots, n, \tag{10.5.19}$$

where $0 < e' < \prod_{\alpha=1}^{\rho_1} m_{i_\alpha}$ for $\rho_1 \geq 1$, $e' = 0$ for $\rho_1 = 0$, and m_j can be any one of the information moduli not corresponding to the erasures.

(ii) $M \leq X + E < 2M$. In this case,

$$\overline{Y} = X + E - M = X + e' \frac{M}{\prod_{\alpha=1}^{\rho_1} m_{i_\alpha}} - M$$

$$\equiv X + \left(e' - \prod_{\alpha=1}^{\rho_1} m_{i_\alpha}\right) \frac{M}{\prod_{\alpha=1}^{\rho_1} m_{i_\alpha}}, \tag{10.5.20}$$

and

$$\Delta_r \equiv \left(e' m_j - m_j \prod_{\alpha=1}^{\rho_1} m_{i_\alpha}\right) \frac{M}{m_j \prod_{\alpha=1}^{\rho_1} m_{i_\alpha}} \pmod{m_{k+r}}$$

$$r = 1, 2, \ldots, n - k - \rho_2, \ r \neq q, \tag{10.5.21}$$

$$\Delta_q \equiv \left(e' - \prod_{\alpha=1}^{\rho_1} m_{i_\alpha}\right) \frac{M}{\prod_{\alpha=1}^{\rho_1} m_{i_\alpha}} - e_{k+q} \pmod{m_{k+q}}$$

$$1 \leq q \leq n - k - \rho_2, \tag{10.5.22}$$

$$\Delta_r \equiv \left(e' - \prod_{\alpha=1}^{\rho_1} m_{i_\alpha}\right) \frac{M}{\prod_{\alpha=1}^{\rho_1} m_{i_\alpha}} - e_{k+r} \pmod{m_{k+r}}$$

$$k + r = n - \rho_2 + 1, \ldots, n, \tag{10.5.23}$$

where $0 < e' < \prod_{\alpha=1}^{\rho_1} m_{i_\alpha}$ for $\rho_1 \geq 1$, $e' = 0$ for $\rho_1 = 0$, and m_j can be any one of the information moduli not corresponding to the erasures.

The procedure to determine error location, and error and erasure values is the same as that for Situation (A). This is with the exception that after no consistent solution is found in Situation (A), we check if any $n - k - \rho_2 - 1$ congruences in either (10.5.7) or (10.5.8) have a consistent solution (say $e_j^{(1)} < m_j \prod_{\alpha=1}^{\rho_1} m_{i_\alpha}$ for (10.5.7) and $e_j^{(2)} < m_j \prod_{\alpha=1}^{\rho_1} m_{i_\alpha}$ for (10.5.8)) which is equivalent to zero modulo m_j. Let us say that either of the two conditions is satisfied when $j = l$ and the only inconsistent congruence corresponds to the $(k+s)$th digit. Then the $(k+s)$th digit is declared to be in error. The correct value of the erroneous digit is,

$$\widehat{x}_{k+s} \equiv y_{k+s} + \Delta_s - e_l^{(1)} \frac{M}{m_l \prod_{\alpha=1}^{\rho_1} m_{i_\alpha}} \quad (\text{mod } m_{k+s}), \qquad (10.5.24)$$

if $e_l^{(1)}$ is the consistent solution and

$$\widehat{x}_{k+s} \equiv y_{k+s} + \Delta_s + (m_l \prod_{\alpha=1}^{\rho_1} m_{i_\alpha} - e_l^{(2)}) \frac{M}{m_l \prod_{\alpha=1}^{\rho_1} m_{i_\alpha}} \quad (\text{mod } m_{k+s}),$$
$$(10.5.25)$$

if $e_l^{(2)}$ is the consistent solution.

It is clear from (10.5.16–10.5.18, 10.5.7) that if one error takes place in the $(k+q)$th digit such that $X + E < M$, then except for the qth congruence, (10.5.7) has a consistent solution $e_j^{(1)} = e'm_j \equiv 0 \pmod{m_j}$ and $s = q$. Similarly, from (10.5.20–10.5.22, 10.5.8), if one error takes place in the $(k+q)$th digit such that $M \leq X + E < 2M$, then except for the qth congruence, (10.5.8) has a consistent solution $e_j^{(2)} = e'm_j \equiv 0 \pmod{m_j}$ and $s = q$. However, it remains to be shown that if more than one error takes place, then neither (10.5.7) nor (10.5.8) has $n - k - \rho_2 - 1$ congruences having a consistent solution which are equivalent to zero modulo m_j, $1 \leq j \leq k$, $j \neq i_\alpha$, $\alpha = 1, 2, \ldots, \rho_1$. Under the same constraint on the form of the moduli as stated in Theorem 10.9, if more than one residue digit are received in error, then neither (10.5.7) or (10.5.8) has exactly $n - k - \rho_2 - 1$ consistent solutions which are equivalent to zero modulo m_j, $1 \leq j \leq k$, $j \neq i_\alpha$, $\alpha = 1, 2, \ldots, \rho_1$. The proof of this statement is similar to the proof of Theorem 10.9. Based on the concepts developed above and under the assumption that the moduli satisfy the condition in Theorem 10.9 or in Lemma 10.1, the algorithm to correct single error and multiple erasures, and simultaneously detect multiple errors is given in Table 10.3. A flow chart for this algorithm is given in Figure 10.2.

EXAMPLE 10.7 Consider the $(9,6)$ RRNS code based on the moduli $m_1 = 23$, $m_2 = 25$, $m_3 = 27$, $m_4 = 29$, $m_5 = 31$, $m_6 = 32$, $m_7 = 41$, $m_8 = 47$ and $m_9 = 53$, where m_7, m_8 and m_9 are the redundant moduli. Clearly, $m_7 m_8 > 2m_5 m_6 - m_5 - m_6$, and therefore, the condition in Lemma 10.1 is satisfied, thereby satisfying the condition in Theorem 10.9 trivially. Let $X = 400,000,000 < \prod_{i=1}^{6} m_i = 446,623,200$, then $\underline{x} = [8\ 0\ 22\ 13\ 25\ 0\ 23\ 14\ 43]$ is a codevector. Assume that one erasure and one error take place in the first and third position, respectively, and the received vector is $[*\ 0\ 23\ 13\ 25\ 0\ 23\ 14\ 43]$. After substituting zero for the

Table 10.3: An Algorithm for Single Error Correction, Multiple Erasure Correction and Multiple Error Detection

Step 1:	According to the received vector with zeros substituted for the erasures, compute the syndromes using BEX as given in Table 6.1 and $\Delta_r \equiv \overline{y}_{k+r} - y_{k+r} \pmod{m_{k+r}}$, $r = 1, 2, \ldots, n - k$.
Step 2:	Check how many syndromes are zero (without including the syndromes corresponding to the erasures in parity).
	(1). If all the syndromes are zero, then the values of erasures information and parity are zero and $\overline{y}_{k+r}$, respectively. Stop.
	(2). If only one syndrome is non-zero, then the corresponding parity digit is in error, and all other residue digits are correct. The values of erasures in the information are zero, and the values of error and erasures in the parity are $\overline{y}_{k+r}$. Correct them and stop.
	(3). If 2 to $\beta + \rho - (\rho_1 + \rho_2)$ syndromes are non-zero, then go to Step 9.
	(4). If $\beta + \rho - (\rho_1 + \rho_2) + 1$ to $d - 1 - \rho_2$ syndromes are non−zero, then go to Step 3.
Step 3:	Let $j = 1$.
Step 4:	If $j = i_\alpha$, $\alpha = 1, 2, \ldots, \rho_1$, then go to Step 8, else go to Step 5.
Step 5:	Perform the single-error/ρ_1-erasure consistency-checking:
	(1). Compute $e_j^{(1)}$ and $e_j^{(2)}$ using (10.5.7) and (10.5.8), respectively.
	(2). Check the consistency of the solutions: $e_j^{(1)} < m_j \prod_{\alpha=1}^{\rho_1} m_{i_\alpha}$ or $e_j^{(2)} < m_j \prod_{\alpha=1}^{\rho_1} m_{i_\alpha}$, where m_{i_α} are the nonredundant moduli corresponding to the erasures in the information, m_j is one of the nonredundant moduli, and $m_j \neq m_{i_\alpha}$. If it is consistent, then go to Step 6. If it is inconsistent, then go to Step 7.
Step 6:	If the consistent solution $e_j^{(1)} \equiv 0 \mod m_j$ (or $e_j^{(2)} \equiv 0 \mod m_j$), then no error occurs. If the consistent solution $e_j^{(1)} \not\equiv 0 \mod m_j$ (or $e_j^{(2)} \not\equiv 0 \mod m_j$), then exactly one error takes place in the jth residue digit. Correct the error and erasures using (10.5.9–10.5.14) and stop.
Step 7:	If exactly $n - k - \rho_2 - 1$ congruences in either (10.5.7) or (10.5.8) have a consistent solution which is equivalent to zero modulo m_j, then exactly one error takes place in the parity for which the solution is not consistent. Correct the error and erasures using (10.5.24, 10.5.25) and (10.5.9, 10.5.11, 10.5.12, 10.5.14), respectively, and stop.
Step 8:	$j = j + 1$, go to Step 4 for $j \leq k$. For $j > k$, go to Step 9.
Step 9:	Declare more than one error detected and stop.

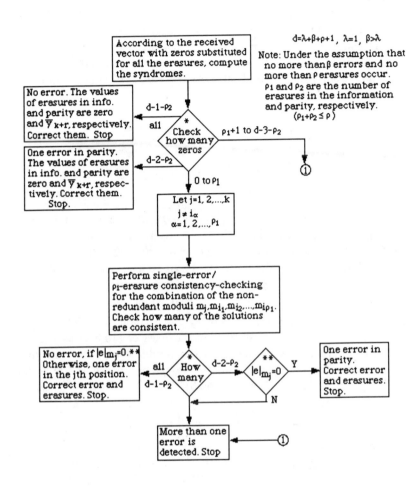

* Without including the syndromes corresponing to the
  erasures in the parity.
** Denote e≡0 (mod m_j) by $|e|_{m_j}$=0, where e is a consistent solution.

Figure 10.2: A Decoder Flowchart for Single Error and Multiple Erasure
Correction and Multiple Error Detection

erasure, the received vector becomes [0 0 23 13 25 0 23 14 43]. Based on the information part [0 0 23 13 25 0] and BEX, we compute the syndrome digits $\Delta_1 \equiv 32 \pmod{41}$, $\Delta_2 \equiv 20 \pmod{47}$ and $\Delta_3 \equiv 12 \pmod{53}$. Following the decoding algorithm, we check the consistency for $j = 1, 2, \ldots, k$, based on (10.5.7) and (10.5.8). It is seen that when $j = 3$, $e_3^{(2)} = 217 < m_1 m_3 = 621$ is a consistent solution to (10.5.8). An error is declared in the third position, and from (10.5.12) and 10.5.14) the values of erasure and error digits are computed as

$$\hat{x}_1 \equiv -e_3^{(2)} \frac{M}{m_1 m_3} \pmod{m_1} \equiv 8,$$

$$\hat{x}_3 \equiv y_3 - e_3^{(2)} \frac{M}{m_1 m_3} \pmod{m_3} \equiv 22.$$

Now, assume an erasure in the first position and an error in the ninth position and let the received vector be [* 0 22 13 25 0 23 14 44]. After substituting zero for the erasure *, the received vector becomes [0 0 22 13 25 0 23 14 44]. Based on the information part [0 0 22 13 25 0] and BEX, we compute the syndrome digits $\Delta_1 \equiv 16 \pmod{41}$, $\Delta_2 \equiv 46 \pmod{47}$ and $\Delta_3 \equiv 44 \pmod{53}$. Following the decoding algorithm, we check the consistency for $j = 2, 3, \ldots, 6$. It is seen that when $j = 2$, $e_2^{(1)} = 1702 > m_1 m_2$ and $e_2^{(2)} = 350 < m_1 m_2$, where $e_2^{(2)} \equiv 0 \pmod{m_2}$. Thus, $e_2^{(2)}$ is a consistent solution to the first two congruences of (10.5.8) as the only inconsistent congruence corresponds to the modulus m_9. So, the ninth residue is declared in error, the estimated values of erasure and erroneous digit being given by (10.5.12) and (10.5.25),

$$\hat{x}_1 \equiv -e_2^{(2)} \frac{M}{m_1 m_2} \pmod{m_1} \equiv 8,$$

$$\hat{x}_9 \equiv y_9 + \Delta_3 + (m_1 m_2 - e_2^{(2)}) \frac{M}{m_1 m_2} \pmod{m_9} \equiv 43.$$

10.6   Extensions of Previous Algorithms

In this section, we extend the algorithms proposed in Yau and Liu [1973] and Jenkins and Altman [1988] for correcting single error to algorithms for simultaneously correcting multiple erasures and single error. The algorithm in Yau and Liu [1973] for single error correction ($d = n - k + 1 = 3$) is based on the fact that if only one of two syndromes is non-zero, then the corresponding parity residue digit is in error. So, the error can be located and corrected by an iterative checking process. It successively selects two

residues (corresponding to the two largest moduli) as redundant residues,
computes the syndromes, and then checks if only one of two corresponding
syndromes is non-zero. By Theorem 10.3, an RRNS with $d = \rho + 3$ can
simultaneously correct ρ erasures and one error. Since an erasure is an
error whose location is known, we may always put the erasures in parity.
Therefore, by Theorem 10.8, a procedure for simultaneously correcting ρ
erasures and one error is the same as that stated above.

The algorithm in Jenkins and Altman [1988] for locating single residue
digit error is based on the properties of modulus projection and MRC.
The modulus m_i-projection of X in an (n, k) RRNS code, denoted by X_i,
is defined in (9.6.1). The mixed radix representation of X_i is defined in
(9.6.2). This algorithm consists in checking if the redundant mixed radix
digits are all zero for each modulus m_i-projection, $i = 1, 2, \ldots, n$. Since the
residue representation of m_i-projection X_i can be treated as if one erasure
takes place in the ith residue digit of the original residue representation of
X, the value of erasure can be obtained by BEX. For the case that both
erasures and error occur, we extend the above algorithm in the following.

The M_Ψ-projection of X, denoted by X_Ψ, is defined as

$$X_\Psi \equiv X \pmod{\frac{MM_R}{M_\Psi}}, \tag{10.6.1}$$

where $M_\Psi = m_j \prod_{\alpha=1}^{\rho_1+\rho_2} m_{i_\alpha}$, $\Psi = \{j, i_1, i_2, \ldots, i_{\rho_1+\rho_2}\}$; $\rho_1 + \rho_2 \leq \rho$, and
$i_1, i_2, \ldots, i_{\rho_1+\rho_2}$ are the erasure locations. The mixed radix representation
of X_Ψ is

$$X_\Psi = \sum_{\substack{l=1 \\ l \notin \Psi}}^{n} a_l \prod_{\substack{r=1 \\ r \notin \Psi}}^{l-1} m_r. \tag{10.6.2}$$

In Chapter 9, it was shown that an illegitimate projection resulting from
ρ or fewer errors can not be smaller than the smallest non-zero number
represented by (10.6.2) with all the nonredundant mixed radix digits set to
zero. Therefore, by checking only the redundant mixed radix digits, it can
be determined whether or not the projection is legitimate. If all the redundant mixed radix digits are zero, the projection is legitimate; otherwise, it
is illegitimate. Based on the above analysis, a procedure for simultaneously
correcting ρ erasures and one error can be outlined as follows:

Step 1: Based on the received residue vector, compute the mixed radix
digits and check if all the redundant mixed radix digits are zero. If
yes, then declare no error and stop. Otherwise, go to next step.

Step 2: Compute the mixed radix digits of M_Ψ-projection for $j = 1, 2, \ldots,$
n, $j \neq i_1, i_2, \ldots, i_{\rho_1+\rho_2}$. Check if all the redundant mixed radix digits

are zero. If yes, then declare the jth residue digit in error, correct the erasures and error by BEX and stop. Otherwise, go to next step.

Step 3: Declare more than one error detected. Stop.

10.7   Computational Complexity Analysis

The algorithm developed here for simultaneously correcting single error and multiple erasures is superior to the algorithms in Yau and Liu [1973] and Jenkins and Altman [1988] from a computational complexity point of view. Based on the pipelined hardware architecture of MRC technique for BEX as described in Jenkins [1983] and Shenoy and Kumaresan [1989], the comparison of our algorithm with that in Yau and Liu [1973] and Jenkins and Altman [1988] in terms of the requirement of MADD and MMULT is shown in Table 10.4. We should point out that the expression in columns 1 and 2 of Table 10.4 is obtained by extending the algorithms in Yau and Liu [1973] and Jenkins and Altman [1988] as described in the previous section. In addition, the new algorithms employ the nonredundant residue digits to perform the BEX while the algorithms in Yau and Liu [1973] and Jenkins and Altman [1988] have to find correct residue digits first and then perform BEX to obtain the correct values of erroneous or erased residue digits. Therefore, these algorithms require fewer multiplexers in computational elements as compared to the previous algorithm in Yau and Liu [1973] and Jenkins and Altman [1988]. However, the algorithms in Yau and Liu [1973] and Jenkins and Altman [1988] remain valid even for the case when the moduli do not satisfy the conditions stated in Theorems 10.6, 10.7 and 10.9. Furthermore, these new algorithms do not require error-correcting lookup table while the algorithms in Watson [1965] and Su and Lo [1990] for single error correction require error-correcting lookup table. The algorithms in Watson [1965] Su and Lo [1990] may become impractical for simultaneously correcting single error and multiple erasures and detecting multiple errors because of large memory space requirement for the lookup table.

NOTES
This is perhaps the first time that algorithms are being described for simultaneous single/multiple erasures correction, multiple errors detection, and single error correction. Thus far most literature is focused on error correction alone. This includes Watson [1965], Mandelbaum [1972], Barsi and Maestrini [1973], Mandelbaum [1976], Yau and Liu [1973], Jenkins and Altman [1988] and Su and Lo [1990]. In this chapter, the concept of the

Table 10.4: MMULT and MADD required for Single Error and Single Erasure Correction $d = 2\lambda + \rho + 1, d = 4, \lambda, \rho = 1$

Yau and Liu Algorithm [1973]	Jenkins and Altman Algorithm [1988]	New Algorithm	Modular Operations
$\frac{1}{4}k^3 + \frac{5}{4}k^2 + \frac{3}{2}k - 3$	$\frac{1}{2}k^3 + \frac{3}{2}k^2 + 3k$	$\frac{1}{2}k^2 + \frac{35}{2}k - 17$	MMULT
$\frac{1}{4}k^3 + \frac{5}{4}k^2 + \frac{5}{2}k - 1$	$\frac{1}{2}k^3 + \frac{3}{2}k^2 + 3k$	$\frac{1}{2}k^2 + \frac{29}{2}k - 8$	MADD

erasure correction in the RRNS is introduced from a coding theory point of view. The relationship between the minimum distance and the error detection and correction and erasure correction capability is derived. This also leads to computationally efficient algorithms for the cases of interest.

This chapter is based on the research paper Sun and Krishna [1993].

10.8 Bibliography

[10.1] F. Barsi and P. Maestrini, "Error Correcting Properties of Redundant Residue Number Systems," *IEEE Transactions on Computers*, vol. C-22, pp. 307-315, 1973.

[10.2] H.L. Garner, "The Residue Number System," *IRE Transactions on Electronic Computers*, vol. EC-8, pp. 140-147, 1959.

[10.3] W.K. Jenkins, "The Design of Error Checker for Self-Checking Residue Number Arithmetic," *IEEE Transactions on Computers*, vol. C-32, pp. 388-396, 1983.

[10.4] W.K. Jenkins and E.J. Altman, "Self-Checking Properties of Residue Number Error Checkers Based on Mixed Radix Conversion," *IEEE Transactions on Circuits and Systems*, vol. 35, no. 2, pp. 159-167, 1988.

[10.5] D. Mandelbaum, "Error Correction in Residue Arithmetic," *IEEE Transactions on Computers*, vol. C-21, pp. 538-545, 1972.

[10.6] D. Mandelbaum, "On a Class of Arithmetic Codes and a Decoding Algorithm," *IEEE Transactions on Information Theory*, vol. IT-22, pp. 85-88, 1976.

[10.7] A.P. Shenoy and R. Kumaresan, "Fast Base Extension Using a Redundant Modulus in RNS," *IEEE Transactions on Computers*, no. 2, pp. 292-297, 1989.

[10.8] C.-C. Su and H.-Y. Lo, "An Algorithm for Scaling and Single Residue Error Correction in Residue Number Systems," *IEEE Transactions on Computers*, no. 8, pp. 1053-1064, 1990.

[10.9] R.W. Watson, "Error Detection and Correction and Other Residue Interlacing Operations in a Redundant Residue Number System," *Ph.D. Dissertation*, University of California at Berkeley, 1965.

[10.10] S.S.-S. Yau and Y.-C. Liu, "Error Correction in Redundant Residue Number Systems," *IEEE Transactions on Computers*, vol. C-22, pp. 5-11, 1973.

[10.11] J.-D. Sun and H. Krishna, "Fast Algorithms for Erasures and Error Correction in Redundant Residue Number Systems," *Circuits, Systems, and Signal Processing*, vol. 12, pp. 503-531, 1993.

Chapter 11

Multiple Error Control in RNS-PC

In Chapter 7, a coding theory approach to error control in RNS-PC was described. It was used in Chapter 8 to derive a fast algorithm for single error correction in RNS-PC.

In this chapter, we extend further the coding theory approach to error control in RNS-PC, and derive computationally efficient algorithms for simultaneously correcting multiple errors and detecting multiple errors and additive overflow. It is found that the computational complexity of these algorithms for correcting double residue errors is $O(n^2)$ while the computational complexity of the algorithms derived by extending the previously known algorithm is $O(n^4)$, n being the number of moduli in the RNS-PC.

11.1 Single Error Correction: Continued

Recall from Chapter 7 that an RNS-PC with $A > max\{2\prod_{i=1}^{\beta+1} m_{j_i}\}$ can simultaneously correct single errors and detect the presence of β ($\beta > 1$) or fewer errors. The algorithm for correcting single errors and simultaneously detecting multiple errors and additive overflow is the same as that in Section 8.3. However, it is be shown that if more than one but up to β errors take place, then there is no consistent solution. Under an additional constraint on the form of the moduli, this result is established in the following theorem.

THEOREM 11.1 *If the moduli and code generator A of an RNS-PC are such that there do not exist integers n_j, n_B; $0 \le n_j < m_j$, $0 \le n_B < M_B =$*

299

$\prod_{\alpha=1}^{\beta} m_{i_\alpha}$, $1 \le i_\alpha \le n$, $j \ne i_\alpha$, *that satisfy*

$$n_j M_B + n_B m_j = A \qquad (11.1.1)$$

or

$$m_j M_B + n_j M_B + n_B m_j = A, \qquad (11.1.2)$$

then for more than one but up to β residue digits received in error, there is no consistent solution to any one of (8.3.9), (8.3.10) or (8.3.11).

PROOF. Assume that c $(1 < c \le \beta)$ errors has occurred. For $X + E < M_T$ and $X = \overline{X}$, if $e_j^{(1)} = e < m_j$ then comparing (7.5.6) to (8.3.9) we obtain

$$(e \prod_{\alpha=1}^{c} m_{i_\alpha} - e' m_j) \frac{M_T}{m_j \prod_{\alpha=1}^{c} m_{i_\alpha}} \equiv 0 \pmod{A}.$$

This congruence can not hold as $|e \prod_{\alpha=1}^{c} m_{i_\alpha} - e' m_j| < m_j \prod_{\alpha=1}^{c} m_{i_\alpha} < A$. Similarly, for $X + E < M_T$ and $X = \overline{X}$, if $e_j^{(2)} = e < m_j$ then comparing (7.5.6) to (8.3.10) we obtain

$$(e' m_j + (m_j - e) \prod_{\alpha=1}^{c} m_{i_\alpha}) \frac{M_T}{m_j \prod_{\alpha=1}^{c} m_{i_\alpha}} \equiv 0 \pmod{A}.$$

This congruence can not hold as $(e' m_j + (m_j - e) \prod_{\alpha=1}^{c} m_{i_\alpha}) < 2m_j \prod_{\alpha=1}^{c} m_{i_\alpha} < A$ for $c < \beta$, and (11.1.1) holds for $c = \beta$. For $X + E < M_T$ and $X = \overline{X}$, if $e_j^{(3)} = e < m_j$, then comparing (7.5.6) to (8.3.11) we obtain

$$(e' m_j + (2m_j - e) \prod_{\alpha=1}^{c} m_{i_\alpha}) \frac{M_T}{m_j \prod_{\alpha=1}^{c} m_{i_\alpha}} \equiv 0 \pmod{A}.$$

This congruence can not hold as $e' m_j + (2m_j - e) \prod_{\alpha=1}^{c} m_{i_\alpha} < A$ for $c < \beta$, and (11.1.2) holds for $c = \beta$. The cases (i) $M_T \le X + E < 2M_T$ and $X = \overline{X}$, (ii) $X + E < M_T$ and $X = \overline{X} - M_T$, and (iii) $M_T \le X + E < 2M_T$ and $X = \overline{X} - M_T$, can also be analyzed in an analogous manner. This proves the theorem.

LEMMA 11.1 *If the moduli and code generator A of an RNS-PC satisfy*

$$A > max\{3m_j M_B - m_j - M_B\}, \qquad (11.1.3)$$

$M_B = \prod_{\alpha=1}^{\beta} m_{i_\alpha}$, $1 \le i_\alpha, j \le n$, *and $j \ne i_\alpha$, then for more than one but up to β residue digits received in error, there is no consistent solution to any one of (8.3.9), (8.3.10) or (8.3.11).*

The proof of Lemma 11.1 is similar to the proof of Lemma 8.1. The algorithm to correct single error and simultaneously detect multiple errors and additive overflow in RNS-PC is given in Table 8.3. A flow-chart for this algorithm is given in Figure 8.2.

EXAMPLE 11.1 Consider an RNS-PC based on the moduli $m_1 = 3$, $m_2 = 5$, $m_3 = 7$, $m_4 = 11$, $m_5 = 13$, $m_6 = 17$, $m_7 = 19$ and $m_8 = 23$, and the code generator $A = 33,263 > 3m_6m_7m_8 - m_6 - m_7m_8 = 21,833$. Let $X = 21,288,320 < M_T = \prod_{i=1}^{8} m_i = 111,546,435$, and $X \equiv 0 \pmod{A}$. Therefore, X is a code-integer in this RNS-PC. The residue representation of X is $\underline{x} = [2\,0\,4\,9\,1\,2\,17\,3]$. Assume that X is altered by a single residue digit error in the first position, and $E \longleftrightarrow \underline{e} = [2\,0\,0\,0\,0\,0\,0\,0]$. Then the received vector is $\underline{y} = [1\,0\,4\,9\,1\,2\,17\,3]$. Based on the received vector and BEX, we compute the syndrome $\Delta \equiv 21,485 \pmod{A}$. Note that $-M_T \pmod{A} \equiv 17,667$ and $\Delta \not\equiv -M_T \pmod{A}$. Following the decoding algorithm, we check the consistency for $j = 1, 2, \ldots, 8$, based on computing $e_j^{(1)}$, $e_j^{(2)}$ and $e_j^{(3)}$ from (8.3.9)–(8.3.11). It is seen that when $j = 1$

$$e_1^{(1)} \equiv \Delta m_1 M_T^{-1} \pmod{A} \equiv 2 < m_1,$$

$$e_1^{(2)} \equiv \Delta m_1 M_T^{-1} + m_1 \pmod{A} \equiv 5 > m_1,$$

$$e_1^{(3)} \equiv \Delta m_1 M_T^{-1} + 2m_1 \pmod{A} \equiv 8 > m_1.$$

Since $e_1^{(1)}$ is a consistent solution, we declare that one error has occurred in the first position and the estimated value of erroneous digit is

$$\widehat{x}_1 \equiv y_1 - e_1^{(1)} \frac{M_T}{m_1} \pmod{m_1} \equiv 2.$$

Now, assume that X is altered by two residue errors, the error vector $\underline{e} = [0\,0\,0\,0\,0\,0\,8\,5]$. Then the received vector becomes $\underline{y} = [2\,0\,4\,9\,1\,2\,6\,8]$. Based on $\underline{y}$ and BEX, we compute the syndrome $\Delta \equiv 28,422 \pmod{A}$. Following the decoding algorithm, we check the consistency for $j = 1, 2, \ldots$, 8. It is found that there is no consistent solution. Therefore, declare more than one error detected.

11.2   Double Error Correction

Once again, an RNS-PC with $A > max\{2\prod_{i=1}^{\beta+2} m_{j_i}\}$, can simultaneously correct double residue errors and detect the presence of β ($\beta > 2$) or fewer errors. Assume that the pth and qth residue digits are in error, then (7.5.5–7.5.12) become

(i) $X + E < M_T$ and $X = \overline{X}$. In this case,

$$\overline{Y} = X + e'\frac{M_T}{m_p m_q}, \quad 0 < e' < m_p m_q, \tag{11.2.1}$$

and

$$\Delta \equiv e'\frac{M_T}{m_p m_q} \pmod{A}. \tag{11.2.2}$$

(ii) $M_T \le X + E < 2M_T$ and $X = \overline{X}$. In this case,

$$\overline{Y} = X + (e' - m_p m_q)\frac{M_T}{m_p m_q}, \quad 0 < e' < m_p m_q, \tag{11.2.3}$$

and

$$\Delta \equiv (e' - m_p m_q)\frac{M_T}{m_p m_q} \pmod{A}. \tag{11.2.4}$$

(iii) $X + E < M_T$ and $X = \overline{X} - M_T$. In this case,

$$\overline{Y} = \overline{X} + (e' - m_p m_q)\frac{M_T}{m_p m_q}, \quad 0 < e' < m_p m_q, \tag{11.2.5}$$

and

$$\Delta \equiv (e' - m_p m_q)\frac{M_T}{m_p m_q} \pmod{A}. \tag{11.2.6}$$

(iv) $M_T \le X + E < 2M_T$ and $X = \overline{X} - M_T$. In this case,

$$\overline{Y} = \overline{X} + (e' - 2m_p m_q)\frac{M_T}{m_p m_q}, \quad 0 < e' < m_p m_q, \tag{11.2.7}$$

and

$$\Delta \equiv (e' - 2m_p m_q)\frac{M_T}{m_p m_q} \pmod{A}. \tag{11.2.8}$$

Given Δ, based on (11.2.1–11.2.8) a procedure to determine p, q (error locations) and e_p, e_q (error values) can be outlined as follows:
For $i = 1, 2, \ldots, n-1$, and $j = i+1, i+2, \ldots, n$, solve the congruences

$$\Delta \equiv e_{ij}^{(1)}\frac{M_T}{m_i m_j} \pmod{A} \tag{11.2.9}$$

$$\Delta \equiv (e_{ij}^{(2)} - m_i m_j)\frac{M_T}{m_i m_j} \pmod{A} \tag{11.2.10}$$

$$\Delta \equiv (e_{ij}^{(3)} - 2m_i m_j)\frac{M_T}{m_i m_j} \pmod{A}. \tag{11.2.11}$$

Then check if one of the following three conditions are satisfied: **Condition (1)** $e_{ij}^{(1)} < m_i m_j$, $e_{ij}^{(2)} > m_i m_j$ and $e_{ij}^{(3)} > m_i m_j$. **Condition (2)** $e_{ij}^{(2)} <$

$m_i m_j$, $e_{ij}^{(1)} > m_i m_j$ and $e_{ij}^{(3)} > m_i m_j$. **Condition (3)** $e_{ij}^{(3)} < m_i m_j$, $e_{ij}^{(1)} > m_i m_j$ and $e_{ij}^{(2)} > m_i m_j$. Let us say that one of the above three conditions is satisfied for $i = f$ and $j = h$. Then the fth and hth residue digits are declared to be in error. The value of errors are

$$e_f \equiv e \frac{M_T}{m_f m_h} \quad (\text{mod } m_f), \qquad (11.2.12)$$

$$e_h \equiv e \frac{M_T}{m_f m_h} \quad (\text{mod } m_h), \qquad (11.2.13)$$

where $e = e_{ij}^{(1)}$ if $e_{ij}^{(1)}$ is a consistent solution to (11.2.9); $e = e_{ij}^{(2)}$ if $e_{ij}^{(2)}$ is a consistent solution to (11.2.10); and $e = e_{ij}^{(3)}$ if $e_{ij}^{(3)}$ is a consistent solution to (11.2.11). Note that if $e \equiv 0 \pmod{m_f}$ or $e \equiv 0 \pmod{m_h}$, then only the hth or fth residue digit is in error. The correct values of the fth and hth residue digits are $(y_f - e_f) \bmod m_f$ and $(y_h - e_h) \bmod m_h$, respectively, and $\underline{y}$ is decoded to $\widehat{\underline{x}}$, where

$$\begin{aligned}
\widehat{\underline{x}} &= [\widehat{x}_1 \, \widehat{x}_2 \, \ldots \, \widehat{x}_n] \\
\widehat{x}_j &= y_j, \, j = 1, 2, \ldots, n, \, j \neq f, h \\
\widehat{x}_f &\equiv y_f - e_f \pmod{m_f} \\
\widehat{x}_h &\equiv y_h - e_h \pmod{m_h}.
\end{aligned} \qquad (11.2.14)$$

It is clear from (11.2.1), (11.2.2) and (11.2.9) that if two errors occur in the pth and qth residue digits such that $X + E < M_T$ and $X = \overline{X}$, then for $i = p$ and $j = q$, $e = e' = e_{ij}^{(1)} < m_i m_j$ is a solution to (11.2.9), and $f = p$ and $h = q$. Similarly, if two errors occur in the pth and qth residue digits such that (1) $M_T \leq X + E < 2M_T$ and $X = \overline{X}$, then for $i = p$ and $j = q$, $e = e' = e_{ij}^{(2)} < m_i m_j$ is a solution to (11.2.10), and $f = p$ and $h = q$; (2) $X + E < M_T$ and $X = \overline{X} - M_T$, then for $i = p$ and $j = q$, $e = e' = e_{ij}^{(2)} < m_i m_j$ is a solution to (11.2.10), and $f = p$ and $h = q$; (3) $M_T \leq X + E < 2M_T$ and $X = \overline{X} - M_T$, then for $i = p$ and $j = q$, $e = e' = e_{ij}^{(3)} < m_i m_j$ is a solution to (11.2.11), and $f = p$ and $h = q$. Based on the above analysis, if Condition (2) is satisfied, then we correct errors and recompute the new syndrome Δ'. If $\Delta' \equiv -M_T \pmod{A}$ then an additive overflow is detected. However, it remains to be shown that there is no consistent solution if one of the following three cases happens: Case (i) one error occurs in the pth residue digit and $p \neq i, j$, Case (ii) two errors occur in the pth and qth residue digits and $p \neq i, j$ and/or $q \neq i, j$, and Case (iii) more than two but up to β errors occur. Under an additional constraint on the moduli and code generator A, the result is established in the following theorem.

THEOREM 11.2 *If the moduli and code generator A of an RNS-PC are such that there do not exist integers n_{ij}, n_B, $0 < n_{ij} < m_i m_j$, $0 \leq n_B < M_B = \prod_{\alpha=1}^{\beta} m_{i_\alpha}$, $1 \leq i_\alpha \leq n$, that satisfy either of the following equations*

$$n_B m_i m_j + n_{ij} M_B = A, \qquad (11.2.15)$$

$$m_i m_j M_B + n_B m_i m_j + n_{ij} M_B = A, \qquad (11.2.16)$$

$1 \leq i, j \leq n$, $i \neq j \neq i_\alpha$, then none of (11.2.9), (11.2.10) and (11.2.11) has a consistent solution for any one of the following three cases, Case (i): one error occurs in the pth residue digit and $p \neq i, j$, Case (ii): two errors occur in the pth and qth residue digits and $p \neq i, j$ and/or $q \neq i, j$, and Case (iii) more than two but up to β errors occur.

PROOF. Consider Case (ii), two errors have occurred and $i \neq p, q$ and/or $j \neq p, q$ (for Case (i), one error in the pth position, $p \neq i, j$, can be analyzed in a similar manner). For $X + E < M_T$ and $X = \overline{X}$, if $e_{ij}^{(1)} < m_i m_j$ is a consistent solution to (11.2.9), then comparing (11.2.2) to (11.2.9), we get

$$\frac{M_T}{m_i m_j m_p m_q}(e' m_i m_j - e_{ij}^{(1)} m_p m_q) \equiv 0 \pmod{A},$$

that is, $e' m_i m_j - e_{ij}^{(1)} m_p m_q$ is a multiple of A. This is not possible as

$$|e' m_i m_j - e_{ij}^{(1)} m_p m_q| < m_i m_j m_p m_q < A.$$

Similarly, if $e_{ij}^{(2)} < m_i m_j$ is a consistent solution to (11.2.10), then comparing (11.2.2) to (11.2.10), we get

$$\frac{M_T}{m_i m_j m_p m_q}\left(e' m_i m_j + (m_i m_j - e_{ij}^{(2)}) m_p m_q\right) \equiv 0 \pmod{A}.$$

This can not hold as $\left(e' m_i m_j + (m_i m_j - e_{ij}^{(2)}) m_p m_q\right) < A$. If $e_{ij}^{(3)} < m_i m_j$ is a consistent solution to (11.2.11), then comparing (11.2.2) to (11.2.11), we get

$$\frac{M_T}{m_i m_j m_p m_q}\left(e' m_i m_j + (2 m_i m_j - e_{ij}^{(3)}) m_p m_q\right) \equiv 0 \pmod{A}.$$

This can not hold as $\left(e' m_i m_j + (2 m_i m_j - e_{ij}^{(3)}) m_p m_q\right) < A$. The cases for (i) $M_T \leq X + E < 2 M_T$ and $X = \overline{X}$, (ii) $X + E < M_T$ and $X = \overline{X} - M_T$, and (iii) $M_T \leq X + E < 2M_T$ and $X = \overline{X} - M_T$ can also be analyzed in an analogous manner.

Now, consider Case (iii) and assume that c $(2 < c \leq \beta)$ errors occur. For $X + E < M_T$ and $X = \overline{X}$, if $e_{ij}^{(1)} < m_i m_j$ then comparing (7.5.6) to (11.2.9) we obtain

$$(e_{ij}^{(1)} \prod_{\alpha=1}^{c} m_{i_\alpha} - e' m_i m_j) \frac{M_T}{m_i m_j \prod_{\alpha=1}^{c} m_{i_\alpha}} \equiv 0 \pmod{A}.$$

This congruence can not hold as

$$|e_{ij}^{(1)} \prod_{\alpha=1}^{c} m_{i_\alpha} - e' m_i m_j| < m_i m_j \prod_{\alpha=1}^{c} m_{i_\alpha} < A.$$

Similarly, for $X + E < M_T$ and $X = \overline{X}$, if $e_{ij}^{(2)} < m_i m_j$ then comparing (7.5.6) to (11.2.10) we obtain

$$(e' m_i m_j + (m_i m_j - e_{ij}^{(2)}) \prod_{\alpha=1}^{c} m_{i_\alpha}) \frac{M_T}{m_i m_j \prod_{\alpha=1}^{c} m_{i_\alpha}} \equiv 0 \pmod{A}.$$

This congruence can not hold as

$$(e' m_i m_j + (m_i m_j - e_{ij}^{(1)}) \prod_{\alpha=1}^{c} m_{i_\alpha}) < 2 m_i m_j \prod_{\alpha=1}^{c} m_{i_\alpha} < A, \ c < \beta;$$

and (11.2.15) holds for $c = \beta$. For $X + E < M_T$ and $X = \overline{X}$, if $e_{ij}^{(3)} < m_i m_j$, then comparing (7.5.6) to (11.2.11) we obtain

$$(e' m_i m_j + (2 m_i m_j - e_{ij}^{(3)}) \prod_{\alpha=1}^{c} m_{i_\alpha}) \frac{M_T}{m_j \prod_{\alpha=1}^{c} m_{i_\alpha}} \equiv 0 \pmod{A}.$$

This congruence can not hold as $e' m_i m_j + (2 m_i m_j - e_{ij}^{(3)}) \prod_{\alpha=1}^{c} m_{i_\alpha} < A$ for $c < \beta$, and (11.2.16) holds for $c = \beta$. The cases (i) $M_T \leq X + E < 2M_T$ and $X = \overline{X}$, (ii) $X + E < M_T$ and $X = \overline{X} - M_T$, and (iii) $M_T \leq X + E < 2M_T$ and $X = \overline{X} - M_T$, can also be analyzed in an analogous manner. This proves the theorem.

LEMMA 11.2 *If the moduli and code generator A of an RNS-PC satisfy*

$$A > max\{3 m_i m_j M_B - m_i m_j - M_B\}, \tag{11.2.17}$$

$M_B = \prod_{\alpha=1}^{\beta} m_{i_\alpha}$, $1 \leq i,j,i_\alpha \leq n$, $i \neq j \neq i_\alpha$, *then none of* (11.2.9), (11.2.10) *and* (11.2.11) *has a consistent solution for either one of the three*

following cases, Case (i): one error occurs in the pth residue digit and
$p \neq i,j$, Case (ii): two errors occur in the pth and qth residue digits and
$p \neq i,j$ and/or $q \neq i,j$, and Case (iii) more than two but up to β errors
occur.

The proof of the Lemma 11.2 is similar to the proof of Lemma 8.1. Based on the concepts developed above and under the assumption that the moduli m_i, $i = 1, 2, \ldots, n$, and the code generator A satisfy the conditions in Theorem 11.2 (or Lemma 11.2), the algorithm to simultaneously correct double errors and detect multiple errors and additive overflow in RNS-PC is given in Table 11.1. A flow chart for this algorithm is given in Figure 11.1.

An RNS-PC with $A > max\{\prod_{i=1}^{4} m_{j_i}\}$ can correct double residue errors. Under an additional constraint on the moduli and code generator A, the above decoding algorithm can be used for simultaneously correcting double residue errors and detecting additive overflow. The result is established in the following theorem and lemma. The proofs of Theorem 11.3 and Lemma 11.3 are similar to the proofs of Theorem 11.2 and Lemma 11.2.

THEOREM 11.3 *If the moduli and code generator A of an RNS-PC are such that there do not exist integers n_{pq}, n_{ij}, $0 < n_{pq} < m_p m_q$, $0 < n_{ij} < m_i m_j$, that satisfy either one of the following equations*

$$n_{pq} m_i m_j + n_{ij} m_p m_q = A, \qquad (11.2.18)$$

$$m_p m_q m_i m_j + n_{pq} m_i m_j + n_{ij} m_p m_q = A, \qquad (11.2.19)$$

$1 \leq i,j \leq n$, $i \neq j \neq p \neq q$, then none of (11.2.9), (11.2.10) and (11.2.11) has a consistent solution for either one of the following cases, Case (i): if one error occurs in the pth residue digit and $p \neq i,j$, and Case (ii): two errors occur in the pth and qth residue digits and $p \neq i,j$ and/or $q \neq i,j$.

LEMMA 11.3 *If the moduli and code generator A of an RNS-PC satisfy*

$$A > max\{3m_p m_q m_i m_j - m_i m_j - m_p m_q,\} \qquad (11.2.20)$$

$1 \leq i,j \leq n$, $i \neq j \neq p \neq q$, then none of (11.2.9), (11.2.10) and (11.2.11) has a consistent solution for either one of the following cases, Case (i): If one error occurs in the pth residue digit and $p \neq i,j$, and Case (ii): Two errors occur in the pth and qth residue digits and $p \neq i,j$ and/or $q \neq i,j$.

EXAMPLE 11.2 Consider an RNS-PC based on the moduli $m_1 = 3$, $m_2 = 5$, $m_3 = 7$, $m_4 = 11$, $m_5 = 13$, $m_6 = 17$, $m_7 = 19$ and $m_8 = 23$,

Table 11.1: An Algorithm for Correcting Double Errors and Simultaneously Detecting Multiple Errors and Additive Overflow

Step 1:	According to the received vector, compute the syndrome Δ using the BEX method as given in Table 7.1.
Step 2:	If $\Delta \equiv 0 \pmod{A}$, then declare that no error has occurred and stop. If $\Delta \equiv -M_T \pmod{A}$, then declare that no error has occurred and an additive overflow is detected and stop. Otherwise, go to next step.
Step 3:	Let $i = 1$.
Step 4:	Let $j = i + 1$.
Step 5:	Perform the double-error consistency-checking for the combination of the moduli m_i and m_j: (1). Compute $e_{ij}^{(1)}, e_{ij}^{(2)}, e_{ij}^{(3)}$ using (11.2.9), (11.2.10), (11.2.11). (2). Check the consistency of the solutions, that is, $\quad e_{ij}^{(1)} < m_i m_j$ or $e_{ij}^{(2)} < m_i m_j$ or $e_{ij}^{(3)} < m_i m_j$. If it is consistent, then go to Step 7. If it is inconsistent, then go to next step.
Step 6:	Let $j = j + 1$, go to Step 5 for $j \leq n$. For $j > n$, $i = i + 1$ go to Step 4 for $i \leq n - 1$. For $i = n$, go to Step 8.
Step 7:	Declare two errors in the ith and jth position. Correct them using (11.2.12), (11.2.13), (11.2.14). If $e_{ij}^{(3)} < m_i m_j$, then declare an additive overflow detected and stop. If $e_{ij}^{(2)} < m_i m_j$, then recompute the new syndrome Δ' as in Step 1. If $\Delta' \equiv -M_T \pmod{A}$, then also declare an additive overflow detected and stop. Otherwise, stop.
Step 8:	Declare more than two errors detected and stop.

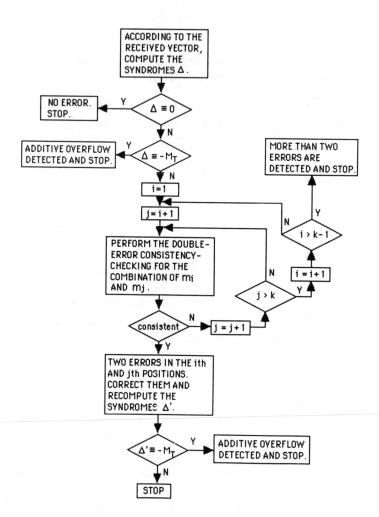

Figure 11.1: A Decoder Flowchart for Double Errors and Simultaneous Multiple Errors and Additive Overflow Detection

and the code generator $A = 289,078 > 3m_5 m_6 m_7 m_8 - m_5 m_6 - m_7 m_8 = 289,073$. Let $X = 5,781,560 < M_T = \prod_{i=1}^{8} m_i = 111,546,435$, and $X \equiv 0 \pmod{A}$. Therefore, X is a code-integer in this RNS-PC. The residue representation of X is $\underline{x} = [2\ 0\ 1\ 4\ 5\ 13\ 12\ 4]$. Assume that X is altered by two residue errors in the seventh and eighth positions, and $E \longleftrightarrow \underline{e} = [0\ 0\ 0\ 0\ 0\ 0\ 8\ 5]$. Then the received vector is $\underline{y} = [2\ 0\ 1\ 4\ 5\ 13\ 1\ 9]$. Based on the received vector and BEX, we compute the syndrome digit $\Delta \equiv 56,474 \pmod{A}$. Note that $-M_T \equiv 37,673 \pmod{A}$. Following the decoding algorithm, we check the consistency for $i = 1, 2, \ldots, 7$, $j = i+1, i+2, \ldots, 8$, based on computing $e_{ij}^{(1)}$, $e_{ij}^{(2)}$ and $e_{ij}^{(3)}$ from (11.2.9–11.2.11). We get

$$e_{78}^{(1)} \equiv \Delta m_7 m_8 M_T^{-1} \pmod{A} \equiv 212 < m_7 m_8,$$

$$e_{78}^{(2)} \equiv \Delta m_7 m_8 M_T^{-1} + m_7 m_8 \pmod{A} \equiv 649 > m_7 m_8,$$

$$e_{78}^{(3)} \equiv \Delta m_7 m_8 M_T^{-1} + 2 m_7 m_8 \pmod{A} \equiv 1086 > m_7 m_8.$$

Since $e_{78}^{(1)} \equiv 212 \pmod{A} < m_7 m_8$ is a consistent solution, we declare that two errors have occurred in the seventh and eighth positions. The estimated values of erroneous residue digits are given by

$$\widehat{x}_7 \equiv y_7 - e_{78}^{(1)} \frac{M_T}{m_7 m_8} \pmod{m_7} \equiv 12,$$

$$\widehat{x}_8 \equiv y_8 - e_{78}^{(1)} \frac{M_T}{m_7 m_8} \pmod{m_8} \equiv 4.$$

EXAMPLE 11.3 In the RNS-PC of Example 11.2, let $X \equiv \overline{X} \pmod{M_T}$ and $\overline{X} = X_1 + X_2$, where $X_1 = 109,849,640 \leftrightarrow \underline{x}_1 = [2\ 0\ 5\ 10\ 4\ 9\ 0\ 7]$, $X_2 = 110,138,718 \leftrightarrow \underline{x}_2 = [0\ 3\ 4\ 8\ 1\ 2\ 12\ 21]$. Then $X = 108,441,923 \leftrightarrow \underline{x} = [2\ 3\ 2\ 7\ 5\ 11\ 12\ 5]$. Based on the vector $\underline{x}$ and BEX, we compute the syndrome $\Delta \equiv 37,673 \pmod{A}$. Since $\Delta \equiv -M_T \pmod{A}$, we declare that an additive overflow is detected.

Now, assume that X is altered by two residue errors in the seventh and eighth positions, and $E \longleftrightarrow \underline{e} = [0\ 000\ 0\ 08\ 5]$. Then the received vector is $\underline{y} = [2\ 32\ 75\ 111\ 10]$. Based on the received vector $\underline{y}$ and BEX, we compute the syndrome $\Delta \equiv 131,820 \pmod{A}$. Following the decoding algorithm, we check the consistency for $i = 1, 2, \ldots, 7$, $j = i+1, i+2, \ldots, 8$, based on computing $e_{ij}^{(1)}$, $e_{ij}^{(2)}$ and $e_{ij}^{(3)}$ from (11.2.9–11.2.11). We get

$$e_{78}^{(1)} \equiv \Delta m_7 m_8 M_T^{-1} \pmod{A} \equiv 288,416 > m_7 m_8,$$

$$e_{78}^{(2)} \equiv \Delta m_7 m_8 M_T^{-1} + m_7 m_8 \pmod{A} \equiv 288,853 > m_7 m_8,$$

$$e_{78}^{(3)} \equiv \Delta m_7 m_8 M_T^{-1} + 2m_7 m_8 \quad (\text{mod } A) \equiv 212 < m_7 m_8.$$

Since $e_{78}^{(3)} \equiv 212 \,(\text{mod } A) < m_7 m_8$ is a consistent solution, we declare that two errors have occurred in the seventh and eighth positions and an additive overflow is detected. The estimated values of erroneous residue digits are computed to be

$$\widehat{x}_7 \equiv y_7 - e_{78}^{(3)} \frac{M_T}{m_7 m_8} \quad (\text{mod } m_7) \equiv 12,$$

$$\widehat{x}_8 \equiv y_8 - e_{78}^{(3)} \frac{M_T}{m_7 m_8} \quad (\text{mod } m_8) \equiv 5.$$

11.3   Multiple Error Correction

An RNS-PC with $A > max\{\prod_{i=1}^{2\lambda} m_{j_i}\}$, $1 \leq j_i \leq n$, can correct λ residue errors. Assume that c $(1 \leq c \leq \lambda)$ errors occur. The altered residue vector and the derivation of the syndrome are stated in (7.5.1–7.5.12). Given Δ, a procedure to determine error locations and error values can be outlined as follows:

For $j_1 = 1, 2, \ldots, n - \lambda + 1$, $j_2 = j_1 + 1, j_1 + 2, \ldots, n - \lambda + 2, \cdots, j_\lambda = j_{\lambda-1} + 1, j_{\lambda-1} + 2, \ldots, n$, solve the congruences

$$\Delta \equiv e^{(1)} \frac{M_T}{\prod_{i=1}^{\lambda} m_{j_i}} \quad (\text{mod } A), \tag{11.3.1}$$

$$\Delta \equiv (e^{(2)} - \prod_{i=1}^{\lambda} m_{j_i}) \frac{M_T}{\prod_{i=1}^{\lambda} m_{j_i}} \quad (\text{mod } A), \tag{11.3.2}$$

$$\Delta \equiv (e^{(3)} - 2 \prod_{i=1}^{\lambda} m_{j_i}) \frac{M_T}{\prod_{i=1}^{\lambda} m_{j_i}} \quad (\text{mod } A), \tag{11.3.3}$$

to obtain the values $e^{(1)}$, $e^{(2)}$ and $e^{(3)}$. Check if the solutions are consistent, that is, one of the following three conditions is satisfied:
Condition (1) $e^{(1)} < \prod_{i=1}^{\lambda} m_{j_i}$, $e^{(2)} > \prod_{i=1}^{\lambda} m_{j_i}$ and $e^{(3)} > \prod_{i=1}^{\lambda} m_{j_i}$;
Condition (2) $e^{(2)} < \prod_{i=1}^{\lambda} m_{j_i}$, $e^{(1)} > \prod_{i=1}^{\lambda} m_{j_i}$ and $e^{(3)} > \prod_{i=1}^{\lambda} m_{j_i}$;
Condition (3) $e^{(3)} < \prod_{i=1}^{\lambda} m_{j_i}$, $e^{(1)} > \prod_{i=1}^{\lambda} m_{j_i}$ and $e^{(2)} > \prod_{i=1}^{\lambda} m_{j_i}$.
Let us say that one of the three conditions is satisfied for $j_i = l_i$, $i = 1, 2, \ldots, \lambda$. Then the l_1th, l_2th, $\ldots$, and l_λth residue digits are declared to be in error, the values of errors being

$$e_{l_i} \equiv e \frac{M_T}{\prod_{i=1}^{\lambda} m_{j_i}} \quad (\text{mod } m_{l_i}), \quad i = 1, 2, \ldots, \lambda,$$

where $e = e^{(1)}$ if $e^{(1)} < \prod_{i=1}^{\lambda} m_{j_i}$; $e = e^{(2)}$ if $e^{(2)} < \prod_{i=1}^{\lambda} m_{j_i}$; and $e = e^{(3)}$ if $e^{(3)} < \prod_{i=1}^{\lambda} m_{j_i}$. The received vector $\underline{y}$ is decoded to $\underline{\hat{x}}$, where

$$\begin{aligned}
\underline{\hat{x}} &= [\hat{x}_1\ \hat{x}_2\ \ldots\ \hat{x}_n] \\
\hat{x}_j &= y_j,\ j = 1,2,\ldots,n,\ j \neq l_i,\ i = 1,2,\ldots,\lambda, \\
\hat{x}_{l_i} &\equiv y_{l_i} - e_{l_i}\ (\bmod\ m_{l_i}),\ i = 1,2,\ldots,\lambda.
\end{aligned} \qquad (11.3.4)$$

Note that if $e_{l_i} = 0$, then the l_ith residue digit is correct.

It is clear from (7.5.5), (7.5.6) and (11.3.1) that if $c\ (1 \leq c \leq \lambda)$ errors occur in the positions $\{i_1, i_2, \ldots, i_c\}$ such that $X + E < M_T$ and $X = \overline{X}$, then for $\{i_1, i_2, \ldots, i_c\} \subseteq \{j_1, j_2, \ldots, j_\lambda\}$,

$$e = e^{(1)} = e' \prod_{\substack{i=1 \\ j_i \neq i_1,i_2,\ldots,i_c}}^{\lambda} m_{j_i} < \prod_{i=1}^{\lambda} m_{j_i}$$

is a solution to (11.3.1). Similarly, if $c\ (1 \leq c \leq \lambda)$ errors occur in the positions $\{i_1, i_2, \ldots, i_c\}$ such that
(1) $M_T \leq X + E < 2M_T$ and $X = \overline{X}$, then for $\{i_1, i_2, \ldots, i_c\} \subseteq \{j_1, j_2, \ldots, j_\lambda\}$,

$$e = e^{(2)} = e' \prod_{\substack{i=1 \\ j_i \neq i_1,i_2,\ldots,i_c}}^{\lambda} m_{j_i} < \prod_{i=1}^{\lambda} m_{j_i}$$

is a solution to (11.3.2);
(2) $X + E < M_T$ and $X = \overline{X} - M_T$, then for $\{i_1, i_2, \ldots, i_c\} \subseteq \{j_1, j_2, \ldots, j_\lambda\}$,

$$e = e^{(2)} = e' \prod_{\substack{i=1 \\ j_i \neq i_1,i_2,\ldots,i_c}}^{\lambda} m_{j_i} < \prod_{i=1}^{\lambda} m_{j_i}$$

is a solution to (11.3.2); and
(3) $M_T \leq X + E < 2M_T$ and $X = \overline{X} - M_T$, then for $\{i_1, i_2, \ldots, i_c\} \subseteq \{j_1, j_2, \ldots, j_\lambda\}$,

$$e = e^{(3)} = e' \prod_{\substack{i=1 \\ j_i \neq i_1,i_2,\ldots,i_c}}^{\lambda} m_{j_i} < \prod_{i=1}^{\lambda} m_{j_i}$$

is a solution to (11.3.3).

Based on the above analysis, if Condition (2) is satisfied, then we correct errors and recompute the syndrome, denoted by Δ'. If the new syndrome $\Delta' \equiv -M_T\ (\bmod\ A)$ then an additive overflow is detected. However, it has to be shown that if $\{i_1, i_2, \ldots, i_c\} \not\subseteq \{j_1, j_2, \ldots, j_\lambda\}$, then there is no consistent solution. Under an additional constraint on the moduli and code generator A, the result is established in Theorem 11.4 and Lemma 11.4. The proofs of Theorem 11.4 and Lemma 11.4 are similar to the proofs of Theorem 11.2 and Lemma 11.2.

THEOREM 11.4 *If the moduli and code generator A of an RNS-PC are such that there do not exist integers n_{Λ_1}, n_{Λ_2}, $0 < n_{\Lambda_1} < \prod_{\substack{i=1 \\ 1 \le j_i \le n}}^{\lambda} m_{j_i}$, $0 < n_{\Lambda_2} < \prod_{\substack{\alpha=1 \\ 1 \le i_\alpha \le n}}^{\lambda} m_{i_\alpha}$, that satisfy either one of the following equations*

$$n_{\Lambda_1} \prod_{\substack{\alpha=1 \\ 1 \le i_\alpha \le n}}^{\lambda} m_{i_\alpha} + n_{\Lambda_2} \prod_{\substack{i=1 \\ 1 \le j_i \le n}}^{\lambda} m_{j_i} = A \qquad (11.3.5)$$

$$\prod_{\substack{\alpha=1 \\ 1 \le i_\alpha \le n}}^{\lambda} m_{i_\alpha} \prod_{\substack{i=1 \\ 1 \le j_i \le n}}^{\lambda} m_{j_i} + n_{\Lambda_1} \prod_{\substack{\alpha=1 \\ 1 \le i_\alpha \le n}}^{\lambda} m_{i_\alpha} + n_{\Lambda_2} \prod_{\substack{i=1 \\ 1 \le j_i \le n}}^{\lambda} m_{j_i} = A, \qquad (11.3.6)$$

where $i_1, i_2, \ldots, i_\lambda$ are the positions of errors, then for $\{i_1, i_2, \ldots, i_\lambda\} \not\subseteq \{j_1, j_2, \ldots, j_\lambda\}$, there is no consistent solution to any one of (11.3.1), (11.3.2) and (11.3.3).

LEMMA 11.4 *If the moduli and code generator A of an RNS-PC satisfy*

$$A > max\{3 \prod_{\substack{\alpha=1 \\ 1 \le i_\alpha \le n}}^{\lambda} m_{i_\alpha} \prod_{\substack{i=1 \\ 1 \le j_i \le n}}^{\lambda} m_{j_i} - \prod_{\substack{\alpha=1 \\ 1 \le i_\alpha \le n}}^{\lambda} m_{i_\alpha} - \prod_{\substack{i=1 \\ 1 \le j_i \le n}}^{\lambda} m_{j_i}\}, \qquad (11.3.7)$$

where $i_1, i_2, \ldots, i_\lambda$ are the positions of errors, then for $\{i_1, i_2, \ldots, i_\lambda\} \not\subseteq \{j_1, j_2, \ldots, j_\lambda\}$, there is no consistent solution to any one of (11.3.1), (11.3.2) and (11.3.3).

Based on the concept developed above and under the assumption that the moduli and code generator A satisfy the conditions in Theorem 11.4 (or Lemma 11.4), the algorithm to correct λ residue errors in the RNS-PC is given in Table 11.2.

11.4   Extensions of Previous Algorithms

In this section, we extend the previous algorithms for single error correction and additive overflow detection in Barsi and Maestrini [1974] and Jenkins

Table 11.2: An Algorithm for Correcting λ Errors and Simultaneously Detecting Additive Overflow

Step 1:	According to the received vector, compute the syndrome Δ using BEX given in Table 7.1.
Step 2:	If $\Delta \equiv 0 \pmod{A}$, then declare that no error has occurred and stop. If $\Delta \equiv -M_T \pmod{A}$, then declare that no error has occurred and an additive overflow is detected and stop. Otherwise, go to next step.
Step 3:	For $j_1 = 1, 2, \ldots, n - \lambda + 1$, $j_2 = j_1 + 1, j_1 + 2,$ $\ldots, n - \lambda + 2, \cdots, j_\lambda = j_{\lambda-1} + 1, j_{\lambda-1} + 2, \ldots, n,$ perform the λ-error consistency-checking for the combination of the moduli $m_{j_1}, m_{j_2}, \ldots, m_{j_\lambda}$: (1). Compute $e^{(1)}, e^{(2)}, e^{(3)}$ using (11.3.1), (11.3.2), (11.3.3). (2). Check the consistency of the solutions, $\qquad e^{(1)} < \prod_{i=1}^{\lambda} m_{j_i}$ or $e^{(2)} < \prod_{i=1}^{\lambda} m_{j_i}$ or $e^{(3)} < \prod_{i=1}^{\lambda} m_{j_i}.$ If it is consistent, then go to Step 4. If it is inconsistent for all the combinations of λ moduli, then go to Step 5.
Step 4:	Declare λ errors in the positions $j_1, j_2, \ldots, j_\lambda$. Correct them using (11.3.4). If $e^{(3)} < \prod_{i=1}^{\lambda} m_{j_i}$, then declare an additive overflow detected and stop. If $e^{(2)} < \prod_{i=1}^{\lambda} m_{j_i}$, then recompute the new syndrome Δ' as in Step 1. If $\Delta' \equiv -M_T \pmod{A}$, then also declare an additive overflow detected and stop. Otherwise, stop.
Step 5:	Declare more than λ errors detected and stop.

[1982] to obtain new algorithms for λ-error correction and β-error and additive overflow detection. In Barsi and Maestrini [1974], the algorithm for locating single residue digit error in an RNS-PC is based on the properties of modulus projection. The modulus m_i-projection of X, X_i, is defined in (9.6.1). The residue representation of X_i is the same as the residue representation of X in a reduced RNS with the ith residue digit x_i deleted. Given $X \leftrightarrow [x_1\ x_2\ \ldots\ x_n]$, the number $X^e \leftrightarrow [x_1\ x_2\ \ldots\ x_n\ 0]$, $X^e \equiv 0 \pmod{A}$, is defined as the image of X in the extended RNS of moduli $m_1, m_2, \ldots, m_n, A$. The set of the images of numbers in the given RNS is defined as the image RNS.

Similarly, the m_i-projection X_i^e of X^e is

$$X_i^e \equiv X^e \pmod{\frac{M_T A}{m_i}} \tag{11.4.1}$$

and $X_i^e \leftrightarrow [x_1\ x_2\ \ldots\ x_{i-1}\ x_{i+1}\ \ldots\ x_n\ 0]$. In Barsi and Maestrini [1974] it is shown that if a number $\overline{Y}^e \leftrightarrow [x_1\ x_2\ \ldots\ \overline{x_i}\ \ldots\ x_n 0]$ is derived from a number $X^e \leftrightarrow [x_1\ x_2\ \ldots\ x_i\ \ldots\ x_n\ 0]$ by an error in the ith residue digit, $\overline{x_i} \neq x_i$, and the code generator A satisfies the inequality

$$A > 2m_i m_j \left(1 - \frac{(M_T)\ \mathrm{mod}\ A}{M_T}\right),\ i \neq j,\ 1 \leq j \leq n,$$

then the image $\overline{Y}_j^e > 2[M_T - (M_T)\ \mathrm{mod}\ A]$, $j = 1, 2, \ldots, n$, $j \neq i$. For $j = i$, $\overline{Y}_j^e \leq 2[M_T - (M_T)\ \mathrm{mod}\ A]$ and the number $X \equiv \overline{Y}_j^e \pmod{A}$ is the correct number. If $\overline{Y}_j^e > M_T$, the correct number is in overflow. The mixed radix representation of $\overline{Y}_j^e$ is

$$\overline{Y}_j^e = \sum_{\substack{l=1 \\ l \neq j}}^{n+1} a_l \prod_{\substack{r=1 \\ r \neq j}}^{l-1} m_r. \tag{11.4.2}$$

In Jenkins [1982], based on the fact that the mixed radix system is a weighted number system, the range into which $\overline{Y}_j^e$ falls is determined by testing the size of a_{n+1} for $j = 1, 2, \ldots, n$. The correct residue x_j is obtained through BEX. In the following analysis, these algorithms extended for detecting and correcting multiple residue digit errors and detecting additive overflow. The M_Γ-projection of X^e, denoted by X_Γ^e, is defined by

$$X_\Gamma^e \equiv X^e \pmod{\frac{M_T A}{M_\Gamma}}, \tag{11.4.3}$$

$M_\Gamma = \prod_{\alpha=1}^{\gamma} m_{i_\alpha}$, $\Gamma = \{i_1, i_2, \ldots, i_c, \ldots, i_\gamma;\ i_1 < i_2 < \cdots < i_c < \cdots < i_\gamma\}$, and $\gamma \leq d - 1$. X_Γ^e can also be represented as a reduced residue representation of X^e with the residue digits $x_{i_1}, x_{i_2}, \ldots, x_{i_\gamma}$ deleted. Then, the mixed radix representation of X_Γ^e is

$$X_\Gamma^e = \sum_{\substack{l=1 \\ l \notin \Gamma}}^{n+1} a_l \prod_{\substack{r=1 \\ r \notin \Gamma}}^{l-1} m_r. \tag{11.4.4}$$

It follows from (11.4.3) that the M_Γ-projection of any code-integer $X = X^e$ in an RNS-PC is still a code-integer, that is, $X_\Gamma^e = X^e = X$. The proof of the following theorem is similar to the proof of Theorem 5.1 in Barsi and Maestrini [1973].

THEOREM 11.5 *Let $\overline{Y}^e$ be a number in image RNS and $\overline{Y}_\Gamma^e$ be M_Γ-projection of $\overline{Y}^e$, where $M_\Gamma = \prod_{\alpha=1}^\gamma m_{i_\alpha}$. If $\overline{Y}_\Gamma^e \neq \overline{Y}^e$, the residue representation of $\overline{Y}_\Gamma^e$ uniquely differs from $\overline{Y}^e$ in one or more of residue digits corresponding to the moduli $m_{i_1}, m_{i_2}, \ldots, m_{i_\gamma}$.*

Before considering the range of M_Γ-projection $\overline{Y}_\Gamma^e$ of any illegitimate number $\overline{Y}^e$ in the RNS-PC, we give the following theorems.

THEOREM 11.6 *Let $\overline{Y}^e$ be an illegitimate number in the RNS-PC having minimum distance d. If the M_Γ-projection $\overline{Y}_\Gamma^e$ ($M_\Gamma = \prod_{\alpha=1}^\gamma m_{i_\alpha}$, $1 \leq \gamma < d$), is a code-integer, then there exists only one code-integer X^e differing from $\overline{Y}^e$ in one or more of the residue digits corresponding to the moduli $m_{i_1}, m_{i_2}, \ldots, m_{i_\gamma}$.*

PROOF. Since $\overline{Y}^e \neq \overline{Y}_\Gamma^e$, it is obvious from Theorem 11.5 that the code-integer $X = \overline{Y}_\Gamma^e$ is a solution to the problem. Now assume that there exists two different code-integers X and X', both differing from $\overline{Y}^e$ in one or more of the residue digits corresponding to the moduli $m_{i_1}, m_{i_2}, \ldots, m_{i_\gamma}$. Then

$$X = Y_\Gamma^e + e \frac{M_T A}{\prod_{\alpha=1}^\gamma m_{i_\alpha}},$$

$$X' = Y_\Gamma^e + e' \frac{M_T A}{\prod_{\alpha=1}^\gamma m_{i_\alpha}},$$

$0 \leq e, e' < \prod_{\alpha=1}^\gamma m_{i_\alpha}$. Since $0 \leq X, X' < M_T$, $0 \leq \overline{Y}_\Gamma^e < M_T$, and

$$\frac{M_T A}{\prod_{\alpha=1}^\gamma m_{i_\alpha}} \geq M_T,$$

it follows that $e = e' = 0$ and $X = X'$. This contradicts the original assumption $X \neq X'$. This proves the theorem.

THEOREM 11.7 *In an RNS-PC with minimum distance $d = \lambda + \beta + 1$, an illegitimate number $\overline{Y}^e$ is derived from a code-integer X^e by λ errors corresponding to the moduli $m_{i_1}, m_{i_2}, \ldots, m_{i_\lambda}$, iff the M_Λ-projection $\overline{Y}_\Lambda^e$ is a code-integer and M_P-projection $\overline{Y}_P^e$ is an illegitimate number. Here $M_\Lambda = \prod_{\alpha=1}^{\lambda} m_{i_\alpha}$, $M_P = \prod_{\alpha=1}^{p} m_{j_\alpha}$, $1 \leq p \leq \beta$, and $gcd(M_P, M_\Lambda) \neq \prod_{\alpha=1}^{\lambda} m_{i_\alpha} = M_\Lambda$.*

PROOF. If $\overline{Y}^e$ differs from X^e in the i_1th, i_2th, $\ldots$, i_λth residue digits, then $\overline{Y}^e$ can be expressed as

$$\overline{Y}^e \equiv X^e + e' \frac{M_T A}{\prod_{\alpha=1}^{\lambda} m_{i_\alpha}},$$

where $0 < e' < \prod_{\alpha=1}^{\lambda} m_{i_\alpha}$. By definition,

$$\overline{Y}_\Lambda^e \equiv \overline{Y}^e \pmod{\frac{M_T A}{M_\Lambda}}$$

$$\equiv X^e + e' \frac{M_T A}{\prod_{\alpha=1}^{\lambda} m_{i_\alpha}} \pmod{\frac{M_T A}{M_\Lambda}}$$

$$\equiv X_\Lambda^e < M_T.$$

Therefore, $\overline{Y}_\Lambda^e$ is a code-integer. Since $gcd(M_P, M_\Lambda) \neq \prod_{\alpha=1}^{\lambda} m_{i_\alpha}$, $\overline{Y}_P^e$ can be treated as a number originating from X_P^e affected by λ or less errors corresponding to the moduli $m_{i_1}, m_{i_2}, \ldots, m_{i_\lambda}$, that is,

$$\overline{Y}_P^e = X_P^e + e' \frac{M_T A}{\prod_{\alpha=1}^{p} m_{j_\alpha} \prod_{\alpha=1}^{\lambda} m_{i_\alpha}},$$

where $0 < e' < \prod_{\alpha=1}^{\lambda} m_{i_\alpha}$. Since $A \geq \prod_{\alpha=1}^{p} m_{j_\alpha} \prod_{\alpha=1}^{\lambda} m_{i_\alpha}$, it follows that $\overline{Y}_P^e$ is an illegitimate number. This proves the necessity. The sufficiency is proved by contradiction. Since $\overline{Y}_\Lambda^e$ is a code-integer, by Theorem 11.6, there exists only one code-integer X^e differing from $\overline{Y}^e$ in one or more of residue digits corresponding to the moduli $m_{i_1}, m_{i_2}, \ldots, m_{i_\lambda}$. However, if the code-integer X^e differs from $\overline{Y}^e$ in less than λ residue digits (say p $(p < \lambda)$ residue digits corresponding to the moduli $m_{i_1}, m_{i_2}, \ldots, m_{i_p}$) then, by Theorem 11.6, the M_P-projection $\overline{Y}_P^e$ is a code-integer, where $M_P = \prod_{\alpha=1}^{p} m_{i_\alpha}$ and $gcd(M_P, M_\Lambda) \neq M_\Lambda$. This contradicts the original assumption. This proves the theorem.

    The range of the illegitimate projection $\overline{Y}_\Lambda^e$ of any illegitimate number $\overline{Y}$ is given in the following theorem.

THEOREM 11.8 *Assume that no more than β $(\beta > \lambda)$ residue errors occur, and the moduli and code generator A of an RNS-PC satisfy the condition*

$$A > max\{2 \prod_{\substack{i=1 \\ 1 \le j_i \le n}}^{\lambda+\beta} m_{j_i}\}. \tag{11.4.5}$$

If the M_Λ-projection $\overline{Y}_\Lambda^e$ of any illegitimate number $\overline{Y}^e$ is illegitimate, $\overline{Y}_\Lambda^e$ will be in the range $[2M_T, \frac{M_T A}{M_\Lambda})$, where

$$M_\Lambda = \prod_{\substack{\alpha=1 \\ 1 \le i_\alpha \le n}}^{\lambda} m_{i_\alpha}.$$

PROOF. The illegitimate projection $\overline{Y}_\Lambda^e$ can be treated as a number originating from X_Λ affected by β or fewer errors corresponding to the moduli $m_{j_1}, m_{j_2}, \ldots, m_{j_\beta}$. Then, $\overline{Y}_\Lambda^e$ can be expressed as

$$\overline{Y}_\Lambda^e \equiv X_\Lambda + e' \frac{M_T A}{\prod_{\alpha=1}^{\lambda} m_{i_\alpha} \prod_{\alpha=1}^{\beta} m_{j_\alpha}} \quad (\text{mod } \frac{M_T A}{M_\Lambda}),$$

$0 < e' < \prod_{\alpha=1}^{\beta} m_{j_\alpha}$. To get the smallest value of $\overline{Y}_\Lambda^e$, let $X_\Lambda = 0$ and $e' = 1$. Since $A \ge 2 \prod_{\alpha=1}^{\lambda} m_{i_\alpha} \prod_{\alpha=1}^{\beta} m_{j_\alpha}$, it follows that $\overline{Y}_\Lambda^e > 2M_T$. This proves the theorem.

Based on the above analysis, a procedure for correcting λ errors and simultaneously detecting β $(\beta > \lambda)$ errors and additive overflow can be outlined as follows:

Step 1. Based on the received residue vector, compute the syndrome Δ using BEX.

Step 2. If $\Delta \equiv 0 \pmod{A}$, then declare no errors and stop. If $\Delta \equiv -M_T \pmod{A}$, then no error has occurred and an additive overflow is detected. Stop. Otherwise, go to next step.

Step 3. Compute the mixed radix digits corresponding to each of the $\binom{n}{\lambda}$ projections of the image. Note that each of the projections is obtained by deleting λ residue digits from the n residue digits. Let M_Λ denote the product of the moduli corresponding to the deleted digits. Let a_{n+1} be the last mixed radix digit.
(1) If $a_{n+1} \in [0, M_\Lambda)$, that is, $\overline{Y}_\Lambda^e \in [0, M_T)$, then declare that λ errors have occurred in the positions corresponding to the moduli of

M_Λ. Correct them by using BEX and stop.

(2) If $a_{n+1} \in [M_\Lambda, 2M_\Lambda)$, that is, $\overline{Y}_\Lambda^e \in [M_T, 2M_T)$, then declare that λ errors have occurred in the positions corresponding to the moduli of M_Λ and an additive overflow is detected. Correct errors by using BEX and stop.

(3) If $a_{n+1} \in [2M_\Lambda, A)$, that is, $\overline{Y}_\Lambda^e \in [2M_T, M_T A/M_\Lambda)$, for all $\binom{n}{\lambda}$ sets of λ different moduli, then go to next step.

Step 4. Declare more than λ errors detected. Stop.

11.5   Computational Complexity Analysis

Based on the coding theory framework developed in Chapter 7, we described new algorithms for correcting and detecting multiple residue errors and detecting additive overflow in RNS-PC in this chapter. They are superior to the algorithms in Barsi and Maestrini [1974] and Jenkins [1982] from the computational complexity point of view. Based on the MRC method for BEX, the comparison of our algorithm with that in Barsi and Maestrini [1974] and Jenkins [1982] in terms of requirement of MMULT and MADD is shown in Table 11.3. We point out that for double-error correction the expression in column 2 of Table 11.3 is obtained by extending the algorithms as described in the previous section. However, the algorithms in Barsi and Maestrini [1974] and Jenkins [1982] remain valid even for the case when the moduli and code generator do not satisfy the conditions stated in Theorems 11.1–11.4.

Table 11.3: MMULT and MADD Required for Single/Double Errors Correction and Additive Overflow Detection

♯ Error Correction	Previous Algorithms	New Algorithm	Modular Operations
1	$\frac{1}{2}n^3 + \frac{3}{2}n - 2$	$n^2 + 2n - 1$	MMULT
$d = 3$	$\frac{1}{2}n^3 + \frac{3}{2}n - 2$	$n^2 + 3n - 1$	MADD
2	$\frac{1}{4}n^4 - n^3 + \frac{7}{4}n^2 + 2n - 5$	$\frac{3}{2}n^2 + \frac{1}{2}n$	MMULT
$d = 5$	$\frac{1}{4}n^4 - n^3 + \frac{7}{4}n^2 + 2n - 5$	$2n^2$	MADD

NOTES

Our effort has been to expose and exploit the underlying mathematical structure present in the RNS in order to derive computationally efficient algorithms for error control. In this regard, we have addressed both single error and the multiple error correction cases. All the algorithms are based on modulo arithmetic. This, we feel, is a very attractive feature from the VLSI implementation point of view.

This chapter is based on the research reported in Sun and Krishna [1993].

11.6 Bibliography

[11.1] F. Barsi and P. Maestrini, "Error Correcting Properties of Redundant Residue Number Systems," *IEEE Transactions on Computers*, vol. C-22, pp. 307-315, 1973.

[11.2] F. Barsi and P. Maestrini, "Error Detection and Correction by Product Codes in Residue Number Systems," *IEEE Transactions on Computers*, vol. C-23, pp. 915-924, 1974.

[11.3] W.K. Jenkins, "Failure Resistant Digital Filters Based on Residue Number System Product Code," *IEEE International Conference on Acoustics, Speech, and Signal Processing*, pp. 60-63, 1982.

[11.4] J.D. Sun and H. Krishna, "On Theory and Fast Algorithms for Error Correction in Residue Number System Product Codes," *IEEE Transactions on Computers*, vol. 43, pp. 840-852, 1993.

Appendix A

Computational Complexity Analysis

Under the assumption that BEX is based on the MRC approach, we compare the computational complexity of the previously known algorithms in Yau and Liu [1973] and Jenkins and Altman [1988] in terms of modular multiplications (MMULT) and modular additions (MADD) with the computational complexity of the algorithm described in Section 8.1 for single error correction in RRNS.

A.1 Yau and Liu [1973] algorithm

Compute $(\frac{k}{2}+1)$ syndrome sets (assume k even), that is, $\{|\Delta|_{m_{k+2}}, |\Delta|_{m_{k+1}}\}$, $\{|\Delta|_{m_k}, |\Delta|_{m_{k-1}}\}, \ldots, \{|\Delta|_{m_{k-2t}}, |\Delta|_{m_{k-2t-1}}\}, \ldots$, and $\{|\Delta|_{m_2}, |\Delta|_{m_1}\}$. For each set:

Step 1. Compute the mixed radix digits

$$\frac{1}{2}(k-1)k = \frac{1}{2}k^2 - \frac{1}{2}k.$$

Step 2. Obtain $|X|''_{m_{k-2t}}$ and $|X|''_{m_{k-2t-1}}$ from the mixed radix digits.

$$(d-1)(k-1) = 2k - 2 \quad \text{for} \quad d = 3.$$

Step 3. Compute

$$\begin{cases} |\Delta|_{m_{k-2t}} = ||X|''_{m_{k-2t}} - |X|'_{m_{k-2t}}|_{m_{k-2t}} \\ |\Delta|_{m_{k-2t-1}} = ||X|''_{m_{k-2t-1}} - |X|'_{m_{k-2t-1}}|_{m_{k-2t-1}} \end{cases}$$

$$d - 1 = 2 \quad \text{for} \quad d = 3 \quad (\text{MADD only}).$$

** The total number of MMULT

$$(\frac{k}{2} + 1)(\frac{1}{2}k^2 - \frac{1}{2}k + 2k - 2) = \frac{1}{4}k^3 + \frac{5}{4}k^2 + \frac{1}{2}k - 2.$$

** The total number of MADD

$$(\frac{k}{2} + 1)(\frac{1}{2}k^2 - \frac{1}{2}k + 2k - 2 + 2) = \frac{1}{4}k^3 + \frac{5}{4}k^2 + \frac{3}{2}k.$$

A.2   Jenkins and Altman [1988] algorithm

Step 1. Successively compute the mixed radix digits for each of the n projections required for error locations.

$$n[\frac{1}{2}(n-1)(n-2)]$$

$$= \frac{1}{2}n(n-1)(n-2)$$

$$= \frac{1}{2}(k+2)(k+1)k \quad \text{for} \quad d = n - k + 1 = 3$$

$$= \frac{1}{2}k^3 + \frac{3}{2}k^2 + k.$$

Step 2. Obtain the correct residue digit from the mixed radix digits.

$$n - 2 = k \quad \text{for} \quad d = 3.$$

** The total number of MMULT/MADD

$$\frac{1}{2}k^3 + \frac{3}{2}k^2 + 2k.$$

A.3   New algorithm

Step 1. Compute only one syndrome set $\{\Delta_1, \Delta_2\}$.
    (a). Compute the mixed radix digits

$$\frac{1}{2}(k-1)k = \frac{1}{2}k^2 - \frac{1}{2}k.$$

    (b). Obtain $\bar{y}_{k+1}$ and $\bar{y}_{k+2}$ from the mixed radix digits.

$$(d-1)(k-1) = 2k - 2 \quad \text{for} \quad d = 3,$$

(c). Compute

$$\begin{cases} \Delta_1 \equiv \overline{y}_{k+1} - y_{k+1} & (\text{mod } m_{k+1}) \\ \Delta_2 \equiv \overline{y}_{k+2} - y_{k+2} & (\text{mod } m_{k+2}) \end{cases}$$

$$d - 1 = 2 \quad \text{for} \quad d = 3 \quad (\text{MADD only}).$$

Step 2. Decoding procedure for $j = 1, 2, \ldots, k$

$$\begin{cases} e_j^{(1,1)} \equiv \Delta_1 \left(\dfrac{M}{m_j} \right)^{-1} & (\text{mod } m_{k+1}) \\ e_j^{(1,2)} \equiv \Delta_2 \left(\dfrac{M}{m_j} \right)^{-1} & (\text{mod } m_{k+2}) \\ e_j^{(2,1)} \equiv (e_j^{(11)} + m_j) & (\text{mod } m_{k+1}) \\ e_j^{(2,2)} \equiv (e_j^{(12)} + m_j) & (\text{mod } m_{k+2}) \end{cases}$$

$$(d - 1)k = 2k \quad \text{for} \quad d = 3.$$

Step 3. Compute the error value and obtain the correct residue digit

$$\begin{cases} e_j = e_j' \dfrac{M}{m_j} & (\text{mod } m_i) \\ \widehat{x}_j \equiv (y_j - e_j) & (\text{mod } m_j). \end{cases}$$

** The total number of MMULT

$$\frac{1}{2}k^2 - \frac{1}{2}k + 2k - 2 + 2k + 1 = \frac{1}{2}k^2 + \frac{7}{2}k - 1,$$

** The total number of MADD

$$\frac{1}{2}k^2 - \frac{1}{2}k + 2k - 2 + 2 + 2k + 1 = \frac{1}{2}k^2 + \frac{7}{2}k + 1.$$

Appendix B

Complexity Analysis: Continued

Under the assumption that BEX is based on the MRC approach, we compare the computational complexity of the new algorithms with that of the extension of the algorithm in Jenkins and Altman [1988] in terms of MMULT and MADD for double error correction in RRNS as described in Chapter 9.

B.1 Extended Jenkins and Altman [1988] algorithm

<u>Step 1</u>. Successively compute the mixed radix digits for each of the $\begin{pmatrix} n \\ 2 \end{pmatrix}$ projections required for error locations.

$$\begin{pmatrix} n \\ 2 \end{pmatrix} [\frac{1}{2}(n-3)(n-2)]$$

$$= \frac{1}{4}n(n-1)(n-2)(n-3)$$

$$= \frac{1}{4}(k+4)(k+3)(k+2)(k+1) \quad \text{for } d = n - k + 1 = 5$$

$$= \frac{1}{4}k^4 + \frac{10}{4}k^3 + \frac{35}{4}k^2 + \frac{50}{4}k + 6.$$

<u>Step 2</u>. Obtain the correct residue digits from the mixed radix digits.

$$2(n-3) = 2(k+1) \quad \text{for } d = 5.$$

** The total number of MMULT/MADD

$$\frac{1}{4}k^4 + \frac{10}{4}k^3 + \frac{35}{4}k^2 + \frac{58}{4}k + 8.$$

B.2   New algorithm

<u>Step 1</u>. Compute only one syndrome set $\{\Delta_1, \Delta_2, \Delta_3, \Delta_4\}$.
    (a). Compute the mixed radix digits

$$\frac{1}{2}(k-1)k = \frac{1}{2}k^2 - \frac{1}{2}k.$$

    (b). Obtain $\bar{y}_{k+1}, \bar{y}_{k+2}, \bar{y}_{k+3}$ and $\bar{y}_{k+4}$ from the mixed radix digits.

$$(d-1)(k-1) = 4k - 4 \quad \text{for} \quad d = 5.$$

    (c). Compute $\Delta_r \equiv \bar{y}_{k+r} - y_{k+r} \pmod{m_{k+r}}, \quad r = 1, 2, 3, 4$.

$$d - 1 = 4 \quad \text{for} \quad d = 5 \text{ (MADD only)}.$$

<u>Step 2</u>. Decoding procedure for $i = 1, 2, \ldots, k-1, \quad j = i+1, i+2, \ldots, k$
    (a).

$$\begin{cases} e^{(1,r)} \equiv \Delta_r \left(\dfrac{M}{m_i m_j}\right)^{-1} & \pmod{m_{k+r}}, \quad r = 1, 2, 3, 4 \\ e^{(2,r)} \equiv e^{(1,r)} + m_i m_j & \pmod{m_{k+r}}, \quad r = 1, 2, 3, 4. \end{cases}$$

    (b).

$$\begin{cases} e_{ij}^{(1)} \equiv \sum_{r=1}^{2} e^{(1,r)} T_r' \frac{m_{k+1} m_{k+2}}{m_{k+r}} \pmod{m_{k+1} m_{k+2}} \\ e_{ij}^{(2)} \equiv \sum_{r=1}^{2} e^{(2,r)} T_r' \frac{m_{k+1} m_{k+2}}{m_{k+r}} \pmod{m_{k+1} m_{k+2}} \\ \text{where } T_r' \frac{m_{k+1} m_{k+2}}{m_{k+r}} \equiv 1 \pmod{m_{k+r}}, \quad r = 1, 2. \\ \text{Check} \begin{cases} \Delta_r \equiv e_{ij}^{(1)} \frac{M}{m_i m_j} \pmod{m_{k+r}}, \quad r = 3, 4 \\ \Delta_r \equiv (e_{ij}^{(2)} - m_i m_j) \frac{M}{m_i m_j} \pmod{m_{k+r}}, \quad r = 3, 4, \end{cases} \end{cases}$$

$$\begin{cases} e_{ij}^{(1)} \equiv \sum_{r=3}^{4} e^{(1,r)} T_r'' \frac{m_{k+3} m_{k+4}}{m_{k+r}} \pmod{m_{k+3} m_{k+4}} \\ e_{ij}^{(2)} \equiv \sum_{r=3}^{4} e^{(2,r)} T_r'' \frac{m_{k+3} m_{k+4}}{m_{k+r}} \pmod{m_{k+3} m_{k+4}} \\ \text{where } T_r'' \frac{m_{k+3} m_{k+4}}{m_{k+r}} \equiv 1 \pmod{m_{k+r}}, \quad r = 3, 4. \\ \text{Check} \begin{cases} \Delta_r \equiv e_{ij}^{(1)} \frac{M}{m_i m_j} \pmod{m_{k+r}}, \quad r = 1, 2 \\ \Delta_r \equiv (e_{ij}^{(2)} - m_i m_j) \frac{M}{m_i m_j} \pmod{m_{k+r}}, \quad r = 1, 2. \end{cases} \end{cases}$$

$$\binom{k}{2}[4+4(d-1)] = 10k(k-1) \quad \text{MMULT} \quad \text{for} \quad d=5.$$

$$\binom{k}{2}[4+2(d-1)] = 6k(k-1) \quad \text{MADD} \quad \text{for} \quad d=5.$$

<u>Step 3</u>. Compute the error values and obtain the correct residue digits

$$\begin{cases} e_i \equiv e' \frac{M}{m_i m_j} \quad (\text{mod } m_i) \\ e_j \equiv e' \frac{M}{m_i m_j} \quad (\text{mod } m_j), \end{cases}$$

$$\begin{cases} \widehat{x}_i \equiv y_i - e_i \quad (\text{mod } m_i) \\ \widehat{x}_j \equiv y_j - e_j \quad (\text{mod } m_j). \end{cases}$$

** The total number of MMULT

$$(\frac{1}{2}k^2 - \frac{1}{2}k) + (4k-4) + (10k^2 - 10k) + 2 = \frac{42}{4}k^2 - \frac{26}{4}k - 2.$$

** The total number of MADD

$$(\frac{1}{2}k^2 - \frac{1}{2}k) + (4k-4) + 4 + (6k^2 - 6k) + 2 = \frac{26}{4}k^2 - \frac{10}{4}k + 2.$$

Index